METHODS IN MOLECULAR BIOLOGY™

For further volumes:
http://www.springer.com/series/7651

Yeast Genetic Networks

Methods and Protocols

Edited by

Attila Becskei

Institute of Molecular Life Sciences, University of Zurich, Zurich, Switzerland

Humana Press

Editor
Attila Becskei
Institute of Molecular Life Sciences
University of Zurich
Zurich
Switzerland
attila.becskei@imls.uzh.ch

ISSN 1064-3745 e-ISSN 1940-6029
ISBN 978-1-61779-085-0 e-ISBN 978-1-61779-086-7
DOI 10.1007/978-1-61779-086-7
Springer New York Dordrecht Heidelberg London

Library of Congress Control Number: 2011923964

Printed on acid-free paper

Humana press is a part of Springer Science+Business Media (www.springer.com)

Preface

A gene changes the activity of the genes it interacts with. The entirety of these effects in a set of genes represents the dynamical behavior of a gene network. The analysis of this behavior can reveal how a network stabilizes the expression level of its components against perturbations, how it specifies the range of signaling intensity and frequency that can be efficiently transmitted in a pathway, or how it induces gene expression to oscillate. Regulation of gene expression – a major determinant of gene activity – occupies a central place in molecular biology. A detailed mechanistic description of the processes involved, methods for highly quantitative measurements, and an array of biotechnological tools are available to understand, to measure and to control gene expression. These favorable conditions explain why yeast genetic networks attracted the attention of many scientists in the nascent field of molecular systems biology. The book *Yeast Genetic Networks: Methods and Protocols* covers approaches to the systems biological analysis of small-scale gene networks in yeast.

Gene expression is primarily determined by how activators and repressors bound to promoters set the level of mRNA production and how quickly the produced mRNA decays. Part I of the book discusses the methods to analyze gene expression quantitatively: identification of promoter regulatory functions, measurement of mRNA production rates, inference of mRNA decay rates based on mRNA production rates, and detection of oscillatory patterns in gene expression. Furthermore, approaches are presented how to control and analyze signaling in genetic networks by implementing self-regulatory synthetic networks and by using microfluidics to dynamically modulate the intensity of external signals.

Part II is a collection of mathematical and computational tools to analyze stochasticity, adaptation, sensitivity in signal transmission, and oscillations in gene expression.

Control of genetic circuits by synthetic elements and dynamical external stimulation are carefully designed for specific purposes. On the other hand, natural genetic variations in a species provide a gratuitous form of control of genetic networks. While the potential to explore the behavior of networks by natural mutations is more restricted, they offer the advantage of identifying the naturally occurring gene variants that shape the behavior of networks. In Part III, methods are presented how to use the tools of quantitative genetics to identify genes that regulate stochasticity and oscillations in gene expression.

Genetic variations are even larger among related fungal species and evolution can shed a different light on network behavior. Thus, Part IV outlines the analysis of conserved gene expression systems and networks in different fungal species: the galactose network in *Kluyveromyces lactis*, and transcriptional silencing is described in *Candida glabrata*. While the former two species are close relatives of the baker's yeast, more diverged pathogenic fungi, *Candida albicans* and *Cryptococcus neoformans* were also included, to emphasize the medical aspects of fungal systems biology.

In summary, *Yeast Genetic Networks: Methods and Protocols* contains a broad range of resources of significant value to both novices and experienced researchers.

Zurich, Switzerland — *Attila Becskei*

Preface

A gene changes the activity of the genes it interacts with. The entirety of these effects in a set of genes represents the dynamical behavior of a gene network. The analysis of this behavior can reveal how a network stabilizes the expression level of its components against perturbations, how it specifies the range of signaling intensity and frequency that can be efficiently transmitted in a pathway, or how it induces gene expression to oscillate. Regulation of gene expression – a major determinant of gene activity – occupies a central place in molecular biology. A detailed mechanistic description of the processes involved, methods for highly quantitative measurements, and an array of biotechnological tools are available to understand, to measure, and to control gene expression. These favorable conditions explain why yeast genetic networks attracted the attention of many scientists in the nascent field of molecular systems biology. The book *Yeast Genetic Networks: Methods and Protocols* covers approaches to the systems biological analysis of small-scale gene networks in yeast.

Gene expression is primarily determined by how activators and repressors bound to promoters set the level of mRNA production and how quickly the produced mRNA decays. Part I of the book discusses the methods to analyze gene expression quantitatively: identification of promoter regulatory functions, measurement of mRNA production rates, inference of mRNA decay rates based on mRNA production rates, and detection of oscillatory patterns in gene expression. Furthermore, approaches are presented how to control and analyze signaling in genetic networks by implementing self-regulatory synthetic networks and by using microfluidics to dynamically modulate the intensity of external signals.

[illegible] adaptation [illegible] activity in signal transmission [illegible] gene expression.

Control of genetic circuits by synthetic elements and dynamical external stimulation [illegible]

[illegible]

Whereas [illegible] largest among related fungal species [illegible] a different light on network behavior. Thus, Part IV outlines the analysis of conserved gene expression systems and networks in different fungal species, the galactose network in *Kluyveromyces lactis*, and transcriptional silencing [illegible]. While the former [illegible] pathogenic fungi, *Candida albicans* and *Cryptococcus neoformans* was also included to emphasize the medical aspects of fungal systems biology.

In summary, *Yeast Genetic Networks: Methods and Protocols* contains a broad range of resources of significant value to both novices and experienced researchers.

Zurich, Switzerland Attila Becskei

Contents

Contributors

Rhys M. Adams · *UT M. D. Anderson Cancer Center, Houston, TX, USA*
Paula Alepuz · *Facultad de Ciencias Biológicas, Departmento de Bioquímica y Biología Molecular, Universitat de València, Burjassot, Spain*
Alexander Anders · *Institut für Biologie, Martin-Luther-Universität Halle-Wittenberg, Halle, Germany*
Gábor Balázsi · *UT M. D. Anderson Cancer Center, Houston, TX, USA*
Attila Becskei · *Institute of Molecular Life Sciences, University of Zurich, Zurich, Switzerland*
Scott A. Berceli · *Department of Surgery, University of Florida, Gainesville, FL, USA*
Arthur Berg · *Center for Statistical Genetics, Pennsylvania State University, Hershey, PA, USA*
Karin D. Breunig · *Institut für Biologie, Martin-Luther-Universität Halle-Wittenberg, Halle, Germany*
Irene Castaño · *Instituto Potosino de Investigación Científica y Tecnológica, San Luis Potosí, SLP, Mexico*
Alejandro De Las Peñas · *Instituto Potosino de Investigación Científica y Tecnológica, San Luis Potosí, SLP, Mexico*
Maurizio Del Poeta · *Department of Biochemistry, Medical University of South Carolina, Charleston, SC, USA*
Tormod Drengstig · *Department of Electrical Engineering and Computer Science, University of Stavanger, Stavanger, Norway*
Steffen Fehrmann · *Laboratoire de Biologie Moléculaire de la Cellule Ecole Normale Superieure de Lyon, Lyon, France*
Jacqueline Garcia · *Department of Biochemistry, Medical University of South Carolina, Charleston, SC, USA*
José García-Martínez · *Facultad de Ciencias Biológicas, Sección de Chips de DNA-S.C.S.I.E, Universitat de València, Burjassot, Spain*
Pascal Hersen · *Department of Molecular and Cellular Biology, School of Engineering and Applied Sciences, Harvard University, Cambridge, MA, USA*
Denes Hnisz · *Max F. Perutz Laboratories, Christian Doppler Laboratory for Infection Biology, Campus Vienna Biocenter, Vienna, Austria*
Alexandra Jauhiainen · *Department of Mathematical Statistics, Chalmers University of Technology and University of Gothenburg, Göteborg, Sweden*
Carl Hirschie Johnson · *Department of Biological Sciences, Vanderbilt University, Nashville, TN, USA*
Alejandro Juárez-Reyes · *Instituto Potosino de Investigación Científica y Tecnológica, San Luis Potosí, SLP, Mexico*
Jaehee Kim · *Department of Statistics, Duksung Women's University, Seoul, South Korea*

THOMAS KJOSMOEN · *Department of Electrical Engineering, University of Stavanger, Stavanger, Norway; Department of Computer Science, University of Stavanger, Stavanger, Norway*
KARL KUCHLER · *Max F. Perutz Laboratories, Christian Doppler Laboratory for Infection Biology, Campus Vienna Biocenter, Vienna, Austria*
NING LI · *Department of Epidemiology and Biostatistics, University of Florida, Gainesville, FL, USA*
JULIA MARÍN-NAVARRO · *Departmento de Biotecnología, Instituto de Agroquímica y Tecnología de Alimentos, Paterna, Spain*
MEGAN N. MCCLEAN · *Lewis-Sigler Institute for Integrative Genomics, Princeton University, Princeton, NJ, USA*
JOAQUÍN MORENO · *Facultad de Ciencias Biológicas, Departmento de Bioquímica y Biología Molecular, Universitat de València, Burjassot, Spain*
DMITRY NEVOZHAY · *UT M. D. Anderson Cancer Center, Houston, TX, USA*
VICENT PELECHANO · *Facultad de Ciencias Biológicas, Departmento de Bioquímica y Biología Molecular, Universitat de València, Burjassot, Spain*
JOSÉ E. PÉREZ-ORTÍN · *Facultad de Ciencias Biológicas, Departmento de Bioquímica y Biología Molecular, Universitat de València, Burjassot, Spain*
ASFIA QURESHI · *Department of Biochemistry, Medical University of South Carolina, Charleston, SC, USA*
SHARAD RAMANATHAN · *Department of Molecular and Cellular Biology, School of Engineering and Applied Sciences, Harvard University, Cambridge, MA, USA*
J. BRIAN ROBERTSON · *Department of Biological Sciences, Vanderbilt University, Nashville, TN, USA*
PRASUNA RATNA · *Institute of Molecular Life Sciences, University of Zurich, Zurich, Switzerland*
PETER RUOFF · *Faculty of Science and Technology, Centre for Organelle Research, University of Stavanger, Stavanger, Norway*
JOHN H. SCHWACKE · *Department of Biochemistry, Medical University of South Carolina, Charleston, SC, USA*
KELLIE J. SIMS · *Department of Biochemistry, Medical University of South Carolina, Charleston, SC, USA*
ARPITA SINGH · *Department of Biochemistry, Medical University of South Carolina, Charleston, SC, USA*
PER SUNNERHAGEN · *Department of Cell and Molecular Biology, Lundberg Laboratory, University of Gothenburg, Gothenburg, Sweden*
YI TAO · *Key Lab of Animal Ecology and Conservational Biology, Centre for Computational and Evolutionary Biology, Institute of Zoology, Chinese Academy of Sciences, Beijing, China*
CHUNFA TONG · *Center for Statistical Genetics, Pennsylvania State University, Hershey, PA, USA*
MICHAEL TSCHERNER · *Max F. Perutz Laboratories, Christian Doppler Laboratory for Infection Biology, Campus Vienna Biocenter, Vienna, Austria*
ZHONG WANG · *Center for Statistical Genetics, Pennsylvania State University, Hershey, PA, USA*

RONGLING WU · *Center for Statistical Genetics, Pennsylvania State University, Hershey, PA, USA*
GAËL YVERT · *Laboratoire de Biologie Moléculaire de la Cellule, Ecole Normale Superieure de Lyon, Lyon, France*
XIU-DENG ZHENG · *Key Lab of Animal Ecology and Conservational Biology, Centre for Computational and Evolutionary Biology, Institute of Zoology, Chinese Academy of Sciences, Beijing, China*

Part I

Experimental Analysis of Signalling in Gene Regulatory Networks

Chapter 1

Global Estimation of mRNA Stability in Yeast

Julia Marín-Navarro, Alexandra Jauhiainen, Joaquín Moreno, Paula Alepuz, José E. Pérez-Ortín, and Per Sunnerhagen

Abstract

Turnover of mRNA is an important level of gene regulation. Individual mRNAs have different intrinsic stabilities. Moreover, mRNA stability changes dynamically with conditions such as hormonal stimulation or cellular stress. While accurate methods exist to measure the half-life of an individual transcript, global methods to estimate mRNA turnover have limitations in terms of resolution in time and precision. We describe and compare two complementary approaches to estimating global transcript stability: (1) direct measurement of decay rates; (2) indirect estimation of turnover from determination of mRNA synthesis rates and steady-state levels. Since the two approaches have distinct strengths yet confer different cellular perturbations, it is valuable to consider results obtained with both methods. The practical aspects of the chapter are written from a yeast perspective; the general considerations hold true for all eukaryotes, however.

Key words: 1-10-Phenanthroline, Microarray, Exponential decay, Transcription

1. Introduction

Regulation of gene products occurs on multiple levels, from initiation of transcription to post-translational modifications. The post-transcriptional level, which starts once a primary transcript has been formed, consists of several steps, including mRNA modification, transport, translation, and eventual degradation. All of these steps can be subject to regulation following, e.g. stress or hormonal stimulation. In this chapter, we describe existing methods to study mRNA turnover rates on a global scale. The abundance of an mRNA species is determined by the rates of its production (transcription) and its decay. However obvious, this relation is many times ignored, and changes in steady-state levels of a transcript are often taken to imply regulation at the level of transcription initiation. The extent of regulation at the level of mRNA stability is increasingly becoming appreciated.

Attila Becskei (ed.), *Yeast Genetic Networks: Methods and Protocols*, Methods in Molecular Biology, vol. 734, DOI 10.1007/978-1-61779-086-7_1, © Springer Science+Business Media, LLC 2011

Quite precise methods for estimating the stability of individual mRNA species under physiologically relevant conditions exist, such as promoter shut-off followed by direct observation of transcript decay. By contrast, methods for global estimation of mRNA stability have limitations regarding resolution in time as well as the physiological disturbances that are imposed on the cell by the respective experimental techniques.

Two principally different approaches will be described. In the first, direct measurement of mRNA decay following arrest of transcription, RNA polymerase II is inactivated either by mutation (e.g. using the temperature-sensitive *rpb1-1* allele in *Saccharomyces cerevisiae* (1)), or by chemical inhibitors. Both techniques suffer from the physiological impact of the necessary temperature shock or the side effects of the chemical, respectively. In an important array-based study, global estimates of mRNA stability using five different RNA pol II inhibitors (1-10-phenanthroline, thiolutin, 6-azauracil, ethidium bromide, and cordycepin) or an *rpb1-1* allele were directly compared (2). It was concluded that there was good agreement between the estimates obtained by different methods, with the inhibitor 1-10-phenanthroline showing the best fit with the RNA pol II mutant. However, the study identified groups of mRNAs specifically affected by one or several inhibitors, which should consequently be excluded from the analysis. Another concern about this traditional approach is that of temporal resolution. If we want to study fundamental decay rates and to estimate the changes in mRNA stability that take place over time in the course of, e.g. a cellular stress response or hormonal stimulation, we may be interested in resolving data points separated by only one or a few minutes. However, the time required for inactivation of a temperature-sensitive allele, or for a chemical inhibitor to penetrate into the cell and fully inactivate its target may be several minutes. In addition, since the half-lives of eukaryotic mRNAs themselves on average are longer than the time course under study, it is intrinsically difficult to obtain data with high resolution in time by direct observation of mRNA decay.

In a second, complementary approach, mRNA decay rates are instead estimated indirectly, from simultaneous measurement of both mRNA amounts (RA) and transcription rates (*TR*). An estimate of *TR*s is achieved by adding labelled RNA precursors to cells permeabilized by treatment with sarcosyl and subsequent hybridisation of the labelled nascent mRNA pool to DNA arrays ("genomic run-on" (GRO); see Chapter 2). Steady-state RA levels are estimated by conventional hybridisation of in vitro labelled mRNA to arrays. Both *TR* and RA data have to be converted to real units (molecules/minute and molecules/cell, respectively) by comparison with external standards in order to determine real mRNA half-lives. A distinct advantage of this approach is that higher resolution in time is possible because the method provides instantaneous determination of *TR* and RA, and so time points in a measurement series as close as only 1 min apart are meaningful. Moreover, the indirect method obviates the drastic perturbations of cell physiology associated with blocking transcription. However, the indirect nature of the estimation introduces additional uncertainty, in particular, when the system is not at steady state (i.e. when transcription rates and/or degradation rates are changing).

In the following, we give an account of practical considerations when estimating mRNA turnover rates with either of these two complementary approaches, both concerning experimentation, data treatment, and analysis.

2. Direct Estimation of mRNA Stability Using Transcriptional Arrest

2.1. Experimental Considerations

When designing an experiment series for the determination of mRNA degradation rates, it is advantageous to include several time points if changing conditions are going to be studied. It has emerged that mRNA stability changes dynamically in the course of stress responses, where early stabilisation of mRNAs required for stress resistance is followed by later destabilisation (3–5). In order to capture these events, therefore, a time course is in order.

It is a good idea to check the in vivo efficiency of the particular RNA pol II inhibitor to be used before large-scale experimentation is commenced. This can be done by, e.g. sampling RNA at various times after the addition of inhibitor and analysing individual genes by Northern blot using probes for transcripts with known half-lives, preferably including at least one reference gene with a slow and one with a rapid decay rate. A 1-10-phenanthroline at a final concentration of 100 ng/ml works well for *S. cerevisiae* (5). This concentration works well also for *Sz. pombe* (Asp et al., in preparation) even though higher concentrations have been reported in the literature. Care should also be taken to store the inhibitor in question to prevent loss of efficacy between experiments. For instance, 1-10-phenanthroline is sensitive to oxidation, and stocks (100 mg/ml in ethanol) should be kept frozen at −20°C in sealed tubes under nitrogen gas.

A typical mRNA stability experiment consists of one sample taken before application of RNA pol II inhibition, which provides the mRNA steady-state levels to be used as a reference. In addition, several samples (usually 2–4) taken after different times after RNA pol II inactivation are included. These will result in one final estimate of the stability for every mRNA, under one set of conditions. Based on our experience, it is not meaningful to incubate yeast cells with 1-10-phenanthroline for a shorter time than 5 min, since it takes this long to achieve full RNA pol II inactivation. If a dynamic event is to be followed, then several time points representing different times after the stimulus in question are needed, each connected with samples representing a series of RNA pol II inactivation times. The total number of arrays needed for stability estimations is thus rather great.

For mRNA stability measurements, yeast cells at a density around 5×10^7/ml (10 ml of culture for *S. cerevisiae*; 20 ml for *Sz. pombe*) are divided into two fractions. To one fraction, the RNA pol II inhibitor is added and incubation is continued. From the other fraction, RNA is prepared and used for the determination of steady-state levels of mRNA species. After different times, samples are taken from the fraction with inhibitor added and RNA prepared by the same method. For convenience, cell samples can be flash frozen in liquid nitrogen and stored at −70°C and RNA prepared at a later time. For array hybridisations, the purified RNA is fluorescence labelled (with or without prior conversion to cDNA). If the two-dye approach is used, then it is convenient to pair samples on arrays representing steady-state levels from different time points of the experiment series with the time = 0 sample, to obtain the steady-state mRNA levels. To obtain stability estimates, the samples taken after different times of RNA pol II inhibition are matched on arrays with the sample taken at the same time point of the experiment but without inhibitor added.

2.2. Microarray Data Processing

All microarray experiments require some kind of normalisation procedure. For two-colour arrays, the purpose is often to remove intensity-dependent trends, and these methods are based on the prerequisite that there is no dependence between log2-ratios (*M*-values) of the two channels to the mean intensities (*A*-values), i.e. that an M/A plot has a cloud centred around zero. The most common normalisation is a loess smoother, used either globally or within print-tip groups. When applying the loess normalisation to arrays in a decay experiment, one should be aware that trends between mRNA length and decay rate will be removed, if such trends exist.

In a typical microarray decay experiment, arrays showing steady-state transcript levels are used as a standard for calculation of decay rates (i.e. from cells treated with some transcriptional inhibitor). The steady-state level arrays can be pre-processed according to standard procedure; however, the decay arrays demand special attention.

If a chemical inhibitor of RNA pol II is used, the levels of particular mRNA groups will be affected for reasons irrelevant to the decay measurement. For instance, 1-10-phenanthroline is a Zn^{2+} chelator, and many genes involved in zinc metabolism will be transcriptionally induced by this compound ((2) and our own observations). If known, such genes should be excluded from further analysis.

In each series of treatment with a transcriptional inhibitor, the arrays from different time points exhibit very different orders of magnitude for the *M*-values. Performing global scale normalisation is therefore seldom appropriate and would result in loss of information. A better approach would be to perform scale normalisation (creating, for example the same median-absolute-deviation (MAD) across arrays) within groups of arrays measuring pools within the same transcriptional inhibitor time point across strains and stress conditions. The arrays measuring steady-state levels can also be scale normalised for comparability.

2.3. Modelling mRNA Stability

The simplest model for mRNA decay is an exponential decay model. We assume that we are observing a single mRNA species, with $N(0)$ copies in the steady-state condition. The number of copies over time, $N(t)$, .under no transcription, would follow $N(t) = N(0)\, 2^{(-t/t_{1/2})}$, where $t_{1/2}$ is a the half-life of the mRNA transcript, often referred to in the literature. Ideally, in a decay experiment of a competitive fashion, the wanted quantity is $N(t)/N(0)$, and since transformations on a log2 scale often is used, we would have $\log 2\,(N(t)/N(0)) = -t/t_{1/2}$.

Unfortunately, this quantity is never observable in practice. Noise is added to the experiments, and the normalisation methods and/or hybridisation schemes cause a shift of the *M*-values of each decay time point. To extract approximate half-lives for the mRNA species, some transformation of the data is required. For the different mRNA microarray decay studies reported in the literature, several normalisation methods have been employed. In some cases, external spike-in controls have been used, for example in microarray studies using *Escherichia coli* or *Halobacterium salinarium* (6, 7). In these studies, the number of external controls was 64 and only one, respectively. Other studies have employed a more computational approach to deduce the decay rates of transcripts. In a study using the archaeon *Sulfolobus* (8), the arrays were loess normalised, followed by the assumption that around 10% of the transcripts were stable. The decay profiles were afterwards adjusted to fit this assumption. Another approach is to assume a mean half-life for the transcripts, and then adjust the decay profiles to match this half-life (2). However, whatever normalisation and decay profile adjustment scheme is employed, it comes with a price in the form of extra assumptions that need to be made on the data.

Alternatively, instead of computing half-lives (which is difficult), the possibility to rely on the strength of multi-parallel (if such are made) is present, to detect differences in half-lives between time series. Systematic errors in parallel decay series (from different stress conditions for example) will be similar, and are likely to cancel when comparing decay slopes between series.

By choosing not to transform the data, the extra assumptions are avoided, however, the global behaviour over each time series is assumed to be unchanged. The quantities which then are compared between time series (e.g. stress conditions) are stability indices, which may be positive or negative compared to a median transcript.

2.4. Statistical Analysis

To estimate the stability indices from a decay experiment, a linear model is adopted to the *M*-values at each time point, with an origin at zero. The slopes for each decay profile are estimated via least-squares, and can be done in, e.g. the open source statistical software R or with Microsoft Excel. Differences in decay indices between parallel time series can be tested using different versions of two-sample *t*-tests. A possibility is to use moderated *t*-tests (9), in which the problem with spurious small variances, due to the small number of replicates, is circumvented.

3. Indirect Determination of mRNA Stability from Transcription Rate and RNA Amount Data

In cases where experimental determination of mRNA decay rate is not feasible or convenient, there is still the possibility of an indirect estimation whenever both mRNA amount and synthesis rate are known. We shall consider two different situations. In the first instance, the cells, under more or less constant environmental conditions, are assumed to keep the unchanged mRNA levels in a dynamical steady-state (i.e. synthesis equals decay). In a second scenario, there is a cell response to an environmental shift leading to relatively fast changes in mRNA levels and steady-state conditions cannot be assumed.

3.1. Estimating mRNA Stability Under Steady-State Conditions

The mRNA concentration (m) is thought to be established as a balance between a zero-order transcription rate (TR) and a first order decay rate with kinetic constant k_{D}. Therefore, the rate of mRNA change is written as:

$$\frac{\mathrm{d}m}{\mathrm{d}t} = TR - k_{\mathrm{D}} \cdot m \tag{1}$$

Under steady-state conditions, m does not vary (i.e. $\mathrm{d}m/\mathrm{d}t = 0$). Thus,

$$TR = k_{\mathrm{D}} \cdot m$$

and

$$k_{\mathrm{D}} = \frac{TR}{m} \tag{2}$$

According to Eq. 2, k_D can be calculated as the ratio of *TR* to *m* determined at a steady state (see Notes 1 and 2). k_D is related to the mRNA half-life ($t_{1/2}$) by

$$t_{1/2} = \frac{\ln 2}{k_D} \approx \frac{0.693}{k_D} \quad (3)$$

which allows mRNA decay to be expressed as a half-life (see Note 3). This procedure has been applied for the indirect estimation of mRNA half-lives of yeast cells growing under steady-state conditions in glucose and galactose media (10).

3.2. Estimating mRNA Stability Under Non-Steady State Conditions

3.2.1. Background

In many interesting biological instances the levels of relevant mRNAs are changing with time. This is the habitual case after imposing a stress or an environmental shift to the cell culture, which results in an adaptation of the gene expression pattern to the new situation. Under these circumstances the steady-state relation between k_D, *TR*, and *m* (Eq. 2) does not hold (at least, transitorily). Moreover, shifts in mRNA levels must result from changes in transcription rate, decay rate, or both. Consequently, for a detailed description of the process, the time course of k_D, *TR*, and *m* should be monitored. It is currently possible to make a point-wise simultaneous measurement of *TR* and *m*, which may be frequently repeated (typically every few minutes) along the experiment, for a whole set of yeast genes by means of the GRO technique (see Chapter 2). Since Eq. 1 must hold at any time, it is still possible to find a relation to infer k_D from the instantaneous values of *TR* and *m* determined by GRO.

If *TR* values are sampled frequently enough, a linear variation between successive time points might be assumed. Under these circumstances, the following expression relating the experimental values of *TR* (TR_1 and TR_2) and *m* (m_1 and m_2) determined a consecutive time points (t_1 and t_2) with k_D has been demonstrated to hold (11):

$$\begin{aligned}&[(TR_2 - TR_1)/(t_2 - t_1)] - TR_2 \cdot k_D + m_2 \cdot {k_D}^2 \\ &\quad = [[(TR_2 - TR_1)/(t_2 - t_1)] - TR_1 \cdot k_D + m_1 \cdot {k_D}^2] \cdot exp \\ &\qquad - [k_D(t_2 - t_1)]\end{aligned} \quad (4)$$

Here, k_D represents an average value of the decay constant in between t_1 and t_2 (11). Equation 4 may be used to calculate k_D values for each time interval in between successive GRO sampling time points. However, Eq. 4 cannot be analytically solved for k_D and, therefore, a numerical approach should be considered. A relatively simple spreadsheet program, like the VBA "Marmor" program for Microsoft Excel (given in Appendix), can be used to perform this calculation. Indeed, this procedure has been already employed to estimate global changes in yeast mRNA stability from GRO data obtained under oxidative and hyperosmotic stress (3, 4). In the following sections, we describe how to prepare, load, and use this program.

3.2.2. Basic Features of the "Marmor" Program

This program uses two separate Microsoft Excel books named "Calk" and "Data." The actual program is written as two Visual Basic for Applications (VBA) macros inserted in "Calk." The first macro operates sequentially, gene by gene, in a three-step cycle: (1) it transfers the data of a particular gene from the "Data" book to the "Calk" book, (2) it runs the second macro, which actually performs the k_D calculation for each pair of consecutive time points for the given gene, and (3) it transfers the resulting k_D values back to the "Data" book, proceeding to the next gene. Technically, k_D is calculated by means of a bisection algorithm which approaches the solution up to a specified degree of precision.

3.2.3. Soft- and Hardware Requirements

The program was originally written for Microsoft Excel 2002 but will run in later versions (such as the current Excel from Microsoft Office 2007). Running of the program (at the yeast genomic scale) requires a personal computer with a 2-GHz (or faster) processor and at least 512 MB of RAM memory. Typically, calculation of the k_Ds (to a <0.0001/min error) for seven GRO time points on the whole yeast genome (about 6,000 genes) takes some 3 min.

3.2.4. Preparing the Excel Books

1. Open a new Microsoft Excel book (to be saved with the name "Data").
2. On sheet 1 of "Data" type in letters (see Fig. 1):
   - "Data book" on cell A1
   - "Data sheet" on cell A2
   - "Calc book" on cell D1
   - "Calc sheet" on cell D2
   - "# time points" on cell G1
   - "# of genes" on cell G2
   - "time" on cell B4
   - "Gene number" on cell A5
   - "Gene name" on cell B5
3. End saving changes in "Data."
4. Open another new Microsoft Excel book (to be saved with the name "Calk").
5. On sheet 1 of "Calk" type in letters (see Fig. 2):
   - "minimum m" on cell A1
   - "precision" on cell A2
   - " # time points =" on cell D1

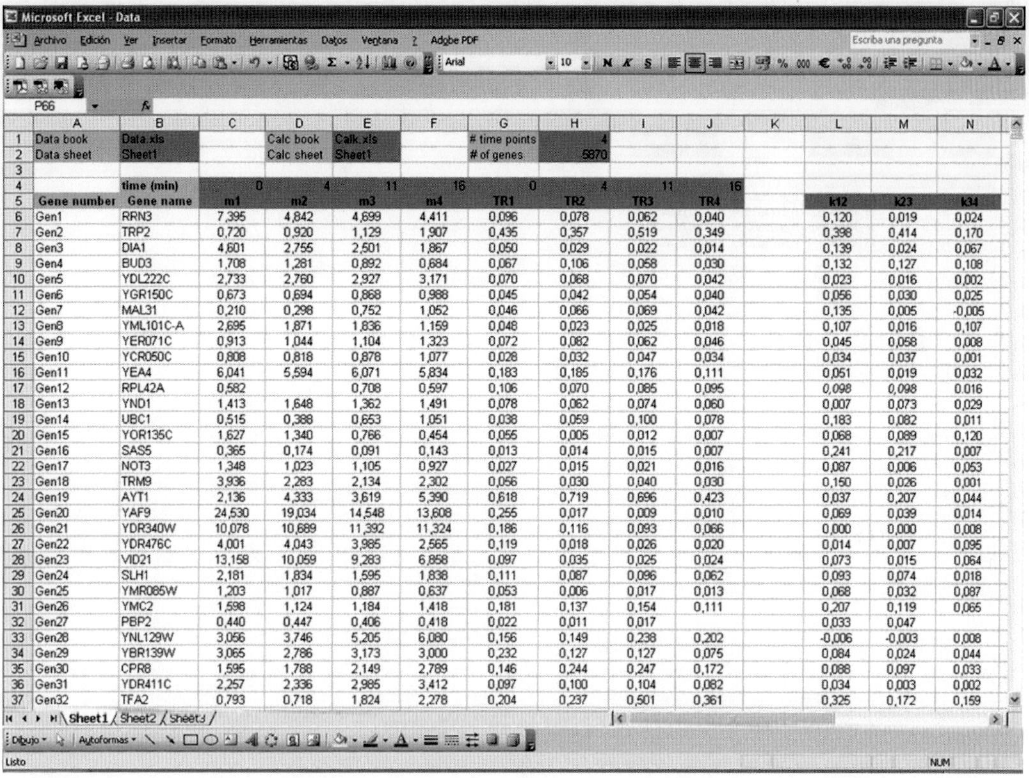

	A	B	C	D	E	F	G	H	I	J	K	L	M	N
1	Data book	Data.xls		Calc book	Calk.xls		# time points	4						
2	Data sheet	Sheet1		Calc sheet	Sheet1		# of genes	5870						
3														
4		time (min)	0	4	11	16	0	4	11	16				
5	Gene number	Gene name	m1	m2	m3	m4	TR1	TR2	TR3	TR4		k12	k23	k34
6	Gen1	RRN3	7,395	4,842	4,699	4,411	0,096	0,078	0,062	0,040		0,120	0,019	0,024
7	Gen2	TRP2	0,720	0,920	1,129	1,907	0,435	0,357	0,519	0,349		0,398	0,414	0,170
8	Gen3	DIA1	4,601	2,755	2,501	1,867	0,050	0,029	0,022	0,014		0,139	0,024	0,067
9	Gen4	BUD3	1,708	1,281	0,892	0,684	0,067	0,106	0,058	0,030		0,132	0,127	0,108
10	Gen5	YDL222C	2,733	2,760	2,927	3,171	0,070	0,068	0,070	0,042		0,023	0,016	0,002
11	Gen6	YGR150C	0,673	0,694	0,868	0,988	0,045	0,042	0,054	0,040		0,056	0,030	0,025
12	Gen7	MAL31	0,210	0,298	0,752	1,052	0,046	0,066	0,069	0,042		0,135	0,005	-0,005
13	Gen8	YML101C-A	2,695	1,871	1,836	1,159	0,048	0,023	0,025	0,018		0,107	0,016	0,107
14	Gen9	YER071C	0,913	1,044	1,104	1,323	0,072	0,082	0,062	0,046		0,045	0,058	0,008
15	Gen10	YCR050C	0,808	0,818	0,878	1,077	0,028	0,032	0,047	0,034		0,034	0,037	0,001
16	Gen11	YEA4	6,041	5,594	6,071	5,834	0,183	0,185	0,176	0,111		0,051	0,019	0,032
17	Gen12	RPL42A	0,582		0,708	0,597	0,106	0,070	0,085	0,095		0,098	0,098	0.016
18	Gen13	YND1	1,413	1,648	1,362	1,491	0,078	0,062	0,074	0,060		0,007	0,073	0,029
19	Gen14	UBC1	0,515	0,388	0,653	1,051	0,038	0,059	0,100	0,078		0,183	0,082	0,011
20	Gen15	YOR135C	1,627	1,340	0,766	0,454	0,055	0,005	0,012	0,007		0,068	0,089	0,120
21	Gen16	SAS5	0,365	0,174	0,091	0,143	0,013	0,014	0,015	0,007		0,241	0,217	0,007
22	Gen17	NOT3	1,348	1,023	1,105	0,927	0,027	0,015	0,021	0,016		0,087	0,006	0,053
23	Gen18	TRM9	3,936	2,283	2,134	2,302	0,056	0,030	0,040	0,030		0,150	0,026	0,001
24	Gen19	AYT1	2,136	4,333	3,619	5,390	0,618	0,719	0,696	0,423		0,037	0,207	0,044
25	Gen20	YAF9	24,530	19,034	14,548	13,608	0,255	0,017	0,009	0,010		0,069	0,039	0,014
26	Gen21	YDR340W	10,078	10,689	11,392	11,324	0,186	0,116	0,093	0,066		0,000	0,000	0,008
27	Gen22	YDR476C	4,001	4,043	3,985	2,565	0,119	0,018	0,026	0,020		0,014	0,007	0,095
28	Gen23	VID21	13,158	10,059	9,283	6,858	0,097	0,035	0,025	0,024		0,073	0,015	0,064
29	Gen24	SLH1	2,181	1,834	1,595	1,838	0,111	0,087	0,096	0,062		0,093	0,074	0,018
30	Gen25	YMR085W	1,203	1,017	0,887	0,637	0,053	0,006	0,017	0,013		0,068	0,032	0,087
31	Gen26	YMC2	1,598	1,124	1,184	1,418	0,181	0,137	0,154	0,111		0,207	0,119	0,065
32	Gen27	PBP2	0,440	0,447	0,406	0,418	0,022	0,011	0,017			0,033	0,047	
33	Gen28	YNL129W	3,056	3,746	5,205	6,080	0,156	0,149	0,238	0,202		-0,006	-0,003	0,008
34	Gen29	YBR139W	3,065	2,786	3,173	3,000	0,232	0,127	0,127	0,075		0,084	0,024	0,044
35	Gen30	CPR8	1,595	1,788	2,149	2,789	0,146	0,244	0,247	0,172		0,088	0,097	0,033
36	Gen31	YDR411C	2,257	2,336	2,985	3,412	0,097	0,100	0,104	0,082		0,034	0,003	0,002
37	Gen32	TFA2	0,793	0,718	1,824	2,278	0,204	0,237	0,501	0,361		0,325	0,172	0,159

Fig. 1. Screen of the "Data" book. *Light-grey* fields contain permanent instructions of the program. *Dark-grey fields* denote Excel location and numerical parameters that may vary from experiment to experiment. Therefore, they have to be changed as needed for each data set. The figure shows data and results of an experiment with four time points. Numerical data are arranged in columns C–J (from row 6 downwards). Calculated k_D results are displayed in columns I–N (also from row 6).

- "gene number =" on cell D2
- " time" on cell A4
- "*m*" on cell B4
- "*TR*" on cell C4
- "*k*" on cell D4

6. End saving changes in "Calk." If you are using the Microsoft Office 2007 version of Excel, you should choose to save in the "book containing macros" format, which will automatically affix the extension ".xlsm".

3.2.5. Recording the Macros

1. Open the Microsoft Excel book "Calk."
2. Open the Visual Basic Editor screen (i.e. Go to Tools → Macro → Visual Basic Editor or, if you are using Office 2007, click on the Developer tab and then on the Visual Basic icon). If the Developer tab is not visible in the Office 2007, you must previously activate it by clicking on the

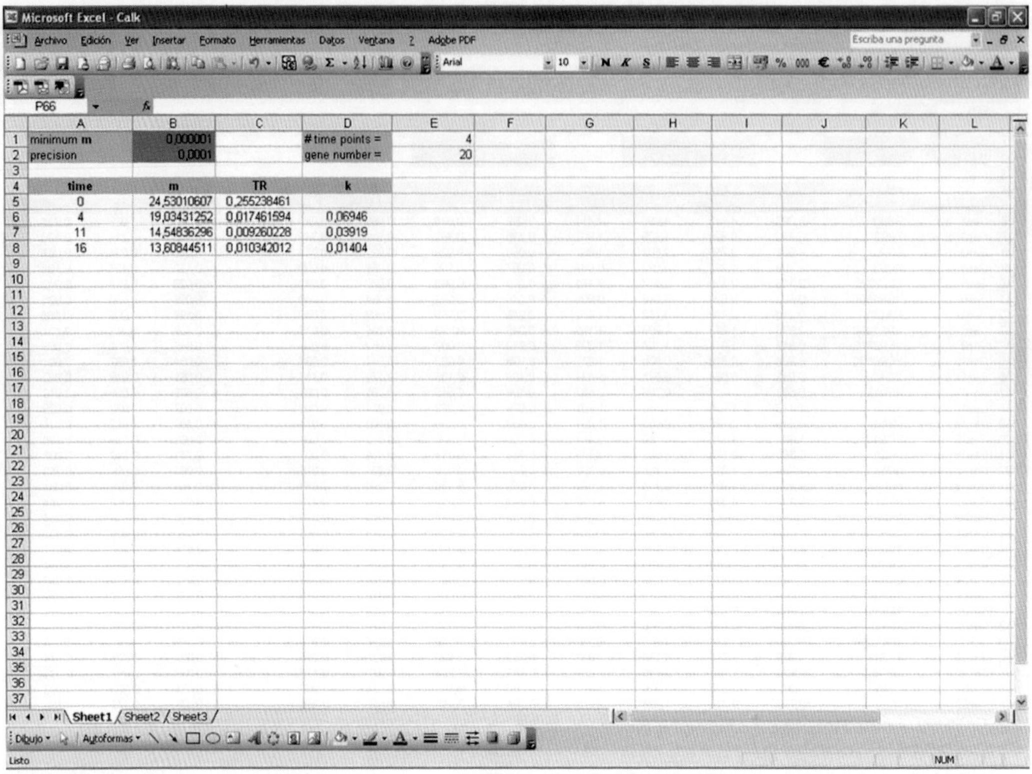

Fig. 2. Screen of the "Calk" book. *Light-grey* fields contain permanent instructions of the program. *Dark-grey fields* denote numerical parameters that may vary from experiment to experiment. Therefore, they have to be changed as needed for each data set. The figure shows *m* and *TR* data (columns B and C from row 5) for a single gene (number 20) taken at four time points (column A), and the corrresponding k_D results (column D). At running the program, each gene (in numbering order) has its data imported into this "Calk" book sheet and, after performing the calculation, its k_D results exported back to the "Data" book. Cells E1 and E2 display automatically the number of time points and the number of the gene being currently processed, respectively.

Microsoft Office Button → Excel Options, and selecting "Show Developer tab in the ribbon."

3. Once in the Visual Basic Editor screen select on the left panel "VBA project (Calk.xls)" and on the upper menu go to Insert → Module.
4. You will see that Module 1 is created in the Module folder within "VBA project (Calk.xls)" and an empty white panel will open on the right side (if not so, open Module 1 by double-clicking on the corresponding icon on the left panel). Copy carefully all lines given in the Appendix under "Macro 1" on this right side panel (see Note 4).
5. Save changes in Calk. If you are using the Microsoft Office 2007 version of Excel, you will be asked to save in the "book

containing macros" format, which will automatically affix the extension ".xlsm".

6. Repeat step 3 to create now Module 2.
7. Open Module 2 and copy carefully all lines given in the Appendix under "Macro 2" as in step 4 (see Note 4).
8. Save changes in Calk.
9. Go back to the "Calk" book in order to assign a shortcut key to Macro 1. Go to Tools → Macro → Macros (or directly click on the Macros icon if you are using Office 2007). In this window, select Macro 1 and click on Options: now select "Ctrl + t" as shortcut key.
10. You will not strictly need a shortcut key for Macro 2 since it will be automatically called from Macro 1. However, you may select a shortcut key (e.g. select "Ctrl + k" as in step 9) just in case you want to run the Macro 2 separately (see Note 5).
11. Save changes in Calk.

3.2.6. Running the Program

Some parameters have to be previously filled in (on dark-shaded cells of Figs. 1 and 2) as indicated below. Afterwards, the data will be introduced in the "Data" book before starting the program. Let us suppose that the data consist in $\boldsymbol{n}$ time points (pairs of $\boldsymbol{m}$ and *TR* values) for $\boldsymbol{N}$ genes.

1. Open the "Calk" book and type in the following cells of sheet 1 (Fig. 2):

   Cell B1: Enter a number which is lower than the sensitivity of experimental detection for $\boldsymbol{m}$ (e.g. 0.000001). This is necessary because the program does not admit 0 as a plausible value for $\boldsymbol{m}$ and will replace all 0s in the $\boldsymbol{m}$ data by this number.

   Cell B2: Enter a number expressing the maximum error allowable in the numerical calculation of k_D (e.g. 0.0001). The program will approach the solution through iterative steps until closer than this limit deviation value.

2. Without closing "Calk," open the "Data" book and type in the following cells of sheet 1 (Fig. 1):

   Cell B1: Enter the name (including file extension) of the Excel book that will contain the data. Initially, it will be "Data.xls" (or "Data.xlsx" if you are using Office 2007 version) (see Note 6).

   Cell B2: Enter the name of the sheet where the data will be pasted (e.g. "Sheet 1").

   Cell E1: Enter the name (including file extension) of the Excel book containing the program macros. Initially, it will be "Calk.xls" (or "Calk.xlsm" if you are using Office 2007 version) (see Note 6).

Cell E2: Enter the name of sheet in "Calk" where the k_D will be calculated (e.g. Sheet 1).

Cell H1: Enter the number of genes (i.e. N).

Cell H2: Enter the number of time points (i.e. n).

Row 4: Columns 3 to $(3 + n - 1)$: enter the times corresponding to the n successive time points (introduce only numbers; the units may be specified in cell B4).

Row 4: Columns $(3 + n)$ to $(3 + 2n - 1)$: repeat times (this is optional).

3. Label consecutively the cells of column A from row 6 to row $(N + 5)$ as "Gene 1" to "Gene N." You may also label the cells of column B with the names of the genes (Fig. 1).
4. In row 5, label columns 3 to $(3 + n - 1)$ as "m1" to "mn"; columns $(3 + n)$ to $(3 + 2n - 1)$ as "TR1" to "TRn" and columns $(3 + 2n + 1)$ to $(3 + 3n)$ as "k12" to "$k(n - 1)n$."
5. Paste the data corresponding to each gene (rows 6 to $N + 5$) and each time point between columns 3 and $(3 + 2n - 1)$ (see Notes 2 and 7). The value in each cell must correspond to its "coordinates" as read in column A (or B) and row 5.
6. Making sure that the "Calk" book is open, start the program from the data screen (i.e., "Data" book, sheet 1) with Ctrl + t.

During the program run you will see the "Calk" book (sheet 1) and you will be able to monitor the advance of the calculation through cell E2, which will display the number of the gene being currently processed. If you wish to abort the calculation at any time, use the Esc key. At the end of the run, k_D results will be printed in the "Data" book, on the rows assigned to the corresponding genes, between columns $(3 + 2n + 1)$ and $(3 + 3n)$ (i.e. a void column is left between the data and the results) (Fig. 1) (see Notes 8–10).

4. Notes

1. A sound use of Eq. 2 requires that the amount of the particular mRNA considered does not vary significantly under the study conditions. This should be experimentally tested. Although steady state may apply to most mRNAs under stable environmental conditions (e.g. exponentially growing yeast cells in standard YPD medium (12)), under certain circumstances

some mRNAs may vary in an oscillatory fashion (even in a constant medium) not reaching a true steady state (13, 14).

2. *TR* and *m* should be expressed in units that cancel down appropriately (e.g. if *m* is in molecules/cell and *TR* in molecules/cell/min, you will get k_D in per minute). See Chapter 2 on how to transform the GRO raw data to absolute units. Whenever *TR* and *m* are determined on a per cell basis and the cells undergo division in the particular conditions of the study, the calculated k_D includes an additive term due to the mRNA dilution into the dividing cells. Therefore,

$$k_D = k_{Dv} + k_{Dg}$$

where k_{Dv} is the dilution rate due to cell division and k_{Dg} is the proper degradation rate of the mRNA. k_{Dv} can be estimated from the cell doubling time of the culture (t_D), as

$$k_{Dv} = \frac{\ln 2}{t_D} \approx \frac{0.693}{t_D}$$

Thus, if t_D is known, the contribution of cell division can be subtracted from k_D to obtain the net rate of mRNA degradation (k_{Dg}). For short half-life mRNAs this correction may be negligible, but for stable mRNAs the dilution rate can make a significant contribution to k_D.

On the other hand, if *TR* and *m* are calculated on a culture volume basis [e.g. *m* in molecules/(ml of culture) and *TR* in molecules/(ml of culture)/min], the influence of cell division is directly offset and the calculated k_D reflects exclusively the degradation rate.

3. Although mRNA half lives can be calculated in a straightforward manner from the k_Ds using Eq. 3, it seems advisable to keep using k_D values for quantitative comparisons and gene clustering because the mathematical transformation to half-lives may amplify substantially any associated error. This is especially relevant for k_Ds close to zero (i.e. stable mRNAs).

4. Instead of typing the lengthy macro instructions you may download them from the following URL: http://scsie.uv.es/chipsdna/chipsdna-e.html#datos. Lines beginning by an apostrophe (‘) in both macros are not strictly needed for running the program and may be deleted. These lines are just comments, but they may be helpful if someone wants to learn what the program is doing at each step.

5. Macro 2 may be run by itself with this shortcut key, thereby calculating k_D for the time, *m* and *TR* values directly introduced in columns A–C of “Calk” (Fig. 2). You may want to run Macro 2 separately to process single gene data or to check for possible errors at program transcription or modification.

Otherwise, Macro 2 is always automatically called by Macro 1 to solve for k_D when needed.

6. You can save the "Data" book with a different name, or copy and paste the pattern of sheet 1 from the "Data" book (Fig. 1) into another sheet from the same book, in order to introduce a new data set. Running the program with this new data set requires only that entries in cells B1 and B2 containing the current book and sheet name, respectively. This sheet should be activated (i.e. on screen) and the "Calk" book should be open when starting the program.

   Similarly, entries in cells E1 and E2 allow to process the data with programs introduced in other books and/or sheets (different from "Calk" "Sheet 1") whenever they exist. This is especially convenient if the macro instructions are modified to fit particular requirements, thereby creating program variants which may be saved in different books. By selecting book and sheet in cells E1 and E2 you may choose an adequate program variant to manage the data.

7. Avoid blank spaces before data numbers. Microsoft Excel may use either a comma (,) or a dot (.), to separate decimal parts depending on the particular configuration of the program (default configurations may also vary between versions for different countries). Make sure that data values are pasted into the sheet in an acceptable number format.

8. In some instances, the program may return a negative value of k_D (see, for example cell N12 in Fig. 1). Obviously, negative values of a kinetic constant make no physical sense, but the message behind this result is that the program, at solving Eq. 4, has found "too much" mRNA at t_2 (i.e. too high m_2) for what was expected from the initial m_1 and the transcription rates TR_1 and TR_2, assuming a linear time course between them. If k_D is "weakly" negative (i.e. near zero) and occurs eventually in single genes, the negative value is most likely a result of experimental error (overestimation of m_2 or underestimation of the *TR*s at the particular time point). Since these errors affecting GRO values appear to be randomly distributed between genes and time points, they are usually averaged out when considering a mean value of a relatively high number of genes (such as in gene clusters). Conversely, whenever significantly negative values persist after this averaging, this strongly suggests that the postulate of a linear progression between TR_1 and TR_2 does not actually hold. Indeed, a prominent negative value of k_D for the interval between t_1 and t_2 indicates that the *TR* value peaked between TR_1 and TR_2. Frequent negative values for many genes (or even clusters) are a clear symptom of excessively separated time points. In these cases, the cultures should be sampled for GRO more often in order to follow the time

course of *TR* and *m* with enough detail as to approach the linear postulate.

9. Occasionally, some values of *TR* and/or *m* may be missing for certain genes and/or time points because of experimental failures. In that instance, you may leave blank cells. Note that the program distinguishes "blank" from "0" with a totally different meaning ("0" means "nothing" while "blank" means "unknown"). Whenever an interspersed value of *TR* or *m* is given as a blank, the program looks for the next time point for which both data are available and calculates k_D for the whole interval between the nearest consecutive fully documented time points, disregarding any intermediate incomplete pair of values. Consequently, it gives the same value of k_D for all intermediate time points encompassed by this interval. To highlight this special circumstance, these values are printed in *italics*. For example, in Fig. 1 the second time point (4 min) is missing from gene 12 (cell D17). As a result, the program calculates k_D for the interval between the first and third time point (from 0 to 11 min) giving a value of 0.098 which is printed in *italics* both under k_{12} and k_{23} (cells L17 and M17). In case that the missing data are at the beginning or the end of the time series, the program will accordingly leave blank cells corresponding to the initial or final intervals for which k_D cannot be determined. For example, the missing value of *TR* at the fourth (and last) time point of gene 27 (cell J32) produces a blank for k_{34} of the same gene (cell N32) in Fig. 1.
10. The program calculates k_D through an iterative method. Initial trials executing the "Marmor" program with experimental data (3, 4) have revealed that a k_D value within a precision of 0.000001/min is usually achieved in less than 20 iterations. However, in order to prevent the program to get stalled between two time points (*i* and *j*) by an inconsistent data set, "Marmor" will stop the calculation after performing 1,000 iterations without reaching the required precision. The message "2 many" (meaning too many iterations) will be printed in the corresponding k_{ij} cell before resuming with the next point.

Acknowledgments

Work in the authors' laboratories is supported by grants from the Spanish MEC (BIO2007-67708-C04-02) and MiCInn (BFU2009-11965, BFU2008-02114, BFU2007-67575-C03-01/BMC), and by the Swedish Research Council (2007-5460).

Appendix: The MARMOR Program

Macro 1

```
Sub Macro1()
'
' Macro1 Macro
'
' Direct acces: CTRL+t
' Assigning cell values to variables
'
  Numgen = Cells(2, 8) + 5
  Nombre = Cells(1, 2)
  Hoja = Cells(2, 2)
  Namek = Cells(1, 5)
  Sheetk = Cells(2, 5)
  tp = Cells(1, 8)
  K1 = 2 * tp + 3
  K2 = K1 + tp + 1
  For A = 6 To Numgen
      For B = K1 To K2
       Cells(A, B) = ""
       Cells(A, B).Font.Italic = False
      Next B
  Next A
'
' Preparing Calculation book
'
  Workbooks(Namek).Worksheets(Sheetk).Activate
  Workbooks(Namek).Worksheets(Sheetk).Cells(1, 5) = tp
  tpkk = tp + 6
  For A = 5 To tpkk
     Workbooks(Namek).Worksheets(Sheetk).Cells(A, 1) = ""
     Workbooks(Namek).Worksheets(Sheetk).Cells(A, 2) = ""
     Workbooks(Namek).Worksheets(Sheetk).Cells(A, 3) = ""
     Workbooks(Namek).Worksheets(Sheetk).Cells(A, 4) = ""
  Next A
  For g = 6 To Numgen
   Ngen = g - 5
   Workbooks(Namek).Worksheets(Sheetk).Cells(2, 5) = Ngen
'
' Pasting m's and TR's on Calculation book
'
      MTR = 0
      For J = 1 To tp
      t = J + 2
      M = J + 2
      TR = J + 2 + tp
      If Workbooks(Nombre).Worksheets(Hoja).Cells(g, M) <> "" _
      And Workbooks(Nombre).Worksheets(Hoja).Cells(g, TR) <> ""
Then
               MTR = MTR + 1
               MTRc = MTR + 4
               Workbooks(Namek).Worksheets(Sheetk).Cells(MTRc, 1) =
_
          Workbooks(Nombre).Worksheets(Hoja).Cells(4, t)
               Workbooks(Namek).Worksheets(Sheetk).Cells(MTRc, 2) =
_
          Workbooks(Nombre).Worksheets(Hoja).Cells(g, M)
```

```
                    Workbooks(Namek).Worksheets(Sheetk).Cells(MTRc, 3) =
_
            Workbooks(Nombre).Worksheets(Hoja).Cells(g, TR)
        End If
        Next J
If MTR > 1 Then
      Ndata = MTR + 5
      tp0 = tp + 5
'
' Deleting remaining rows from previous data sets
'
For A = Ndata To tp0
      Workbooks(Namek).Worksheets(Sheetk).Cells(A, 1) = ""
      Workbooks(Namek).Worksheets(Sheetk).Cells(A, 2) = ""
      Workbooks(Namek).Worksheets(Sheetk).Cells(A, 3) = ""
      Workbooks(Namek).Worksheets(Sheetk).Cells(A, 4) = ""
Next A
Application.Run "Macro2"
'
' Pasting kDs from Calculation book to Data book
'
Select Case MTR
       Case tp
         tp0 = tp + 4
         tpf = 2 * tp + 4
         For A = 6 To tp0
            Workbooks(Nombre).Worksheets(Hoja).Cells(g, tpf) = _
            Workbooks(Namek).Worksheets(Sheetk).Cells(A, 4)
            tpf = tpf + 1
         Next A
       Case Is < tp
         CINM = 3
         CINTR = 3 + tp
         CK = 2 * tp + 4
  Do While Workbooks(Nombre).Worksheets(Hoja).Cells(g, CINM) = "" Or
_
         Workbooks(Nombre).Worksheets(Hoja).Cells(g, CINTR) = ""
         Workbooks(Nombre).Worksheets(Hoja).Cells(g, CK) = ""
                   CINM = CINM + 1
                   CINTR = CINTR + 1
                   CK = CK + 1
  Loop
         tp2 = tp + 2
         K = 6
         For A = CINM To tp2
           B = A + tp
         If Workbooks(Nombre).Worksheets(Hoja).Cells(g, A) <> "" And
_
          Workbooks(Nombre).Worksheets(Hoja).Cells(g, B) <> "" Then
                Workbooks(Nombre).Worksheets(Hoja).Cells(g, CK) = _
                Workbooks(Namek).Worksheets(Sheetk).Cells(K, 4)
                CK = CK + 1
                K = K + 1
         Else
                Kant = K - 1
                CKant = CK - 1
                Workbooks(Nombre).Worksheets(Hoja).Cells(g, CK) = _
                Workbooks(Namek).Worksheets(Sheetk).Cells(Kant, 4)
    Workbooks(Nombre).Worksheets(Hoja).Cells(g, CK).Font.Italic =
True
```

```
Workbooks(Nombre).Worksheets(Hoja).Cells(g, CKant).Font.Italic =
True
                CK = CK + 1
        End If
        Next A
End Select
End If
Next g
    Workbooks(Nombre).Worksheets(Hoja).Activate

End Sub
```

Macro 2

```
Sub Macro2()
'
' Macro2 Macro
'
'
' Calculating the number of time points with full data
'
    tpk = Cells(1, 5) + 5
    For I = 5 To tpk
        If Cells(I, 2) = "" Then
            N = I - 1
            Exit For
        End If
    Next I
'
' Correcting the minimum messenger level
'
    L = Cells(1, 2)
    For I = 5 To N
        If Cells(I, 2) = 0 Then Cells(I, 2) = L
    Next I
'
' Evaluation of Case 1 (steady state conditions)
'
    For I = 6 To N
        Cont2 = -1
        Iant = I - 1
        m1 = Cells(Iant, 2)
        m2 = Cells(I, 2)
        TR1 = Cells(Iant, 3)
        TR2 = Cells(I, 3)
        If m1 = m2 And TR1 = TR2 Then
            Cells(I, 4) = TR1 / m1
            GoTo Proximo
            Else: GoTo Posibilidad2
        End If
'
' Case 2 is non-steady state conditions
'
Posibilidad2:
    Dt = Cells(I, 1) - Cells(Iant, 1)
    p = (TR2 - TR1) / Dt
     Select Case m2
        Case Is < (m1 + (((TR1 + TR2) / 2) * Dt))
            ki = 0
            If TR1 <> 0 Then ks = TR1 / m1 Else ks = 0.001 / m1
            GoTo Posibilidad3A
        Case Is = (m1 + (((TR1 + TR2) / 2) * Dt))
            Cells(I, 4) = 0
            GoTo Proximo
```

```
        Case Is > (m1 + (((TR1 + TR2) / 2) * Dt))
            ki = 0
            If TR1 <> 0 Then ks = -(TR1 / m1) Else ks = -0.001 / m1
            GoTo Posibilidad3B
    End Select
'
' Case 3A handles k > 0
'
' Searching for a high limit when k > 0
'
Posibilidad3A:
    Am = p - TR2 * ks + m2 * (ks ^ 2)
    Bm = (p - TR1 * ks + m1 * (ks ^ 2)) * Exp(-Dt * ks)
    Ds = Bm - Am
    Select Case Ds
        Case Is >= 0
            ks = 2 * ks
            GoTo Posibilidad3A
        Case Is < 0
            GoTo Posibilidad3AA
    End Select
'
' Iterations between low and high limits (with k > 0)
'
Posibilidad3AA:
        Cont2 = Cont2 + 1
         If Cont2 > 1000 Then
             Cells(I, 4) = "2 many"
             GoTo Proximo
         End If
        K = (ki + ks) / 2
        If (ks - ki) < Cells(2, 2) Then
            Cells(I, 4) = K
            GoTo Proximo
        End If
        A = p - TR2 * K + m2 * (K ^ 2)
        B = (p - TR1 * K + m1 * (K ^ 2)) * Exp(-(Dt * K))
        D = B - A
        Select Case D
            Case Is = 0
                Cells(I, 4) = K
                GoTo Proximo
            Case Is > 0
                ki = K
                GoTo Posibilidad3AA
            Case Is < 0
                ks = K
                GoTo Posibilidad3AA
        End Select
'
' Case 3B handles k < 0
'
' Searching for a low limit with k < 0
'
Posibilidad3B:
    Am = p - TR2 * ks + m2 * (ks ^ 2)
        Control = -Dt * ks
        If Control > 700 Then
            Incidencia = "k < " & ks
            Cells(I, 4) = Incidencia
            GoTo Proximo
        End If
    Bm = (p - TR1 * ks + m1 * (ks ^ 2)) * Exp(-(Dt * ks))
    Ds = Bm - Am
    Select Case Ds
        Case Is <= 0
            ks = 2 * ks
            GoTo Posibilidad3B
        Case Is > 0
```

```
                GoTo Posibilidad3BB
        End Select
    '
    ' Iterations between low and high limit (with k < 0)
    '
    Posibilidad3BB:
            Cont2 = Cont2 + 1
            If Cont2 > 1000 Then
                Cells(I, 4) = "2 many"
                GoTo Proximo
            End If
            K = (ki + ks) / 2
            If (ki - ks) < Cells(2, 2) Then
                Cells(I, 4) = K
                GoTo Proximo
            End If
            A = p - TR2 * K + m2 * (K ^ 2)
            B = (p - TR1 * K + m1 * (K ^ 2)) * Exp(-Dt * K)
            D = B - A
            Select Case D
                Case Is = 0
                    Cells(I, 4) = K
                    GoTo Proximo
                Case Is < 0
                    ki = K
                    GoTo Posibilidad3BB
                Case Is > 0
                    ks = K
                    GoTo Posibilidad3BB
            End Select
    '
    ' Proceed to the next gene
    '
    Proximo:
        Next I

    End Sub
```

References

1. Nonet, M., Scafe, C., Sexton, J., and Young, R. (1987) Eucaryotic RNA polymerase conditional mutant that rapidly ceases mRNA synthesis. *Mol Cell Biol* 7, 1602–11.
2. Grigull, J., Mnaimneh, S., Pootoolal, J., Robinson, M. D., and Hughes, T. R. (2004) Genome-wide analysis of mRNA stability using transcription inhibitors and microarrays reveals posttranscriptional control of ribosome biogenesis. factors. *Mol Cell Biol* **24**, 5534–47.
3. Molina-Navarro, M. M., Castells-Roca, L., Bellí, G., García-Martínez, J., Marín-Navarro, J., Moreno, J., Pérez-Ortín, J. E., and Herrero, E. (2008) Comprehensive transcriptional analysis of the oxidative response in yeast. *J Biol Chem* **283**, 17908–18.
4. Romero-Santacreu, L., Moreno, J., Pérez-Ortín, J. E., and Alepuz, P. (2009) Specific and global regulation of mRNA stability during osmotic stress in *Saccharomyces cerevisiae*. *RNA* **15**, 1110–20.
5. Molin, C., Jauhiainen, A., Warringer, J., Nerman, O., and Sunnerhagen, P. (2009) mRNA stability changes precede changes in steady-state mRNA amounts during hyperosmotic stress. *RNA* **15**, 600–14.
6. Bernstein, J. A., Khodursky, A. B., Lin, P. H., Lin-Chao, S., and Cohen, S. N. (2002) Global analysis of mRNA decay and abundance in *Escherichia coli* at single-gene resolution using two-color fluorescent DNA microarrays. *Proc Natl Acad Sci U S A* **99**, 9697–702.
7. Hundt, S., Zaigler, A., Lange, C., Soppa, J., and Klug, G. (2007) Global analysis of mRNA decay in *Halobacterium salinarum* NRC-1 at single-gene resolution using DNA microarrays. *J Bacteriol* **189**, 6936–44.
8. Andersson, A. F., Lundgren, M., Eriksson, S., Rosenlund, M., Bernander, R., and Nilsson, P. (2006) Global analysis of mRNA stability in the archaeon *Sulfolobus*. *Genome Biol* 7, R99.
9. Smyth, G. K. (2004) Linear models and empirical Bayes methods for assessing differential expression in microarray experiments. *Stat Appl Genet Mol Biol* **3**, Article 3.

10. García-Martínez, J., Aranda, A., and Pérez-Ortín, J. E. (2004) Genomic run-on evaluates transcription rates for all yeast genes and identifies gene regulatory mechanisms. *Mol Cell* **15**, 303–13.
11. Pérez-Ortín, J. E., Alepuz, P. M., and Moreno, J. (2007) Genomics and gene transcription kinetics in yeast. *Trends Genet* **23**, 250–7.
12. Pelechano, V., and Pérez-Ortín, J. E. (2010) There is a steady state transcriptome in exponentially growing yeast cells. *Yeast* **27**, 413–22.
13. Tu, B. P., Kudlicki, A., Rowicka, M., and McKnight, S. L. (2005) Logic of the yeast metabolic cycle: temporal compartmentalization of cellular processes. *Science* **310**, 1152–8.
14. Reinke, H., and Gatfield, D. (2006) Genome-wide oscillation of transcription in yeast. *Trends Biochem Sci* **31**, 189–91.

Chapter 2

Genomic-Wide Methods to Evaluate Transcription Rates in Yeast

José García-Martínez, Vicent Pelechano, and José E. Pérez-Ortín

Abstract

Gene transcription is a dynamic process in which the desired amount of an mRNA is obtained by the equilibrium between its transcription (TR) and degradation (DR) rates. The control mechanism at the RNA polymerase level primarily causes changes in TR. Despite their importance, TRs have been rarely measured. In the yeast *Saccharomyces cerevisiae*, we have implemented two techniques to evaluate TRs: run-on and chromatin immunoprecipitation of RNA polymerase II. These techniques allow the discrimination of the relative importance of TR and DR in gene regulation for the first time in a eukaryote.

Key words: Yeast, *Saccharomyces cerevisiae*, Transcription rate, Functional genomics, ChIP-on-chip, Run-on

1. Introduction

Transcription rate (TR) is the rate at which RNAs are produced as molecules per time unit. Measurement of TRs is not as straightforward as the measurement of mRNA amounts (RA). Even at the individual level, the TR of a given gene has been rarely measured because of the difficulty of quantifying nascent RNA molecules. One possibility of evaluating TR is by measuring the RNA polymerase densities in the transcribed regions of the genes. Since each elongating enzyme has a single nascent RNA molecule, the number of RNA polymerases on a gene reflects the number of RNAs being produced, while density reflects the TR if we assume a constant RNA polymerase speed. RNA polymerase II (Pol II) density can be counted by either the run-on (1) or the chromatin immunoprecipitation (chIP) techniques using specific antibodies (Abs).

The run-on technique can be used in many kinds of eukaryotic cells prior to nuclei isolation (2, 3). However, whole cells can be used only in yeast because sarkosyl detergent permeabilizes

Attila Becskei (ed.), *Yeast Genetic Networks: Methods and Protocols*, Methods in Molecular Biology, vol. 734,
DOI 10.1007/978-1-61779-086-7_2, © Springer Science+Business Media, LLC 2011

cell membranes and allows labeled UTP utilization for RNA synthesis (1). This permits an instantaneous labeling of the physiologically real RNA transcription. We adapted the run-on technique to the genomic scale [genomic run-on (GRO)] using [α-^{33}P]rUTP labeling and nylon macroarray hybridization (Figs. 1 and 2, and ref. 4). Using GRO, the nascent TRs for all the genes of an organism have been calculated for the very first time. Since the experiment includes a parallel RA determination, the mRNA stabilities can be calculated at the genomic scale if considering steady-state conditions (4, 5) or even under non-steady-state conditions (6, 7). This utility of the GRO technique will be discussed in a companion chapter of this book (8). Similar protocols have been used in other eukaryotes, but without a real determination of TRs (2, 3). The GRO technique has also been adapted to massive parallel sequencing technologies but, again, without TR calculation (9).

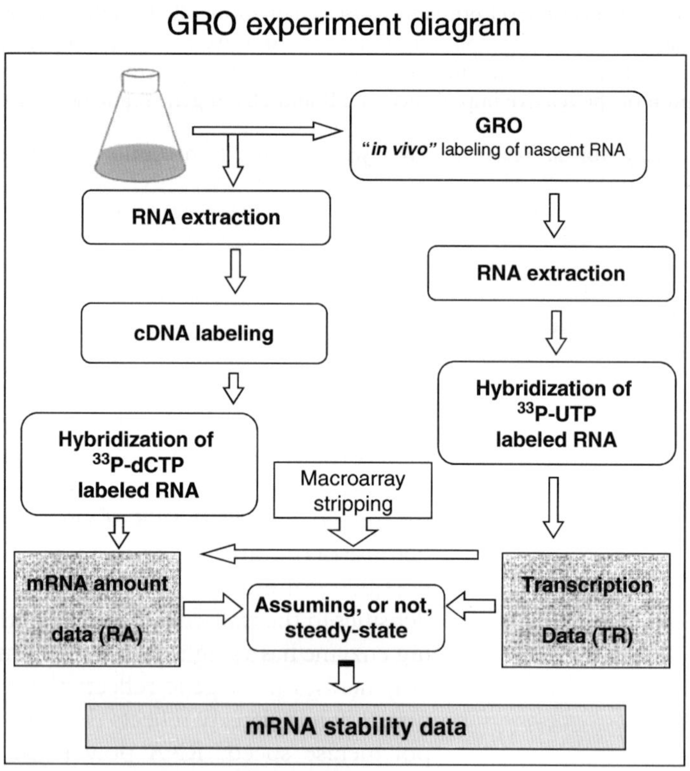

Fig. 1. Genomic run-on protocol for simultaneous TR and RA measurements. Grown cells are subjected to two independent protocols: GRO for nascent RNA labeling (*right*) and direct RNA extraction (*left*). The data from the GRO hybridized macroarrays are used to obtain transcription rates (TR) after normalization and corrections. The data from successive cDNA hybridization onto the same macroarray (after stripping it) are used to obtain mRNA amounts (RA). If one assumes steady-state conditions for mRNA amounts, it is possible to calculate mRNA stability data by dividing RA by TR. If there is no steady-state, a mathematical approximation is also possible see ref. 15.

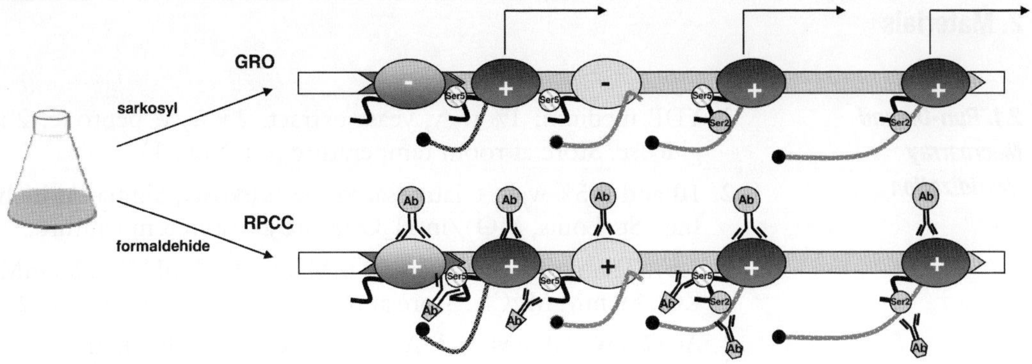

Fig. 2. Comparison of the GRO and RPCC methods. Different forms of Pol II molecules (*ovals*) are bound to a transcription unit (*horizontal rectangle*). Pol II molecules are represented with a CTD tail that can be, or not, modified in Ser5 and/or Ser2 (*dashed circles*) and with or without an mRNA molecule (long string with a *filled circle*, 5′cap). All of them are cross-linkable to the adjacent DNA sequences. If a "general" Ab is used in the RPCC method (such as 8WG16, which recognizes hypophosphorylated molecules, but also others (21) or Ab against tags is added to a Pol II subunit, different forms represented), all the cross-linked Pol II (all kinds of *ovals*) are immunoprecipitated. If specific Abs against the post-translational modifications are used, only those molecules will be precipitated. Run-on, however, only labels true elongating Pol II molecules (*dark ovals*), as well as the other nuclear RNA polymerases (I and III, not shown).

On the other hand, RNA polymerase molecules have been shown to be cross-linked to transiently bound DNA sequences (10, 11). The scaling of this method at the genomic level using DNA chips has been demonstrated for human cells (12) and yeast cells (13–15) using tiling arrays. These studies proved very powerful in terms of the description of the RNA polymerase distribution within the genome and the genes, but they were not used to calculate TRs. However, the use of DNA arrays containing whole ORF probes enables the calculation of an average distribution of Pol II density over the genes. We call this method RNA Polymerase II ChIP-on-chip (RPCC) (Fig. 2). Although the RPCC technique may be used to calculate the TRs in yeast, it is technically more complex than the GRO technique and, moreover, is affected by a higher background due to the unavoidable amplification of co-precipitated nonbound DNA, which is typical of ChIP. This results in a narrower dynamic range than that seen in the GRO technique.

Interestingly, the comparison of RPCC and GRO methods allows the detection and correction of technique-specific biases (V. Pelechano et al., in press). Moreover, the comparison between the presence of Pol II and the elongation activity measured by GRO allows the discovery of biological differences in the way in which the genes are transcribed (16). The RPCC can be done using any antibody that recognizes Pol II. However, the quality of the results depends on the antibody's affinity. We have successfully used Abs against either a tagged Pol II or the different phosphorylation forms of the carboxy terminal domain (CTD) of its largest subunit. Abs against other Pol II subunits may also be used (13, 15).

2. Materials

2.1. Run-On and Macroarray Hybridization

1. YDP medium: 1% w/v, yeast extract, 2% w/v, peptone, 2% glucose. Store at room temperature (see Note 1).
2. 10 and 0.5% w/v, L-laurylsarcosine (sarkosyl, Sigma–Aldrich Inc., St. Louis, MO)/in H_2O. Store at room temperature.
3. 2.5× Transcription buffer: 50 mM Tris–HCl, pH 7.7, 50 mM KCl, 80 mM $MgCl_2$. Store at room temperature (see Note 2).
4. ACG mix (10 mM each ATP, CTP, GTP, Roche, Mannheim, Germany). Store frozen.
5. 0.1 M DTT (Invitrogen, Carlsbad, CA). Store frozen.
6. [α-^{33}P]rUTP (~3,000 Ci/mmol, 10 μCi/μL, PerkinElmer, Waltham, MA). Store at 4°C (see Note 3).
7. Transcription mix: 120 μL of 2.5× Transcription buffer, 16 μL AGC mix, 6 μL 0.1 M DTT, and 16 μL of [α-^{33}P] rUTP. Prepare fresh (see Note 4).
8. LETS buffer: 100 mM LiCl, 10 mM EDTA, 10 mM Tris–HCl, pH 7.5, 0.2% w/v, SDS. Store at room temperature.
9. Acid phenol:chloroform:isoamilic alcohol (125:24:1), equilibrated with water, not buffered. Store at 4°C.
10. 5 M Lithium chloride. Store at room temperature.
11. Hybridization solution: 0.5 M sodium phosphate buffer, 1 mM EDTA, 7% w/v, SDS, pH 7.2, 100 μg/mL sonicated salmon sperm DNA. Do not autoclave. Store at room temperature. Add the DNA (stored frozen in 10 mg/mL solution in small aliquots) just before use (see Note 5).
12. Wash buffer I 1× SSC, 0.1% w/v, SDS and wash buffer II 0.5× SSC, 0.1% w/v, SDS. 20× SSC is 300 mM Na citrate, 3 M NaCl, pH 7.0 adjusted with HCl. Store at room temperature (see Note 5).
13. 1 M and 50 mM NaOH. Store at room temperature.
14. Neutralizing buffer: 50 mM Tris–HCl, pH 7.5, 0.1× SSC, 0.1% w/v, SDS. Store at room temperature.
15. Stripping solution: 5 mM sodium phosphate buffer, pH 7.0, 0.1% w/v, SDS. Store at room temperature.
16. Yeast nylon macroarrays. Described in (17).

2.2. cDNA Labeling

1. 5× First Strand Buffer (Invitrogen). Store frozen.
2. 0.1 M DTT (Invitrogen). Store frozen.
3. RNase OUT (Invitrogen). Store frozen.
4. DNase I (RNase free, 10/μL) (Roche). Store frozen.

5. Chloroform (Panreac, Barcelona). Store at room temperature.
6. 3 M Sodium acetate, pH 4.5. Store at room temperature.
7. Random Hexamers (3 μg/μL) (Invitrogen). Store frozen.
8. Oligo dT ($T_{15}VN$) (500 ng/μL). Store frozen.
9. dNTP's mix:16 mM each of dATP, dGTP, dTTP, and 1 mM dCTP. Divide into small aliquots and store frozen.
10. [α-^{33}P]dCTP (~3,000 Ci/mmol, 10 μCi/μL) (PerkinElmer). Store at 4°C.
11. SuperScript II Reverse Transcriptase (200 U/μL) (Invitrogen). Store frozen.
12. 0.5 M EDTA, pH 8.0 buffered with NaOH. Store at room temperature.
13. ProbeQuant G-50 or SR-H300 columns (GE, Niskayuna, NY). G-50 columns at room temperature and SR-H300 columns at 4°C, according to the suppliers.

2.3. Chromatin Immunoprecipitation

1. 37% w/v, formaldehyde solution in H_2O (Sigma–Aldrich). Store at room temperature.
2. 2.5 M Glycine. Store in small autoclaved aliquots at room temperature.
3. TBS buffer: 20 mM Tris–HCl, 140 mM NaCl, pH 7.5.
4. Glass beads, acid-washed and autoclaved (425–600 mm, Sigma–Aldrich). Store at room temperature.
5. 8GW16 antibody (Covance Inc., Berkeley, CA). Store frozen; once thawed, keep at 4°C.
6. Dynabeads® Protein G for immunoprecipitation (Invitrogen). Store at 4°C.
7. 5 mg/mL bovine serum albumin (BSA) in PBS buffer: 140 mM NaCl, 2.7 mM KCl, 10 mM Na_2HPO_4, 1.8 mM KH_2PO_4, pH 7.4. Divide into small aliquots and store frozen.
8. 10 mg/mL yeast tRNA (Applied Biosystems, Austin, TX). Store frozen.
9. Lysis buffer: 50 mM HEPES–KOH, pH 7.5, 140 mM NaCl, 1 mM EDTA, 1% v/v, Triton X-100, 0.1% w/v, sodium deoxycholate, 1 mM phenylmethylsulfonyl fluoride (PMSF), 1 mM benzamidine and one pill of complete protease inhibitor cocktail (Roche) for 50 mL of buffer. Prepare fresh (see Note 6).
10. Wash buffer:10 mM Tris–HCl, pH 8.0, 250 mM LiCl, 0.5% w/v, Nonidet P-40, 0.5% w/v, sodium deoxycholate, 1 mM EDTA, pH 8.0. Prepare fresh.
11. TE: 10 mM Tris–HCl, pH 8.0, 1 mM EDTA. Store at room temperature.

12. Elution buffer: 50 mM Tris–HCl, pH 8.0, 10 mM EDTA, 1% w/v, SDS. Store at room temperature.
13. Proteinase K (Roche) stock solution: 1 mg/mL in water. Store frozen divided into aliquots.
14. QIAquick PCR purification columns (Qiagen, Valencia, CA). Store at room temperature.
15. Neutral phenol:chloroform: Phenol:chloroform:isoamilic alcohol (25:24:1, saturated with 50 mM Tris–HCl, pH 7.5 buffer). Store at 4°C.

2.4. Ligation-Mediated PCR (LM-PCR) DNA Amplification

1. T4 DNA polymerase. Store frozen.
2. T4 DNA ligase. Store frozen.
3. Linkers oJW102 (5′-GCGGTGACCCGGGAGATCTGAATTC) and oJW103 (5′-GAATTCAGATC) (18). The linker oligonucleotides are mixed to a final concentration of 15 μM in the presence of 250 mM Tris–HCl (pH 7.9). The mixture is distributed into 50 μL aliquots and denatured for 5 min at 95°C. Then they are transferred to a 70°C heated block and allowed to cool down slowly to room temperature. Afterward, the block with the tubes is placed at 4°C and allowed to cool down again. The linkers are then stored frozen, and should always be thawed and kept on ice.
4. Glycogen 20 mg/mL (Roche). Store frozen.

2.5. Macroarray Hybridization

1. Hybridization, washing, and stripping solutions are identical to those described for GRO (see Subheading 1).

3. Methods

3.1. Genomic Run-On

3.1.1. Run-On

1. Allow cells to grow to the desired OD_{600} (we normally use 0.4–0.6).
2. Two aliquots of the culture are needed: 50 and 20 mL (corresponding to about 6 × 10^8 and 2.5 × 10^8 cells, respectively). Other volumes may be required if using different cell densities for the transcription rate (TR) and the mRNA amount (RA, see Subheading 3.1.5) measurements, respectively (see Note 4).
3. Cells are pelleted in a 50-mL falcon tube by centrifugation at 2,500 × *g*-force for 3 min.
4. Eliminate the supernatant and resuspend the cells in 5 mL of 0.5% sarkosyl at room temperature (see Note 7).

5. Pellet the cells as before. The aliquot for RA is directly frozen in dry ice (see Note 8) and the TR aliquot is resuspended in 1 mL 0.5% sarkosyl.
6. Transfer resuspended cells into a 1.5-mL tube, and pellet the cells in a microcentrifuge by centrifuging at 3,300 × *g*-force for 30 s. Discard the supernatant and centrifuge again, if necessary, to eliminate any remains of sarkosyl.
7. Resuspend the cells in 120 μL (see Note 4) of RNase-free water. Pre-warm both cells and mix separately at 30°C for 5 min. Add 158 μL of the transcription mix: the final reaction volume should be ~300 μL (see Note 9).
8. Incubate the mix at 30°C for 5 min in a Thermomixer (Eppendorf, Hamburg, Germany), or similar, with 600 rpm agitation (see Note 10).
9. Stop the run-on reaction by adding 1 mL of ice-cold RNase-free water. Recover cells by centrifuging at 3,300 × *g*-force for 1 min and discard the supernatant (which contains the non-incorporated radioactive nucleotide).
10. Start the RNA extraction by resuspending cells in 500 μL of LETS buffer.
11. Transfer the cells resuspended in LETS to a fresh tube containing 500 μL of glass beads and 500 μL of acid phenol: chloroform.
12. Break cells by vortexing tubes three times for 30 s at 5.5 intensity in a Fast-Prep 24 (MP Biomedicals, Solon, OH) (see Note 11).
13. Centrifuge tubes for 5 min at 13,400 × *g*-force to separate the phases, and transfer the upper water phase to a fresh tube. Add one volume of acid phenol:chloroform, mix well by vortexing, and centrifuge as before.
14. Transfer the new upper aqueous phase to a fresh tube and add 0.1 volume of 5 M LiCl and two volumes of cold 96% ethanol. Mix and incubate at −20°C for at least 3 h (see Note 12).
15. Recover the total RNA by centrifugation at 13,400 × *g*-force in a microcentrifuge for 15 min. Discard the supernatant and wash the pellet with 0.7 mL of 70% ethanol. Dry the pellet in a Speed-vac (Thermo Savant, Waltham, MA) for 2–3 min, and dissolve the RNA in 300 μL of RNase-free water (see Note 13).
16. Prepare a 1:100 dilution of the dissolved RNA in H_2O. Quantify the extracted RNA by measuring at A_{260}. A spectrophotometer that is capable of measuring low volumes (as 50 μL) will avoid losses of the valuable material. Use 5 μL of each one from the same dilutions to measure the

radioactivity incorporated into a scintillation counter. The radioactivity obtained ranges of between 0.8 and 3.5×10^7 dpm (see Note 14). All the labeled RNA is used in hybridization.

3.1.2. Hybridization of Run-On Samples

1. Prehybridize the yeast nylon macroarray (17) for a minimum of 30 min at 65°C with 5 mL of hybridization solution in a hybridization tube on a roller oven (see Note 15).
2. Hybridization is performed with fresh hybridization solution by adding the labeled RNA. The volume of fresh hybridization solution may be adjusted to obtain in a hybridization solution of between 1 and 7×10^6 dpm/mL. Allow to hybridize for 20–24 h at 65°C in a roller oven (see Note 15).
3. After hybridization, wash the macroarray once with washing buffer I at 65°C for 10 min, and twice with washing buffer II at 65°C for 10 min (see Note 5).
4. After washing, the membranes are saran-wrap sealed and exposed between 1 and 7 days to an Imaging Plate (Fujifilm BAS IP or similar), depending on the intensity of the signal measured with a Geiger counter (see Note 16).

3.1.3. Analysis of Run-On Hybridized Macroarrays

1. Scan the macroarrays in a suitable phosphorimager (such as a Fujifilm FLA, Fujifilm BAS, GE Storm, or GE Typhoon), with a resolution of at least 50 μm.
2. The macroarray image data are analyzed by using ArrayVision 7.0 (Imaging Research Inc., Ontario, Canada) or by other array analysis softwares. Biological replicates of the experiment should be done. We recommend at least three.
3. Before manipulating the raw data, we use genomic hybridizations to eliminate any differences due to the filter (see Note 17). Thus, each run-on hybridization dataset was divided by the corresponding genomic hybridization dataset done on the same nylon membrane. This procedure also serves to normalize the signals of the different probes, which enables comparable TR results for all the genes.
4. Values for each replicate are corrected by the number of cells used (see Note 18).
5. Hybridization values for each gene probe in each replicate are normalized and averaged by using ArrayStat 1.0 (Imaging Research Inc.), or other statistical array analysis softwares, in order to obtain a sure transcription value per cell for each gene (TR values).
6. Average values from step 5 are corrected for each gene by the percentage of U in each probe-coding strand.

7. RNA polymerase densities reflect transcription rates if we consider they have a constant elongation speed (4). The TR values obtained are, however, in arbitrary units (radioactive intensities). In order to convert them into real rates (i.e., molecules/min) it is necessary to use a reference. We have used the known TR for *HIS3* gene, 0.43 mRNAs/min (19). In this way, knowing the ratio of the radioactive intensities between *HIS3* and a given gene, the real TR can be calculated for that gene. Another possibility is to use the whole set of absolute values for mRNA concentrations (called m or RA) and mRNA half-lives $t_{1/2}$, e.g., that described in ref. 20 to determine a set of indirect TR using the Eqs. 2 and 3 described in the companion chapter (8) and plot it against the arbitrary units set to obtain a conversion factor (V. Pelechano et al., in press). This last method is more robust than the one previously described.

3.1.4. Stripping Run-On Hybridizations

Nylon macroarrays can be used several times (up to ten times in our hands). Therefore, it is necessary to strip them of the radioactive sample before they are reused. They should be stripped even if they are not to be used immediately (see Note 16).

1. Incubate the membrane inside the hybridization tube with 25 mL of 50 mM NaOH at 45°C for 45 min.
2. Wash once with the same volume of neutralizing buffer at 45°C for 15 min.
3. Transfer the filter to a plastic box and perform an additional washing step with boiling stripping solution for 5–10 min with agitation.
4. Membranes can be reused directly or stored after air-drying.

3.1.5. cDNA Labeling: RNA Extraction

A cDNA labeling experiment requires a series of independent protocols that we describe independently (from Subheadings 3.1.5–3.1.10).

Two different procedures can be followed depending on the primer used in the cDNA synthesis: random primers (RP labeling) or oligo d(T) (dT labeling). If RP labeling is used, it is necessary to perform a DNase I digestion of the RNA in order to eliminate any remains of contaminant DNA that co-extracted with the RNA. This is not necessary with dT labeling because it only primes at poly(A)-mRNAs (see Notes 19 and 20).

1. Total RNA is extracted from the 20-mL frozen culture aliquot for mRNA measurements as in an in vivo run-on protocol. The RNA extraction yield is evaluated by A_{260} (see Subheading 3.1.1, steps 2 and 10–16, but also see Note 12).

3.1.6. DNase I Digestion

1. Use a total of around 100 μg of total RNA (to prevent loss after the phenolization and precipitation steps). Dissolve it in 17 μL of H_2O.
2. Add 2 μL of 5× first strand buffer (Invitrogen), 1 μL of RNase OUT (Invitrogen) and 0.6 μL of RNase free-DNase I.
3. Incubate at 37°C for 30 min. Once again, add 0.4 μL of RNase free-DNase I, and incubate under the same conditions for 30 min more.
4. Remove the RNase free-DNase I by extracting once with acid phenol:chloroform and once with chloroform.
5. Precipitate the RNA with 0.1 volume of 3 M sodium acetate, pH 4.8, and 2.5 volumes of 96% ethanol, incubating at −20°C for a minimum of 1 h.
6. Recover the RNA by centrifugation in a microcentrifuge at 13,400 × *g*-force for 15 min. Remove the supernatant and wash with 0.7 mL of 70% ethanol, and centrifuge again at 13,400 × *g*-force for 5 min.
7. Dry the RNA for 1–2 min in a Speed-vac (see Note 13).

3.1.7. Labeling Reaction

1. Take 50 μg of total RNA (DNase I-digested or not, see Note 12) in a volume of 12.3 μL, add 1 μL of RNase OUT and, alternatively, 1.2 μL of random hexamers (3 μg/μL) or 1.2 μL of Oligo d(T) (500 ng/μL), depending on the labeling option. The final volume of that mix must be 14.5 μL.
2. Incubate the mix at 70°C for 10 min and leave at room temperature for 5–10 min. Then place it on ice.
3. To the previous sample, add 6 μL of the 5× first strand buffer, 3 μL of 0.1 M DTT, 1.5 μL of dNTP's mix, 4 μL of [α-^{33}P]-dCTP, and 1 μL of SuperScript II Reverse Transcriptase. The final reaction volume must be 30 μL (see Note 9).
4. Incubate at 42°C for 1 h and stop the reaction by adding 1 μL of 0.5 M EDTA, pH 8.0.
5. Add water to the reaction to a final volume of 50 μL, and eliminate the nonincorporated nucleotides by using ProbeQuant G-50 or SH-300R columns according to the manufacturer's instructions.
6. Estimate the radioactive incorporation by measuring 1 μL in the scintillation counter to calculate the total dpm.

3.1.8. Hybridization of cDNA Samples

1. Perform a prehybridization of the macroarray as for the run-on samples (Subheading 3.1.2, step 1).
2. Denature the labeled sample at 95°C for 5 min and transfer to an ice bath.

3. Add the denatured labeled cDNA sample to the corresponding volume of hybridization solution to obtain a radioactivity concentration ranging between 5 and 10×10^6 dpm/mL.
4. Hybridization, washing, and scanning are performed as previously described (Subheading 3.1.2, steps 2–5).

3.1.9. Stripping cDNA Hybridizations

1. Perform three washes in a dish with boiling stripping solution for 5–10 min in agitation.
2. Membranes can be reused directly or kept air-dried.

3.1.10. Analysis of the cDNA Hybridized Macroarrays

1. The hybridized macroarrays are scanned and the images are analyzed as before with the run-on samples. Biological replicates of the experiment should be done. Again, we recommend at least three.
2. As in Subheading 3.1.3, genomic hybridizations are used for eliminating any differences due to the filter; again, ArrayStat or a similar software was used to normalize and average the cDNA hybridization values (see Note 17).
3. When different conditions are analyzed, normalized, and averaged, the cDNA values are corrected by the combined factor of total RNA per cell (see Note 18) and the proportion of mRNA in the total RNA (see Note 21) in order to obtain the mRNA values per cell (RA values).
4. Average values from step 3 are corrected for each gene by the percentage of G in each probe-coding strand.
5. As for TR values, the RA values obtained are in arbitrary units (radioactive intensities). In order to convert them into real units (molecules/cell) it is necessary to use a reference. We have used the whole set of absolute values for mRNA concentrations described in ref. 20, and plot it against the arbitrary units to obtain a conversion factor, and transform the arbitrary units into real ones.

3.2. RNA Polymerase-ChIP-on-Chip

The first step of this protocol, and the most critical one, is chromatin immunoprecipitation (IP). To obtain reliable and reproducible results, it is important to ensure that the Pol II IP is successful. It is advisable to perform a control PCR to check IP efficiency using a gene that is known to be expressed as a positive control before proceeding to the array hybridization (11, 21).

The genomic RPCC data should be obtained using the IP data that have been normalized by a positive control of the total chromatin (whole cell extract, WCE). A negative control (such as an IP without a specific antibody) is highly variable between different technical replicates due to the low amount of contaminant DNA. Therefore, although it is advisable to perform negative control replicates to discard any nonspecific IP, they are not used to normalize the final IP data.

3.2.1. Chromatin Immunoprecipitation

1. For each IP reaction or for the negative control, 50 mL cells of yeast culture (OD_{600} ~ 0.5) are cross-linked by adding formaldehyde at a final concentration of 1% for 15 min at room temperature. Then the reaction is quenched by the addition of glycine at a final concentration of 125 mM (see Note 22). Cells are washed four times with 30 mL ice-cold TBS buffer, frozen in liquid N_2, and stored at −20°C until use. Samples can be kept several weeks in this stage.
2. Thaw cells on ice and resuspend them in 300 μL lysis buffer. Then, transfer cells to an ice-cold 1.5 mL screw-capped tube with 0.2 mL of glass beads and break them by vortexing at the maximum power for 12 min at 4°C in a Genie 2 vortex (Scientific Industries Inc., Bohemia, NY) or similar.
3. Add 300 μL lysis buffer to the tubes and transfer the lysed cells to a new tube. Sonicate the chromatin at 4°C (see Note 23).
4. Remove the cell debris by centrifugation at 14,000 × *g* at 4°C for 5 min. A 10 μL aliquot of this WCE is kept as a positive control.
5. The magnetic beads with the Ab should be prepared 1 day prior to their use. Beads (50 μL/sample) are washed twice with 600 μL PBS/BSA using a magnet (DynaMag™-2, Invitrogen). Then they are resuspended with 15 μL 8WG16 Ab (2 μg/μL) and 1 μL yeast tRNA as a blocking agent. For a no-Ab negative control, the volume of Ab is changed by an equal volume of PBS/BSA. Beads are kept in a tube rotator overnight at 4°C (Roto-Torque, Cole-Parmer, Vernon Hills, IL). The next day, beads are washed four times with 600 μL PBS/BSA. Afterward, they are resuspended in 30 μL of PBS/BSA and the sonicated chromatin obtained from 50 mL cells (step 4) is added. The samples with the beads are incubated in a rotator for 1.5 h at 4°C (see Note 24). Wash beads twice with 1 mL lysis buffer, twice with 1 mL lysis buffer supplemented with 360 mM NaCl, twice with 1 mL wash buffer, and once with 1 mL TE. In order to elute the samples, beads are resuspended in 50 μL of elution buffer and incubated for 10 min at 65°C under agitation (600 rpm in a Thermomixer). Then 30 μL of eluted sample is recovered and an additional amount of 30 μL of elution buffer is added. Repeat this incubation and recover an additional amount of 30 μL of the eluted sample. It is important in this step to be careful not to touch beads excessively with the tip to avoid contamination or any bead carryover. Raise the final volume of the samples to 300 μL with TE and incubate overnight at 65°C

with agitation (600 rpm in a Thermomixer) to reverse the cross-linking.

6. To digest the proteins, 142.5 μL TE and 7.5 μL proteinase K (to 20 μg/mL) are added to each sample. Incubation is kept at 37°C with agitation (600 rpm) for 1.5 h. Samples are purified using QIAquick PCR purification columns (or similar) with two binding steps and the same column for each sample. The sample is eluted in 50 μL. Up to 5 μL sample should be used in this step to check IP efficiency by performing a standard PCR analysis for an expressed control gene (11, 21). These DNA samples are only stable for a few days at −20°C. For this reason, the rest of the sample should be used as soon as possible for the DNA amplification step (next paragraph).

3.2.2. DNA Amplification by LM-PCR

1. The ends of the DNA molecules are blunted. The entire IP sample is used for this, but only 2 μL of the sample is used for the WCE (4% of the total). The reaction is allowed to proceed for 20 min at 12°C in the presence of 0.6 U of T4 DNA polymerase in its buffer supplemented with 80 μM dNTPs. Then, the sample is extracted twice with neutral phenol:chloroform and precipitated with two volumes of ethanol in the presence of 0.1 volume of sodium acetate and 12 μg of glycogen.
2. Ligate the blunt-ended sample overnight at 16°C using 0.5 U of T4 DNA ligase in a final volume of 50 μL in the presence of the annealed linkers oJW102 and oJW103 (1.5 μM) (18). Precipitate the ligated sample with ethanol and resuspend it in 25 μL of milliQ sterile water.
3. Amplify the sample in a 50-μL PCR mix using 1 μM of oJW102 primer. The PCR program is 2 min at 95°C, 30 (or less) cycles (30 s at 95°C, 30 s at 55°C, and 2 min at 72°C), with a final cycle of 4 min at 72°C. The number of PCR cycles should be tested and kept as low as possible. Precipitate the DNA with ethanol and resuspend it in 50 μL of milliQ water (see Note 25). In this state, the sample can be kept at −20°C for months.

3.2.3. Sample Labeling and Macroarray Hybridization

1. Label the sample by one additional cycle of PCR in the presence of α-[^{33}P]-dCTP. 15 μL of sample containing 1–2 μg of DNA from LM-PCR in 50 μL final volume, including: 1× Taq DNA pol buffer, 2 mM $MgCl_2$, 0.2 mM dATP, dTTP, and dGTP, 25 μM dCTP, 1 μM oJW102, 0.8 μCi α-[^{33}P]-dCTP, and 5 U Taq DNA pol. Denature the mix for 5 min at 95°C, anneal for 5 min at 50°C, and amplify for 30 min at 72°C (see Note 26). Purify the reaction product with a ProbeQuant G-50 column following the manufacturer's

instructions to remove unincorporated [^{33}P]-dCTP and the remaining oligonucleotides.

2. Yeast nylon macroarrays (17) should be used. The conditions for the hybridization, washing scanning, and stripping of the macroarrays are identical to those described for the cDNA experiments (see Subheadings 3.1.8 and 3.1.9).

3.2.4. Analysis of the Hybridized Macroarrays

The macroarray image data are analyzed by ArrayVision 7.0 (Imaging Research Inc.) or other array analysis softwares. Relative immunoprecipitation is computed as the ratio between the IP and WCE samples. Biological replicates of the experiment should be done. We recommend at least three. To compare the RPCC data under the different conditions, the median binding ratio for spots with negligible Pol II binding (e.g., probes for the rRNA) can be taken and arbitrarily set to 0 to normalize the RPCC ratio under these conditions.

4. Notes

1. Although YPD is the most common culture medium, other complete or synthetic media may also be used. We have experienced that the total labeling obtained depends on the culture growth rate. This is mainly due to the high dependence of Pol I + Pol III TR ($\geq$60% of the total TR, our unpublished observations using an inhibitor of Pol II, α-amanitin) on the cells growth rate.
2. All home-made buffers and most solutions are autoclaved at 2 kg/cm^2 for 1 h to inactivate DNases and RNases.
3. We have found that there are important differences in the quality of the [α-^{33}P]rUTP depending on the supplier. We recommend testing the efficiency of incorporation if a different supplier has been used.
4. We have checked different amounts of [α-^{33}P]rUTP. Depending on the cells' run-on efficiency (see Note 1), between 13 and 25 μL may be used, and the water volume needs to be checked to resuspend cells. We also have checked different amounts of cells in the assay affect the total incorporation of ^{33}P. We recommend to use always the same number of cells.
5. This solution allows a faster hybridization and higher signals than that originally described (4). However, it has a greater tendency to cause radioactive stains on the macroarray. Alternatively, you may use: 5× SSC, 5× Denhart's, 0.5% SDS, 100 μg/mL salmon sperm DNA, and hybridize for 40–48 h.

In this case, washing is done once for 20 min with 2× SSC 0.1% SDS, and twice for 30 min with 0.2× SSC 0.1% SDS (4).

6. In order to avoid precipitation when preparing the buffer, all the compounds, except for benzamidine, the PMSF and the protease inhibitor are dissolved first at room temperature. Then benzamidine is added. The solution is mixed and cooled on ice. Finally, all the remaining compounds are added.
7. We have checked that centrifuging cells at 4°C and/or washing them with cold sarkosyl, or cold water (as in ref. 4), induces a cold stress response that affects TR and RA of some genes. By resuspending cells in room temperature sarkosyl this stress is avoided because cells are quickly killed under nonstressing conditions. Alternatively, 10% w/v, sarkosyl stock solution can be added directly to the cells' culture medium to obtain a concentration of 0.5% w/v, then proceed with cell recovery under the same conditions as before.
8. We have observed that the slow freezing of sarkosyl-treated cells causes some RNA degradation. It is recommended that the cells for total RNA extraction are to be frozen in liquid nitrogen or on dry ice.
9. For multiple reactions, prepare a master mix with an excess of 5–10% to compensate minor pipetting inaccuracy.
10. We have checked that longer incubation times do not improve labeling. This coincides with previously described run-on protocols (1). Probably, the run-on reaction is completed in only a few minutes. Temperature, however, has a clear effect on run-on labeling. We have checked that it increases up to 50% from 30°C to 37°C. Traditionally, run-on in *Saccharomyces cerevisiae* is done at 30°C (1) because this is the standard growth temperature for this yeast. Nevertheless, because run-on is an in vitro assay in which yeast cells are dead, there is no obstacle to perform it at a different temperature. Different temperatures could, however, affect the length of the elongation in nascent mRNA which may cause differential effects on genes. Therefore, we recommend using the same temperature for all the experiments.
11. Other RNA extraction methods are also possible, such as hot-phenol or commercial RNA extraction kits. It is important to verify that the selected method is highly efficient because the amount of in vivo labeled RNA can be limiting for sensitive detection.
12. Instead of ethanol, one volume of 5 M LiCl may also be used, followed by incubation as before. This precipitation procedure is more selective for RNA and avoids contaminant DNA. This is not a major problem, however, because DNA is not

labeled. Li^+ ions could inhibit an enzymatic processing of the RNA, such as cDNA synthesis. Therefore, it is not a problem for the RNA sample obtained with the GRO protocol (Subheading 3.1.1), but may be a problem in RNA precipitation in the sample for cDNA labeling (Subheading 3.1.5). For this reason, we recommend a new RNA precipitation using 0.1 volume of 3 M sodium acetate, pH 4.8, and two volumes of 96% ethanol.

13. Over-drying the pellet results in a difficulty to dissolve RNA. For a complete dissolution keep the RNA pellet with water in a Thermomixer at 40°C for about 30 min. Lower temperatures and longer times may also be used. Check the dissolution by carefully inspecting while pipetting.

14. A precise quantification of the incorporated radioactivity is only needed in certain experiments. In such cases, we recommend taking several independent measures of RNA amount and scintillation counting of each one. In other cases, a single measurement or, even a simple Geiger estimation, of the incorporated radioactivity may be more convenient.

15. In order to compare hybridizations between samples it is necessary to use the same volume for all samples although the dpm/mL concentration was different (4). We use cylindrical plastic tubes with a slightly longer diameter circumference than the macroarray width and a slightly longer length than the macroarray length. In this way, the required volume for hybridization is kept at a minimum and the concentration of the radioactive sample is kept at a maximum.

16. We use both a thick plastic base and a saran-wrap cover and heat seal them to avoid macroarray drying which could irreversibly link the radioactivity to the nylon. The thin saran-wrap facing the exposure side of the filter reduces the shielding caused by the plastic on the ^{33}P radioactive emission.

17. Genomic DNA is labeled by random-priming using standard protocols (17). A single sample of labeled genomic DNA is added to the hybridization solution to give between 4 and 7×10^6 dpm/mL, and is divided into aliquots to hybridize all the macroarrays to be used in a given experiment. Intensity values are then corrected for each gene by the percentage of C +G in the probe. In this way, the differences in hybridization will be due only to the differences in the particular macroarrays, and will serve to correct and normalize the GRO and cDNA results.

18. The cell number in each GRO experiment should be very similar to avoid differences in labeling during the run-on.

We estimate the real number of cells used from the amount of RNA obtained after purification (Subheading 3.1.1, step 16). If the amount of RNA per cell is known (this can be obtained from a series of independent RNA purifications from the known amount of cells), the number of cells is derived from it.

19. Depending on the primer used in cDNA labeling, the radioactive label will either be more uniformly distributed along the target ORF (RP labeling) or more concentrated at the 3′-end of the ORF (dT labeling). This should be considered when analyzing the results. We originally used RP labeling (4) because the uniform labeling along the ORF was similar to the distribution of Pol II along the transcribed region. RP-labeled cDNA is, however, less efficient for hybridization than dT labeling. We currently recommend dT labeling for this reason and because of the similar bias toward 3′ as the GRO.
20. The TR calculated by GRO in our macroarrays (17) has a bias with regard to gene length which is due to the 3′-oriented movement of the RNA polymerases during run-on elongation (V. Pelechano et al., in press). This effect can also be seen while doing cDNA labeling using oligo d(T) instead of random primers (Fig. 3). It is possible to use RPCC data, which do not show this effect, to correct the GRO dataset (Fig. 3). For most instances, however, such a correction is not necessary because the GRO-determined TR values are to be compared between themselves.
21. mRNA per total RNA can be obtained by doing a dot-blot hybridization of the total RNA samples (maybe those used in the experiment or others obtained from similar cells) with a 5′-labeled oligo-d(T) probe (see refs. 4, 22 for details). This procedure assumes that all the mRNA is polyadenylated.
22. Both the cross-linking and sonication times can affect the final size of the chromatin obtained. It is advisable, therefore, to check that 15 min is the optimum time for cross-linking your samples.
23. The sonication time should be optimized for each condition and the chromatin size checked by running an agarose gel after reversing the cross-linking. We routinely use five pulses of 30 s at a high output (200 W) in a Bioruptor (Diagenode SA, Liège, Belgium), and obtain DNA fragments whose average size is 350 ± 150 bp.
24. For less efficient antibodies, this incubation time could be extended by up to 4 h.
25. A 5-μL aliquot of DNA of the LM-PCR amplified sample should be analyzed in 1.2% w/v, agarose gel to check both

size and PCR efficiency. A similar sized smear of the original chromatin fragments should be seen. Smears reaching a longer size do not represent a problem, but discrete bands in the gel are indicative of very low IP efficiency, thus the IP step should be repeated.

26. For this, we recommend the use of a set of three independent tube incubators, each at a different temperature. That at 72°C may be a Thermomixer at 600 rpm agitation.

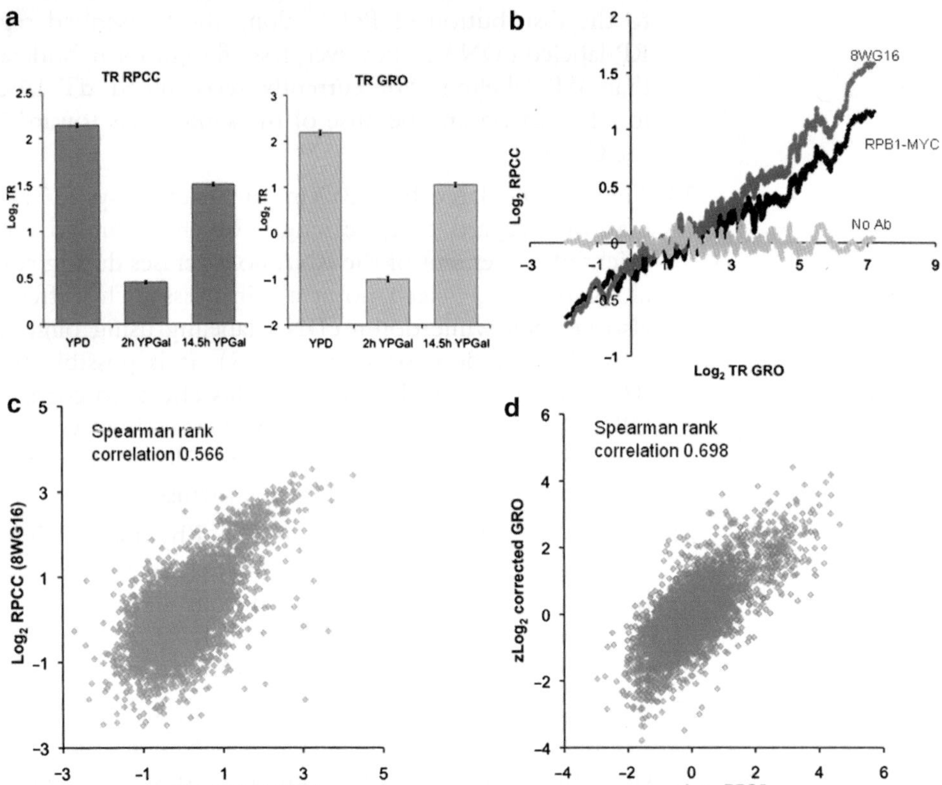

Fig. 3. Correlation between the RPCC and GRO measurements (16). There is a general correlation between the amount of Pol II present in the genes detected by RPCC and the transcriptional density measured by GRO. (**a**) Average transcription rate for all the genome using either RPCC or GRO under three different conditions (exponentially grown in YPD, nongrowing cells after 2 h of changing them to YPGal, and exponentially growing cells in YPGal 14.5 h after a change of medium; refs. 4, 16). The 95% confidence interval of the median is presented. (**b**) General correlation between the RPCC datasets using two different antibodies (Rpb1-Myc in *black*, and 8WG16 which recognizes the hypophosphorylated CTD in *dark gray*) toward the GRO TR. A negative control using a mock IP (NA, No-antibody IP in *light gray*) is also shown. All the curves represent the smoothness of the data using the average values for a sliding window of 100 genes. All the values are presented in arbitrary units. (**c**) A comparison is made between RPCC using two different antibodies: one against the total Rpb1-tagged RNA pol II (Rpb1-Myc) and another against hypophosphorylated CTD (8WG16). Spearman's rank correlation coefficient is shown. (**d**) Comparison between RPCC (8WG16) and GRO using standardized values (*Z*-scores) Spearman's rank correlation coefficient is shown.

Acknowledgments

We wish to thank Priyanka Palit, Toni Jordán, and Fany Carrasco for their help in optimizing the GRO protocol, and also Sebastián Chávez and Paula Alepuz for critically reviewing the manuscript. This work has been supported by grants BFU2007-67575-CO3-01/BMC from the Spanish Ministry of Education and Science and by grant ACOMP/2009/368 from the Generalitat Valenciana (Valencian Regional Government) awarded to JEP-O.

References

1. Hirayoshi, K. and Lis, J. T. (1999) Nuclear run-on assays: assessing transcription by measuring density of engaged RNA polymerases. *Methods Enzymol.* **304**, 351–62.
2. Fan, J., Yang, X., Wang, W., Wood, W. H. 3rd, Becker, K. G. and Gorospe, M. (2002) Global analysis of stress-regulated mRNA turnover by using cDNA arrays. *Proc. Natl. Acad. Sci. USA* **99**, 10611–10616.
3. Legen, J., Kemp, S., Krause, K., Profanter, B., Herrmann, R. G. and Maier, R. M. (2002) Comparative analysis of plastid transcription profiles of entire plastid chromosomes from tobacco attributed to wild-type and PEP-deficient transcription machineries. *Plant J.* **31**, 171–188.
4. García-Martínez, J., Aranda, A. and Pérez-Ortín, J. E. (2004) Genomic Run-On evaluates transcription rates for all yeast genes and identifies new gene regulatory mechanisms. *Mol. Cell* **15**, 303–313.
5. Pérez-Ortín, J. E., Alepuz, P. and Moreno, J. (2007) Genomics and gene transcription kinetics in yeast. *Trends Genet.* **23**, 250–257.
6. Molina-Navarro, M. M., Castells-Roca, L., Bellí, G., García-Martínez, J., Marín-Navarro, J., Moreno, J., Pérez-Ortín, J. E. and Herrero, E. (2008) Comprehensive transcriptional analysis of the oxidative response in yeast. *J. Biol. Chem.* **283**, 17908–17918.
7. Romero-Santacreu, L., Moreno, J., Pérez-Ortín, J. E. and Alepuz, P. (2009) Specific and global regulation of mRNA stability during osmotic stress in *Saccharomyces cerevisiae*. *RNA* **15**, 1110–1120.
8. Marín-Navarro, J., Jauhiainen, A., Moreno, J., Alepuz, P.M., Pérez-Ortín, P.M. and Sunnerhagen, P. (2010). Global estimation of mRNA stability in yeast. (this book, chapter 1).
9. Core, L. J., Waterfall, J. J. and Lis, J. T. (2008) Nascent RNA sequencing reveals widespread pausing and divergent initiation at human promoters. *Science* **322**, 1845–8.
10. Sandoval, J., Rodríguez, J. L., Tur, G., Serviddio, G., Pereda, J., Boukaba, A., Sastre, J., Torres, L., Franco, L. and López-Rodas, G. (2004) RNA Pol-ChIP: a novel application of chromatin immunoprecipitation to the analysis of real-time gene transcription. *Nucleic Acids Res.* **32**, e88.
11. Alepuz, P. M., de Nadal, E., Zapater, M., Ammerer, G. and Posas, F. (2003) Osmostress-induced transcription by Hot1 depends on a Hog1-mediated recruitment of the RNA Pol II. *EMBO J.* **22**, 2433–2442.
12. Brodsky, A. S., Meyer, C. A., Swinburne, I.A., Hall, G., Keenan, B. J., Liu, X. S., Fox, E. A. and Silver, P. A. (2005) Genomic mapping of RNA polymerase II reveals sites of co-transcriptional regulation in human cells. *Genome Biol.* **6**, R64.
13. Steinmetz, E. J., Warren, C. L., Kuehner, J. N., Panbehi, B., Ansari, A. Z. and Brow, D. A. (2006) Genome-wide distribution of yeast RNA polymerase II and its control by Sen1 helicase. *Mol. Cell* **24**, 735–746.
14. Jasiak, A. J., Hartmann, H., Karakasili, E., Kalocsay, M., Flatley, A., Kremmer, E., Strasser, K., Martin, D. E., Soding, J. and Cramer, P. (2008) Genome-associated RNA polymerase II includes the dissociable Rpb4/7 subcomplex. *J. Biol. Chem.* **283**, 26423–26427.
15. Venters, B. J. and Pugh, B.F. (2009) A canonical promoter organization of the transcription machinery and its regulators in the *Saccharomyces* genome. *Genome Res.* **19**, 360–371.
16. Pelechano, V., Jimeno-González, S., Rodríguez-Gil, A., García-Martínez, J., Pérez-Ortín, J. E. and Chávez, S. (2009) Regulon-specific control of transcription elongation across the yeast genome. *PLoS Genet.* **5**, e1000614.
17. Alberola, T. M., García-Martínez, J., Antúnez, O., Viladevall, L., Barceló, A., Ariño, J., Pérez-Ortín, J. E. (2004) A new set of DNA

macrochips for the yeast *Saccharomyces cerevisiae*: features and uses. *Int. Microbiol.* 7, 199–206.

18. Ren, B., Robert, F., Wyrick, J. J., Aparicio, O., Jennings, E. G., Simon, I., Zeitlinger, J., Schreiber, J., Hannett, N., Kanin, E., Volkert, T. L., Wilson, C. J., Bell, S.P. and Young, R. A. (2000) Genome-wide location and function of DNA binding proteins. *Science* **290**, 2306–2309.
19. Iyer, V. and Struhl, K. Absolute mRNA levels and transcriptional initiation rates in *Saccharomyces cerevisiae* (1996). *Proc. Natl. Acad. Sci. USA* **93**, 5208–5212.
20. Wang, Y., Liu, C.L., Storey, J.D., Tibshirani, R.J., Herschlag, D. and Brown. P.O. (2002). Precision and functional specificity in mRNA decay. *Proc. Natl. Acad. Sci. USA* **99**, 5860–5865.
21. Komarnitsky, P., Cho, E. J. and Buratowski, S. (2000) Different phosphorylated forms of RNA polymerase II and associated mRNA processing factors during transcription. *Genes Dev.* **14**, 2452–2460.
22. Radonjic, M., Andrau, J. C., Lijnzaad, P., Kemmeren, P., Kockelkorn, T. T., van Leenen, D., van Berkum, N. L. and Holstege, F.C. (2005). Genome-wide analyses reveal RNA polymerase II located upstream of genes poised for rapid response upon *S. cerevisiae* stationary phase exit. *Mol. Cell* **18**, 171–183.

Chapter 3

Construction of *cis*-Regulatory Input Functions of Yeast Promoters

Prasuna Ratna and Attila Becskei

Abstract

Promoters contain a large number of binding sites for transcriptional factors transmitting signals from a variety of cellular pathways. The promoter processes these input signals and sets the level of gene expression, the output of the gene. Here, we describe how to design genetic constructs and measure gene expression to deliver data suitable for quantitative analysis. Synthetic genetic constructs are well suited to precisely control and measure gene expression to construct *cis*-regulatory input functions. These functions can be used to predict gene expression based on signal intensities transmitted to activators and repressors in the gene regulatory region. Simple models of gene expression are presented for competitive and noncompetitive repressions. Complex phenomena, exemplified by synergistic silencing, are modeled by reaction–diffusion equations.

Key words: *Saccharomyces cervisiae*, Ssn6, Sir3, Sum1, Transcriptional interference, Estradiol

1. Introduction

The expression of eukaryotic genes requires the binding of transcriptional activators to their promoters and the subsequent recruitment of the RNA polymerase. Further tuning of expression is attained by inhibitory processes, exemplified by transcriptional repression and transcriptional interference (1, 2). The dependence of gene expression on its regulators is described by *cis*-regulatory functions (3).

The lack of precise knowledge of interactions between transcriptional factors and unknown cellular components and the pleiotropic effects of signaling events hamper the identification of *cis*-regulatory input functions. Synthetic genes that mimic the functioning of biological systems are designed to study this interplay and to give a quantitative insight into gene expression. Synthetic gene systems combine components from different natural systems (4, 5). The advantages of using synthetic gene networks

Attila Becskei (ed.), *Yeast Genetic Networks: Methods and Protocols*, Methods in Molecular Biology, vol. 734,
DOI 10.1007/978-1-61779-086-7_3,

are that they are inducible, reproducible, and reusable and provide greater control over cellular processes. *Saccharomyces cerevisiae* is an excellent eukaryotic model organism to study molecular mechanisms, because many cellular processes are conserved between yeast and higher eukaryotes.

In *S. cerevisiae*, the major modes of transcriptional inhibition include repression, silencing, and transcriptional interference. The global corepressors Ssn6 and Tup1 are known to be recruited to the promoter by promoter-specific repressors to inhibit transcription. The Ssn6–Tup1 complex generally targets the transcriptional machinery and triggers chromatin modifications. The Ssn6–Tup1 complex inhibits nearly 5% of genes on the yeast genome (6). Silencing is a specialized inhibitory mechanism relying on the spreading of silencing proteins along the chromosome. When they spread to the promoter they interfere with transcriptional initiation. The spreading of SIR proteins, including Sir2, Sir3, and Sir4, induces heterochromatin formation. Silencing occurs mainly at the telomeres, ribosomal DNA arrays, and the mating type loci. Sir2 is an enzyme that deacetylates histones, creating high-affinity binding sites for Sir3, which is thought to facilitate the spreading of silencing complexes (7). Transcriptional interference occurs when transcription initiated at one promoter interferes with the expression driven by another promoter (2, 8).

Here we summarize the principles of how to design genetic constructs, in which gene expression is controlled by synthetic activators and repressors. Subsequently, methods are presented to determine the copy number of the integrated constructs and to measure the expression of reporter genes. Fluorescent reporter genes, GFP and its variants, allow for the measurement of expression in single cells by flow cytometry or microscopy. The β-galactosidase enzymatic reporter gene is typically used for bulk expression assays, but it has the advantage over fluorescence measurements that even weakly induced gene expression can be measured. At the end, equations are presented that can be fit to gene expression data to determine a suitable *cis*-regulatory function.

1.1. Design of Gene Expression Constructs

The gene expression constructs are composed of the following sequence modules: chromosomal targeting sequence, transcriptional terminator, reporter gene, activator binding site, core promoter, and repressor binding site.

Chromosomal targeting sequence. A typically 300–800 bp long DNA sequence from the yeast genome, where the genetic construct is targeted by homologous recombination. For example, the reporter gene can be inserted to euchromatic or heterochromatic regions of the genome. This sequence should have a unique restriction site in the plasmid so that the plasmid can be linearized for transformation. It is placed either upstream or downstream of the gene expression construct comprising all the other modules.

Activator binding sequence. It is placed upstream of the core promoter. Up to a distance of around 1 kb from the transcriptional initiation site, activators are able to activate transcription in yeast (9, 10). We used the *GAL1* UAS, the upstream activating sequence of GAL1, which has four binding sites for Gal4p. The *GAL1* UAS, when fused to core promoters, activates expression with a broad dynamical range. The endogenous Gal4p can be used to activate gene expression but increasing concentrations of galactose often increase the proportion of highly expressing cells rather than the mean expression value (see Note 1).

Alternatively, the synthetic activator GEV can be used that generates a graded response to increasing estradiol concentrations (11). GEV is composed of the Gal4p DNA binding domain, the estradiol receptor, and the VP16 activator domain (11, 12). The expression of GEV was driven by the *MRP7* promoter. With increase in estradiol concentration, the VP16 domain becomes disinhibited. In our experience, estradiol concentrations higher than 500 nM elicit toxic effect, e.g., alterations in the forward and side scatter of the cells in flow cytometric measurements (see Note 2). Strains in which *GAL4* is deleted should be used to avoid interference with gene activation by GEV.

The core promoter includes the TATA box, the transcription initiation site and the promoter sequence up to the start codon. Frequently used strong core promoters are derived from the *CYC1* and *GAL1* promoters.

Repressor binding sequence. We used the *tet* operators to recruit the chimeric tetR-repressor proteins. In general, we obtained more efficient repression by fusion proteins recruited to the *tet* operators than by inserting repressor binding sequences into the promoter constructs. tetR is the bacterial repressor protein that binds to the tet operators in the absence of doxycycline. In eukaryotes, it does not act as a repressor, it merely serves as a DNA-binding domain. For repressor proteins, the binding sites should be inserted between the activator binding sites and the core promoter, or directly upstream of the activator binding sites. The former configuration results in a stronger repression efficiency (11, 13). Silencing proteins and even transcriptional repressors can repress transcription also when they bind downstream of the reporter–transcriptional terminator sequence. We could elicit high-efficiency repression with the tetR–Ssn6, tetR–Sum1, tetR–Sum1-1, and tetR–Sir3 fusion proteins (13). When their expression was driven by the *RET2* promoter, the repression was relieved at doxycycline concentrations higher than 1 μM.

Reporter gene sequence. The ORF (open reading frame and coding region) of the reporter gene (e.g., GFP and lacZ) must be followed by a transcriptional terminator sequence. The expression of fluorescent proteins can be monitored by flow cytometry or microscopy, while lacZ expression is monitored by the enzymatic assay of the lacZ gene product, β-galactosidase.

Transcriptional terminator sequence. It has to be inserted downstream of the reporter gene, but it is highly recommended to insert a terminator upstream of the promoter to prevent transcriptional interference from upstream sequences found in the plasmid. Terminators of the *ACT1*, *ADH1*, and *CYC1* genes (T_{ACT1}, T_{ADH1}, and T_{CYC1}) have been shown to efficiently terminate transcription (14). We observed that the terminator sequence downstream of a reporter gene affects the absolute level of gene expression. Considering GFP expression with T_{CYC1} as unity (1), the expression with T_{ACT1} is 1.3 and with T_{ADH1} is 0.7.

1.2. Determination of the Copy Number of the Integrated Constructs

During transformation, a single copy or multiple copies of the linearized plasmid containing the genetic constructs can integrate into the chromosome. The advantage of having multiple copies is that expression is amplified, increasing the dynamic range of detection. However, multiple copies can affect the functioning of the system when long-range interactions can arise between the individual copies. For example, the Sir and Sum1 proteins recruited to the tet operators can interact synergistically over long distances to suppress expression. In this case, strains with multiple integrations of the reporter gene are not desired. For the above reasons, it is important to assess the copy number of the integrated genes. In practice, PCR can be used to confirm the site and the copy number of the integration up to 2–3 copies. The correct integration can be confirmed by having primers so that one of the primers anneals to the chromosome and the other anneals to the integrated construct. On the other hand, when both primers anneal to the chromosomal regions flanking the integrated construct, the length of the integrated sequence, and hence the copy number, can be determined. However, even two copies of the integrated construct will typically result in a PCR product with a length of more than 10 kb. The drop of the efficiency of PCR product formation with increasing length limits this technique to the detection of low copy numbers. On the other hand, Southern blotting of genomic DNA is not affected by this limitation, although it is a more time-consuming experiment.

A simple way of estimating the copy number of the reporter gene is to measure the GFP fluorescence of a large number (>20) of colonies. This is particularly well suited for flow cytometry that provides highly accurate data in a high-throughput way. It is important to use lower concentration of DNA for transformation to increase the frequency of single-copy integrations. The fluorescence of high-copy integrands are generally whole number multiples of their single-copy counterparts, with a coefficient of variation of around 20–30%. The preferred method for the determination of the copy number of constructs containing the transcriptional activator or repressor genes is Southern blotting.

2. Materials

2.1. Yeast Transformation

1. Lithium acetate Tris buffer (LiAc–Tris buffer): 100 mM lithium acetate in 10 mM Tris–HCl pH 7.5.
2. Wash buffer: 10 mM Tris–HCl pH 7.5.
3. PEG solution: PEG-4000 (polyethylene glycol-4000) dissolved in 1g:1 ml (w:v) ratio in LiAc–Tris buffer. It should be prepared always freshly. Filter sterilize with a help of a syringe.
4. Carrier DNA: Salmon sperm DNA (Sigma) 10 mg/ml. The DNA is denatured for 10 min at 100°C and immediately put on ice. Frequent denaturation is recommended.
5. YPAD: yeast extract 1%, peptone 2%, adenine sulfate 0.002%, and dextrose 2%.
6. Selection plates to select the transformants: yeast nitrogen base (Formedium) 0.69%, glucose 2%, agar 2%, and 100 ml of 10× amino acid drop-out (Formedium) solution in 1 l media.

2.2. Genomic DNA Extraction

1. 1.2 M SCE buffer: 1.2 M Sorbitol, 0.1 M NaCl, 75 mM EDTA, pH 7.0.
2. Lysis buffer (1×): Mix 7.5 ml of water, 1.0 ml of 1 M Tris–HCl (pH 9.7), 1.0 ml of 0.5 M EDTA (pH 8.0), and 0.5 ml of 10% SDS. Store at room temperature.
3. Phosphate-buffered saline (PBS): Prepare 10× stock with 1.37 M NaCl, 27 mM KCl, 100 mM Na_2HPO_4, and 18 mM KH_2PO_4 (adjust to pH 7.4 with HCl if necessary) and autoclave before storage at room temperature. Prepare working solution by the dilution of one part with nine parts water.
4. Lyticase (100 U/μl): Lyticase 10,000 U (Sigma) dissolved in 50% glycerol and 50% PBS. Store at −20°C.
5. Ammonium acetate: prepare 7 M solution in water and adjust pH to 7.0.
6. Wash buffer: 10 mM Tris–HCl pH 7.5.
7. RNase A solution (10 U/μl): RNase A is added to Tris–HCl 10 mM pH 7.5, NaCl 15 mM. Store at −20°C.

8. YPAD: yeast extract 1%, peptone 2%, dextrose 2%, and adenine sulfate 0.002% and autoclave before storage at room temperature.

2.3. Southern Blotting and Detection

1. TBE buffer (5×): 53 g of Tris, 27.5 g of boric acid, 20 ml of 0.5 M EDTA and made up to 1 l with water (adjust pH to 8.0).
2. Agarose gel: 0.5% agarose is prepared in TBE buffer (1×) and heated until completely melted. On cooling 0.5 μg/ml ethidium bromide is added before pouring the gel.
3. Nylon hybridization membrane (e.g., Hybond N^+, Amersham International, Amersham, UK).
4. Depurination buffer: 500 ml of 0.25 M HCl. Store at room temperature.
5. Denaturation buffer: 1,000 ml solution consisting of 1.5 M NaCl and 0.5 M NaOH. Store at room temperature.
6. Neutralization solution: 1,000 ml solution consisting of 1.5 M NaCl and 0.5 M Tris–HCl, pH 7.0. Store at room temperature.
7. Transfer buffer: 1,000 ml solution consisting of 1.5 M NaCl and 0.25 M NaOH.
8. Standard saline citrate (SSC 20×): 3 M NaCl, 0.3 M trisodium citrate, pH 7.0.
9. UV-transparent plastic wrap.
10. Whattman filter paper 3MM.
11. 0.2 M EDTA is used to stop labeling reaction.
12. Maleic acid buffer: 1,000 ml solution comprising 0.1 M maleic acid, 0.15 M NaCl, pH 7.5. Store at room temperature.
13. Washing buffer: 0.3% (v/v) Tween 20 dissolved in maleic acid buffer. Store at room temperature.
14. Detection buffer: Prepare 500 ml solution with 0.1 M Tris–HCl, 0.1 M NaCl, pH 9.5.
15. Antibody solution: Anti-digoxigenin-AP 1:10,000 (75 mU/ml) in blocking solution (Amersham). This solution must be freshly prepared.
16. DIG High Prime DNA Labeling and Detection Starter Kit II (Roche Applied Science).

2.4. Inducer Stocks

1. Estradiol stock: 5 M Estradiol is prepared in 99% ethanol. Store this concentrated stock at −20°C. For each experiment, an intermediate stock is prepared by diluting the concentrated stock to 200 μM in dimethyl sulfoxide (DMSO). Then dilute this stock in the relevant medium at the required final

concentration. If a dilution series is used to prepare the growth medium, the most concentrated solution is 5 μM.

2. Doxycycline stock: 50 M doxycycline stock is made in 50% ethanol. Put an aluminum foil around the microcentrifuge tube to protect it from light. All these stocks can be stored at −20°C.

2.5. Beta Galactosidase CPRG Assay

1. Buffer 1: 2.38 g HEPES, 0.9 g NaCl, 0.065 g L-aspartate, 1 g bovine serum albumin (BSA), and 50 μl Tween 20 and make up to 100 ml with water, pH 7.3. Store at 4°C.
2. Buffer 2: 27.1 mg of CPRG (chlorophenol red-β-D-galactopyranoside) in 20 ml of Buffer 1. It should be prepared fresh, although it can be stored at 4°C up to 2 weeks. Older Buffer 2 turns red.
3. Zinc chloride solution: 100 ml of 3 mM $ZnCl_2$ is prepared in water.

3. Methods

3.1. Yeast Transformation

1. The yeast strain to be transformed is inoculated overnight in YPAD.
2. The overnight inoculum is diluted to an OD at 600 nm of 0.1. The volume of the YPAD media should be $(n + c) \times 5$ ml, where n is number of plasmids to be transformed and c is the number of different strain backgrounds for transformation control. Grow for 4 h.
3. In the meantime, digest the plasmid with a restriction enzyme to linearize it in the integrative sequence of the plasmid for chromosomal integration (see Note 3).
4. Spin the culture and wash with 10 mM Tris–HCl pH 7.5.
5. Resuspend the pellet in $(n + c) \times 5$ ml of LiAc–Tris buffer and keep the centrifuge tube on a rocker for 40 min at room temperature and shake it gently.
6. In the meantime, add 5 μl of carrier DNA to the linearized plasmid.
7. After the incubation, spin the cells at 3,000 ×g for 10 min and remove the supernatant. Resuspend the cell pellet in $(n + c)\times$ 200 μl of LiAc–Tris buffer. Add 100 μl of cell suspension to the plasmid containing solution and incubate for 5 min at room temperature.
8. Add 300 μl of PEG solution and incubate for 5 min at room temperature.
9. Heat shock (42°C) is applied for 15 min.

10. Spin the cells at 10,000 ×*g* for 1 min and remove the supernatant.
11. Resuspend the cell pellet in 1 ml YPAD and grow them for 45 min at 30°C.
12. Spin the cells for 5 min, remove the supernatant, add 100 μl of wash buffer, and plate the cells on the appropriate selection plates.

3.2. Genomic DNA Extraction

1. Prepare the overnight inoculum in YPAD or minimal medium.
2. From the overnight culture, prepare a fresh culture in 5 ml of YPAD or minimal medium. Grow for 3–5 h so that the OD 600 nm is around 0.6–0.8.
3. Spin the cells and wash them with 10 mM Tris–HCl pH 7.5.
4. Resuspend the cells in 150 μl of 1.2 M SCE and add 1 μl of lyticase 100 U/μl. Incubate the cells shaking at 37°C for 60 min (see Note 4).
5. Add 500 μl of lysis buffer to the above cell suspension and incubate at room temperature for 5 min.
6. Add 360 μl of 7 M ammonium acetate pH 7.0 and incubate at room temperature for 10 min.
7. Incubate at 65°C for 10 min and immediately put it on ice for another 10 min.
8. Add 650 μl of chloroform and vortex.
9. Spin the cells for 5 min and transfer the supernatant carefully into a 2 ml microcentrifuge tube.
10. Add 1 ml of isopropanol to the supernatant and incubate at room temperature for 15 min.
11. The extracted genomic DNA is treated with 5 μl RNAase in TE buffer by incubating it at 37°C for 30 min. This helps reducing nonspecific signals in the Southern blot.
12. To the above mixture, add 20 μl of 3 M sodium acetate and 450 μl of ethanol and incubate at −20°C for 60 min.
13. Spin for 10 min and discard the supernatant. Wash the DNA pellet twice with 70% ethanol and dry. During washes care should be taken not to lose the pellet.
14. Dissolve the genomic DNA pellet in 50 μl of 10 mM Tris–HCl pH 7.5 or water (see Note 5).

3.3. Southern Blotting and Detection

The genomic DNA has to be digested with an appropriate choice of restriction enzymes. If the DNA is cut with enzymes recognizing sites within the two genomic sequences flanking of the integrated construct, the length of the inserted DNA, and hence the copy number can be directly measured. This technique is suitable to measure DNA fragments up to 30–40 kb on a 0.4% agarose gel. Alternatively, an additional restriction enzyme can be added that recognizes a site within the integrated construct. In this way, the digestion results in two bands in the case of a single-copy integration. In the case of multiple integrations, a third band appears whose length is equal to the length of the plasmid. In this case, the ratio of the signal intensity measured for the plasmid to that of the flanking segments can be used to determine the number of integrated copies. The procedure involves blotting the membrane with digested genomic DNA, labeling the probe, hybridization of probe to the membrane, and detection of the target sequence with the labeled probe. A protocol using nonradioactive labeling is presented, using the DIG High Prime DNA Labeling and Detection Starter Kit II (Roche Applied Science).

1. Digest the genomic DNA with appropriate restriction enzyme(s). Load the digested genomic DNA, marker DNA, and positive control in a 0.4% agarose gel. Stain with ethidium bromide.
2. The agarose gel is rinsed in distilled water and then depurinated in 0.25 M HCl by slowly shaking on a platform shaker for 30 min at room temperature.
3. Discard the depurination solution and rinse the gel with distilled water. Treat the gel with denaturation solution for 20 min at room temperature.
4. Discard denaturation and rinse the gel with distilled water. Treat the gel with neutralization solution for 20 min at room temperature.
5. Place a stack of 8–10 paper towels. Over this, place 6–8 dry Whatman 3MM papers, two Whatman 3MM papers treated with 2× SSC, nylon membrane pretreated with 2× SSC, the agarose gel (avoid air bubbles while placing), saran wrap with a window slightly smaller than the gel size, two Whatman 3MM papers treated with 2× SSC and two dry Whatman 3MM papers in the order mentioned. The gel should be handled with care that it does not break into pieces. Wrap the whole set of Whatman papers, membrane and gel in saran wrap and make sure that there is no short-circuiting of 20× SSC from the reservoir. Air bubbles between membrane and gel must be avoided, as they can reduce the efficiency of transfer.
6. Form a bridge between the gel and the 20× SSC reservoir with a Whatman 3MM paper, place a glass plate over this to maintain things in place. Leave this transfer apparatus for overnight.

7. Carefully disassemble the set up and wrap the membrane in a UV transparent plastic wrap. Irradiate the membrane on the side with DNA with UV light for 1 min, 1.5 J/cm^2. Membranes could be used immediately for detection or can be stored at 2–8°C (see Note 6).
8. Denature 1 μg of probe DNA (200–1,000 bp) that is specific to the construct integrated into the chromosome, by heating in a boiling water bath for 10 min and quickly chilling on ice. The probe should be highly pure.
9. DIG label the probe with DIG-High prime for 1 h or overnight at 37°C. To stop the reaction, add 0.2 M EDTA.
10. Pre-hybridize the membrane with hybridization solution for 30 min at 37–42°C.
11. Add the DIG-labeled DNA probe and the hybridization solution to the membrane and incubate overnight.
12. Discard hybridization solution with probe and rinse the membrane with washing buffer.
13. Incubate the membrane in blocking solution for 30 min.
14. Discard blocking solution and incubate the membrane in antibody solution for 30 min.
15. Pour off the antibody solution and rinse the membrane twice with wash solution. Improper washing would give spotty background.
16. Incubate the membrane with detection buffer for 5 min and discard the buffer.
17. Spread the detection reagent over the membrane and leave for 5 min at 20–25°C (see Note 7).
18. Expose the membrane to the suitable imager.

3.4. Yeast Growth and Induction of Gene Expression

Cells starting with an OD_{600} of 0.05 are grown in minimal medium supplemented with 2% (w/v) glucose with estradiol and doxycyclin. For β-galactosidase assays, we typically grow the cultures for 3–5 h, while for GFP fluorescence measurements for 4–6 h, because GFP has a maturation time of around 40 min to 1 h.

1. Grow the overnight culture of the strain at 30°C. Measure the OD_{600}.
2. Prepare the media 5 ml each in centrifuge tubes with the appropriate concentrations of estradiol and doxycyclin (see Note 8).
3. Grow the cells for 3–6 h, starting with 0.05 OD_{600}. The samples must be continuously shaken to prevent the cells from settling down during growth. They are grown for 6 h to reach approximately the steady-state expression level.

3.5. Beta-Galactosidase Assay with CPRG

The β-galactosidase enzyme is encoded by the bacterial lacZ gene and converts β-galactosides into monosaccharides. The enzyme is extremely stable, resistant to proteolysis, and easily assayed. CPRG (chlorophenol red-β-D-galactopyranoside) is broken down by β-galactosidase into galactose and the red-colored chlorophenol red, whose absorbance is measured at 595 nm. The detection of lacZ expression by CPRG is ten times more sensitive than by ONPG (*o*-nitrophenyl-β-D-galactopyranoside). In our experience, gene expression can be measured with the combination of lacZ and CPRG to attain the same sensitivity and dynamical range as with quantitative real-time PCR.

1. Pellet 1.5 ml of cells in microcentrifuge tubes at 16,000 × *g* for 1 min. Wash the pellet with cold Buffer 1 (4°C) and spin the cells.
2. Resuspend the cells in 0.3 ml of Buffer 1. Now the concentration factor is 5 because the 1.5 ml is concentrated to 0.3 ml.
3. Remove 0.1 ml and dilute into 1 ml of water to measure the OD at 600 nm.
4. Take 0.1 ml of the remaining 0.3 ml in a screw-capped tube and freeze–thaw for 3–4 times in liquid nitrogen and 37°C water bath, to break open the cells (see Note 9).
5. Add 0.7 ml of cold buffer 2, kept at 4°C, and mix thoroughly by vortexing. Blank reactions have to prepared, as well, by adding 0.7 ml of buffer 2 to 0.1 ml of buffer 1; these will be the blank solutions during spectrophotometric measurements.
6. Transfer the tubes to a water bath kept at 37°C. Start countdown at this time point and stop when the color of the samples changes from yellow to dark red or brown when the reaction should be quenched with 0.5 ml of 3 M $ZnCl_2$. The blank reactions should also be treated in the similar manner as the samples (see Notes 10 and 11).
7. Spin the reactions at 20,000 × *g* for 1 min to pellet cell debris.
8. Transfer supernatant to fresh tubes and measure the OD at 595 nm.
9. Calculate β-galactosidase units with the following formula.

   $$\beta\text{-Galactosidase units} = 1{,}000 \times OD_{595}/(t \times V \times OD_{600})$$

   where t is the time of the reaction from adding buffer 2 to adding $ZnCl_2$ to stop the reaction and $V = 0.1 \times$ concentration factor (here it is 5, step 4).

3.6. Flow Cytometry

Flow cytometry is a technique used to measure fluorescence intensity of single cells in a high-throughput fashion. Cells are carried to the laser intercept in a fluid stream. Here the fluorescent cells scatter laser light and the scattered and fluorescent light are collected by lenses, which are steered to the detectors. These detectors produce electronic signals. With flow cytometry, fluorescence of single cells can be evaluated for large cell samples. While the dynamic range is very large 10^3–10^4, the detection of weak signals is limited by the endogeneous cellular background fluorescence. The protocol below is used for Beckmann Coulter CYTOMICS FC 500 flow cytometry system equipped with the CXP software.

1. After the growth, transfer 1 ml of cells into FACS tubes andkeep the cells on ice. Samples once kept on ice must be measured within 30–45 min.
2. Run the cleaning protocol with 0.5% bleach followed by water. Always clean the flow cytometer before and after use.
3. Before sample acquisition cytometer voltages and gains are adjusted in the Cytometer Control panel. The following parameters were used. For FS (Forward Scatter) Voltage = 790, Gain = 5; for SS (Side Scatter) Voltage = 70, Gain = 1; for FL1 Voltage = 490–550, Gain = 1.
4. With the help of a multicarousel loader, 32 tubes can be loaded and each tube is individually and automatically vortexed before sample acquisition. The flow rate must not exceed 3,000 events per seconds.
5. Mean fluorescence is obtained from the histogram so that the cell population is gated in the linear SS versus linear FS plots. 5–15% of the total cell population is selected that encompasses 20,000–30,000 cells (see Note 12).
6. A control strain without an integrated GFP is used as a fluorescence background control. This background fluorescence is subtracted from each measured fluorescence.

3.7. Fluorescence Microscopy

The fluorescence microscope offers the advantage of being able to detect multiple distinct fluorophores in the same cell and to measure the expression of multiple genes. For example, the pair of GFP-derived fluorescent proteins, YFP and CFP, have similar kinetic properties, allowing the precise comparison of expression levels. The protocol below describes the measurement of fluorescence intensity using the Zeiss Observer. Z1 inverted microscope and the AxioVision 4.6 software to obtain images.

1. After growth, cells are spun down at 4°C and concentrated to 500 μl.
2. One to two microliter of cells are pipetted on the glass slide and covered with cover slip without air bubbles. While taking measurements, care should be taken that cells are not floating and a single layer of cells is formed on the slide.

3. Switch on the microscope and the computer connected to it and set the light flow to the camera.
4. The 63× or 100× objectives with immersion oil are used for imaging.
5. The Multidimensional Acquisition tool helps to capture images using more than one fluorescent channel. Go to WORK AREA or ACQUISITION and click on Multidimensional Acquisition.
6. GFP and DIC channels are chosen. In the case of two color experiments, YFP, CFP, and DIC channels are chosen.
7. The AUTO mode is chosen in the Acquisition and click on MEASURE. This will give a well-exposed image and automatically calculates the exposure time. This step is repeated 3–4 times to get the mean exposure time.
8. Once the exposure time is known, set to FIXED mode, enter value in TIME box and click on MEASURE. Place a new glass slide with the same cell sample onto the lens. In this way, multiple exposures of the same cells can be avoided, which could lead to photobleaching and give erroneous fluorescence measurements. Click on START. This will generate images from all the channels and shows an overlay.
9. Clicking on RUN PROGRAM opens the RUN AUTOMATIC MEASUREMENT PROGRAM window. Choose OPEN IMAGES and unclick automatic. Choose the image file that is just created and click on EXECUTE.
10. *SEGMENTATION window opens, where TOLERANCE and EDGE SIZE can be adjusted. Adjust Color Saturation clicking on ADVANCED. Once adjusted, click on CONTINUE.*
11. INTERACTIVE EXECUTION window opens which allows the measurement of morphological parameters by drawing contours interactively and marking points relatively to user defined coordinate system. One can discard dead cells or cells that are not completely in frame. Note the measurement of gray background and click CONTINUE.
12. Data files of fluorescence values of individual cells are generated as _Regs.CSV extension file. Save the data files and the image files. Collect data for at least 300 cells from each induction condition.
13. Subtract the fluorescence value of individual cells with gray background measured during interactive execution.

3.8. Fitting

The measured expression values, Ex, for each inducer concentration can be used to fit model equations representing *cis*-regulatory input functions. In the simplest form, the input variables represent the inducer concentrations (estradiol and doxycycline).

Alternatively, repression efficiency versus the estimated amount of the activator bound to the promoter can be used to fit the model equation. The activator bound to the promoter, A_P, can be assumed to be linearly proportional to gene expression GA when the repressor is not bound to the promoter. In the case of GEV, GA is the ratio of the expression level at the actual concentration of estradiol to the maximal expression level. Thus, GA corresponds to normalized gene activation, and it offers the advantage of being independent of the fluctuations of the inducer activity. Furthermore, it makes possible to directly compare different promoters.

$$\mathrm{Ex} = w\frac{A_P}{K_D^A + A_P + K_D^A f(R) + \alpha A_P f(R)}$$

where K_D^A is the dissociation constant of the activator binding to the polymerase, w is a proportionality constant, while $f(R)$ is a lumped parameter incorporating the concentration and the dissociation constant of the inhibitor.

This general equation represents different forms of transcriptional inhibition depending on the value of α. Repression is competitive for $\alpha = 0$, noncompetitive for $\alpha = 1$, or noncompetitive with cooperative binding of the activator and repressor for $\alpha > 1$. It is important to note that even without direct competition in the binding of the transcription factors to the promoter, repression can be competitive due to the antagonistic effects of the activator and repressor on the recruitment of the polymerase (Fig. 1).

The repression efficiency is typically expressed as percent repression or fold inhibition. Fold inhibition conveys a better intuitive feeling of changes at strong repression, while percent repression at weak repression. For example, if gene expression is reduced from 200 to 25 U due to the binding of the repressor of the promoter, then expression is reduced by 87.5%, while the corresponding fold inhibition is 8. On the other hand, if expression is reduced from 200 to 150, expression is reduced by 25%, and the corresponding fold inhibition is 1.33. When fold inhibition is plotted on a logarithmic scale, the relative changes are distorted

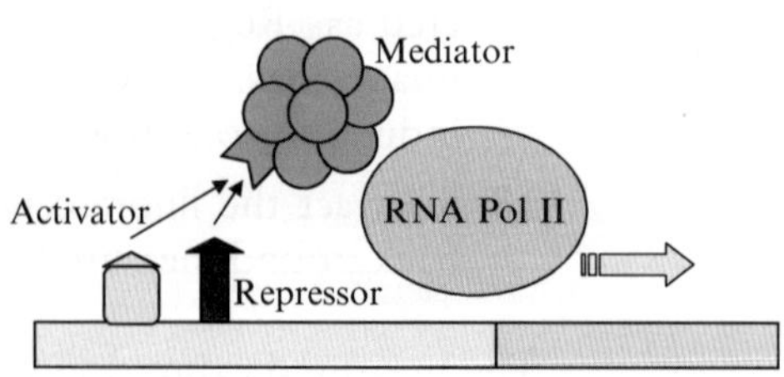

Fig. 1. A hypothetical model for competitive repression in the absence of competitive binding of repressor and activator to the promoter. The activator and the repressor bound to the promoter compete for the recruitment of the polymerase.

at values close to 1 (the value that indicates the absence of repression). This distortion can be circumvented by plotting fold inhibition-1. The fitting is performed by nonlinear regression.

Prokaryotic repression is typically modeled by noncompetitive inhibition. Eukaryotic repressors display both noncompetitive and competitive forms of repression (11, 13).

More complex forms of transcriptional inhibition, such as the synergistic interaction of multiple silencing sites cannot be explained by such simple equilibrium models. In the latter case, reaction–diffusion models can be used that account for the spreading of silencing proteins.

$$\frac{\partial c}{\partial t} = r(c) + s(x) + D_A \frac{\partial}{\partial x}\left(c\frac{\partial c}{\partial x}\right)$$

$$r(c) = L\frac{c^n}{K + c^n} - k_d c + b$$

The changes in the concentration of the silencing protein at a given point of the space–time, $c(x, t)$, are governed by source $s(x)$, reaction $r(c)$, and nonlinear diffusion terms. The nucleation term, $s(x)$, represents the recruitment of the silencing proteins. It is assumed that the autocatalytic association of the silencing proteins is superimposed onto a basal, nonspecific, association, occurring at a rate of b. The former is represented by a Hill function, where L stands for the maximal association rate in the limit of $c \rightarrow \infty$. The dissociation of the silencing proteins is a linear process and occurs at a rate of k_d.

It is assumed that the fold inhibition-1 is directly proportional to the concentration of silencing proteins in the promoter region. The effect of activators on the silencing proteins can be modeled assuming that the activator reduces the spreading of the silencing proteins (i.e., D_A is reduced) or that the activator reduces the affinity of the silencing proteins to the chromatin (i.e., K is reduced).

4. Notes

1. The galactose signal propagates through a network of cascaded feedback loops (15). The *GAL2* and *GAL3* genes, which encode the galactose permease and the galactose signal transducer proteins, respectively, enclose positive feedback loops. In principle, these positive feedback loops can generate binary response. In this case, only the proportion of OFF (weakly expressing) and ON (strongly expressing) cells change when the galactose concentration is varied, and intermediate expression levels are not observed in single cells.

Depending on the strain background, one or the other feedback loop has more pronounced effects. In some strains and growth conditions, the deletion of *GAL2* results in a graded or a mixed graded-binary response to galactose (16). It has to be noted that galactose is taken up by nonspecific hexose transporters, as well (17), so that the resulting graded response can be utilized to precisely control gene activation.

2. There are two opposing factors that determine the optimal copy number of the integrated activator construct containing the GEV. High-copy integrations result in high expression levels and squelching, toxic side effects of highly expressed activators that reduce cell growth. On the other hand, high expression levels require lower estradiol concentrations, reducing the side effects of estradiol. In our experience, intermediated copy numbers of the MRP7 promoter – GEV constructs (2–4 copies) are optimal: they do not display squelching, while expression reaches its maximal level at estradiol concentrations as low as 40–200 nM.
3. A further treatment of the digested plasmid with alkaline phosphates may increase the number of transformants.
4. Incubation with lyticase for longer time results in better lysis. Yeast cells grown for longer time, close to saturation of the culture, give problems with lysis as they have more resistant cell walls. If a larger amount of genomic DNA is required, increase the volume of the culture but not the cell density.
5. Genomic DNA pellet can be dissolved in a larger volume of water and later concentrated. Dissolving the DNA pellet for 1 h at 42°C and then at 4°C overnight would result in complete dissolution.
6. Care should be taken to avoid complete drying out of the membrane as it hinders the binding of antibody. Membranes that would not be used immediately should be drained off any liquid, wrapped in a saran wrap and stored at 4°C.
7. Incubation of membrane with detection reagent for longer period gives false positives. The mentioned time of incubation with detection reagent must be followed strictly.
8. When preparing a series of concentrations of the inducers, it is better to have a single working stock solution to avoid dilution errors.
9. The repeated freezing and thawing can lead to increased pressure within the tube so that the tube may explode. Using screw-capped tubes prevents these explosions.
10. If an intense product color appears quickly, use a diluted sample of disrupted cells. Do not forget to include this dilution factor in the formula.

11. If there is no color development repeat freeze thaw. Lack of lacZ expression due to mutations in the lacZ open reading frame, misintegration of the construct or low transformation efficiency might also be the reasons for no color development. It is always good to have a positive control to ensure that there is no problem with the reagents.
12. We typically gate small cells that have low forward and side-scatter values to exclude large mitotic cells and cell doublets. In this way, the histogram reflects fluorescence distribution of single cells.

References

1. Struhl, K. (1999) Fundamentally different logic of gene regulation in eukaryotes and prokaryotes, *Cell* **98**, 1–4.
2. Sneppen, K., Dodd, I. B., Shearwin, K. E., Palmer, A. C., Schubert, R. A., Callen, B. P., and Egan, J. B. (2005) A mathematical model for transcriptional interference by RNA polymerase traffic in Escherichia coli, *J Mol Biol* **346**, 399–409.
3. Setty, Y., Mayo, A. E., Surette, M. G., and Alon, U. (2003) Detailed map of a cis-regulatory input function, *Proc Natl Acad Sci U S A* **100**, 7702–7707.
4. May, T., Eccleston, L., Herrmann, S., Hauser, H., Goncalves, J., and Wirth, D. (2008) Bimodal and hysteretic expression in mammalian cells from a synthetic gene circuit, *PLoS One* **3**, e2372.
5. Haynes, K. A., and Silver, P. A. (2009) Eukaryotic systems broaden the scope of synthetic biology, *J Cell Biol* **187**, 589–596.
6. Smith, R. L., and Johnson, A. D. (2000) Turning genes off by Ssn6-Tup1: a conserved system of transcriptional repression in eukaryotes, *Trends Biochem Sci* **25**, 325–330.
7. Rusche, L. N., Kirchmaier, A. L., and Rine, J. (2003) The establishment, inheritance, and function of silenced chromatin in Saccharomyces cerevisiae, *Annu Rev Biochem* **72**, 481–516.
8. Buetti-Dinh, A., Ungricht, R., Kelemen, J. Z., Shetty, C., Ratna, P., and Becskei, A. (2009) Control and signal processing by transcriptional interference, *Mol Syst Biol* **5**, 300.
9. Dobi, K. C., and Winston, F. (2007) Analysis of transcriptional activation at a distance in Saccharomyces cerevisiae, *Mol Cell Biol* **27**, 5575–5586.
10. Petrascheck, M., Escher, D., Mahmoudi, T., Verrijzer, C. P., Schaffner, W., and Barberis, A. (2005) DNA looping induced by a transcriptional enhancer in vivo, *Nucleic Acids Res* **33**, 3743–3750.
11. Ratna, P., Scherrer, S., Fleischli, C., and Becskei, A. (2009) Synergy of repression and silencing gradients along the chromosome, *J Mol Biol* **387**, 826–839.
12. Gao, C. Y., and Pinkham, J. L. (2000) Tightly regulated, beta-estradiol dose-dependent expression system for yeast, *Biotechniques* **29**, 1226–1231.
13. Kelemen, J. Z., Ratna, P., Scherrer, S., and Becskei, A. (2010) Spatial epigenetic control of mono- and bistable gene expression, *PLoS Biol* **8**, e1000332.
14. Guo, Z., and Sherman, F. (1995) 3'-end-forming signals of yeast mRNA, *Mol Cell Biol* **15**, 5983–5990.
15. Acar, M., Becskei, A., and van Oudenaarden, A. (2005) Enhancement of cellular memory by reducing stochastic transitions, *Nature* **435**, 228–232.
16. Hawkins, K. M., and Smolke, C. D. (2006) The regulatory roles of the galactose permease and kinase in the induction response of the GAL network in Saccharomyces cerevisiae, *J Biol Chem* **281**, 13485–13492.
17. Wieczorke, R., Krampe, S., Weierstall, T., Freidel, K., Hollenberg, C. P., and Boles, E. (1999) Concurrent knock-out of at least 20 transporter genes is required to block uptake of hexoses in Saccharomyces cerevisiae, *FEBS Lett* **464**, 123–128.

Chapter 4

Luminescence as a Continuous Real-Time Reporter of Promoter Activity in Yeast Undergoing Respiratory Oscillations or Cell Division Rhythms

J. Brian Robertson and Carl Hirschie Johnson

Abstract

This chapter describes a method for generating yeast respiratory oscillations in continuous culture and monitoring rhythmic promoter activity of the culture by automated real-time recording of luminescence. These techniques chiefly require the use of a strain of *Saccharomyces cerevisiae* that has been genetically modified to express firefly luciferase under the control of a promoter of interest and a continuous culture bioreactor that incorporates a photomultiplier apparatus for detecting light emission. Additionally, this chapter describes a method for observing rhythmic (cell cycle-related) promoter activity in small batch cultures of yeast through luminescence monitoring.

Key words: *Saccharomyces cerevisiae*, Luciferase, Bioluminescence, Continuous culture, Bioreactor, Yeast respiratory oscillation

1. Introduction

The bioluminescent reaction catalyzed by the enzyme firefly luciferase has become a useful genetic reporting system for monitoring rhythmic promoter activity in circadian studies of mammals (1, 2), insects (3), plants (4), and filamentous fungi (5). In addition, our work introduced the use of luciferase as a genetic reporter of respiration and cell cycle rhythms in the budding yeast *Saccharomyces cerevisiae* (6). Luciferase from fireflies emits light when the 62-kDa protein catalyzes the oxidation of a bioluminescent substrate "luciferin" (in the presence of O_2, ATP, and Mg^{+2}) into oxyluciferin (and ADP and CO_2) (7). The emitted light is, therefore, an immediate and measurable indication of the enzyme's activity. The relatively short half-life of luciferase (~30 min for destabilized luciferase (6)) allows its expression to dynamically reflect transcription on a faster time scale than longer-lived reporters such

Attila Becskei (ed.), *Yeast Genetic Networks: Methods and Protocols*, Methods in Molecular Biology, vol. 734, DOI 10.1007/978-1-61779-086-7_4,

as βGal, CAT, or GFP (7, 8). Additionally, luciferase does not need excitation from an external light source as does GFP and other fluorescent reporters. Therefore, issues of photobleaching, autofluorescence, phototoxicity, and biological responses to an intense excitatory illumination can be avoided with a luciferase reporter.

Much of our research on the adaptation of the luciferase reporter system to yeast was the study of the yeast respiration oscillation (YRO) in bioreactors (6). A bioreactor (sometimes called a fermentor or chemostat) is a continuous culture apparatus that maintains a microorganism culture in a near steady-state level of exponential growth in which one component of the medium is the growth-limiting factor (9, 10). Within the reactor's vessel, a specified volume of aerated medium sustains yeast growth in much the same way batch growth occurs, but unlike batch growth, the growth environment (including pH, temperature, nutrition, biomass, and metabolic byproducts) is kept relatively constant by continually monitoring and adjusting variables such as pH and temperature in addition to constantly introducing fresh media at a steady rate while removing culture (i.e., media, cells, and byproducts) from the vessel at the same rate. As a result of these conditions, an inoculated culture grows to a concentration that is limited by the depletion of some component(s) of the medium and from that time onward, the growth rate is determined by the rate at which fresh medium is supplied (10).

Under a range of specific conditions of glucose-limited, aerobic continuous culture in bioreactors, spontaneous perturbations of the steady-state can lead to oscillations in various metabolite concentrations in the medium that are sometimes accompanied by (and possibly reinforced by) subpopulations of synchronously dividing cells (11). The most easily observed oscillating metabolite in the continuous culture is the dissolved oxygen (DO) concentration, which reflects the culture alternating between respiro-fermentative metabolism and respiration (Fig. 1) (12–14). We call this phenomenon the yeast respiratory oscillation (YRO), but it also goes by other names including the yeast metabolic cycle (YMC) (13) and the energy metabolism oscillation (EMO) (14). Rhythmic transcription of many genes has been shown to occur at different phases of the YRO using microarrays (13, 15) and northern blots (14), but these methods are time consuming and ultimately limited by the frequency and number of samples taken from the culture. Bioluminescence monitoring of a promoter-coupled luciferase reporter in yeast is a good way to monitor rhythmic transcription continuously over the course of several days as well as to observe transcriptional responses to various treatments in real time.

In addition to having luciferase expressed in a desired strain of yeast, oxygen and luciferin are two requirements for the light

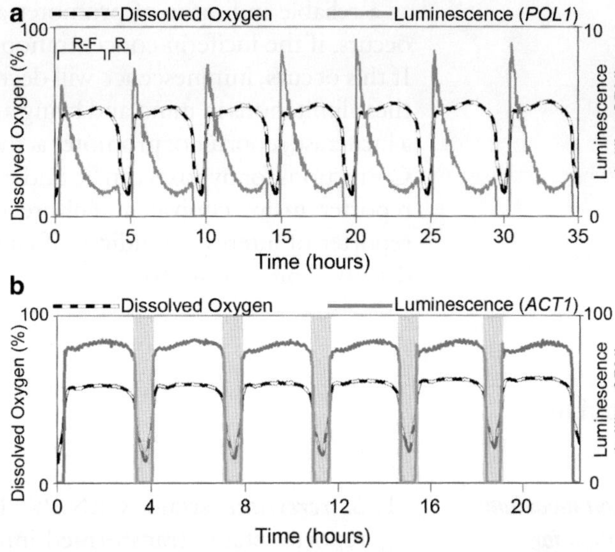

Fig. 1. Examples of the yeast respiratory oscillation monitored by dissolved oxygen (DO) and simultaneously plotted with luminescence measurements from two different luciferase reporters. (**a**) Seven cycles of a yeast respiratory oscillation are shown for 35 continuous hours by monitoring the dissolved oxygen concentration (*dashed black line*) and bioluminescence from yeast transformed with a destabilized luciferase driven by the promoter for *POL1* (a cell cycle-regulated promoter whose activity peaks near the G1/S boundary; *gray line*). This particular culture has a period of about 5 h for the YRO. Dissolved oxygen is measured as percent saturation by atmospheric oxygen of the medium. The brackets above the first oscillation labeled R-F and R show the respiro-fermentative phase and respiration phase of the oscillation, respectively. (**b**) Six cycles of a yeast respiratory oscillation from a separate experiment are shown for 22 continuous hours by monitoring the dissolved oxygen concentration (*dashed black line*) and bioluminescence from yeast transformed with luciferase driven by the promoter for *ACT1* (a constitutive promoter under these conditions, *gray line*). This particular culture has a period of 3.75 h for the YRO. Note that during times of recurring hypoxia (indicated by the *gray highlighted regions*), the luminescence signal drops to nearly zero until adequate oxygen levels return.

emitting reaction that may become limiting during growth (and cause light levels to decrease regardless of luciferase expression). In particular, the researcher must be aware that during times of severe hypoxia, the luminescence signal may not represent the expression level of the promoter coupled to the luciferase gene (as shown during recurring hypoxic periods in Fig. 1, corresponding to the gray highlighted portions of Fig. 1b and the similar but not highlighted portions of Fig. 1a). During these hypoxic "masks," no quantification of promoter activity can be obtained with the luciferase reporter regardless of its expression level. However, we have shown (by immunoblotting with anti-Luc) that levels of luciferase expressed from a constitutive promoter (*ACT1*) remain high during the hypoxic mask (as expected) and once this period of oxygen depletion subsides, luminescence from the reporter returns as a reliable indicator of promoter activity (6). A similar problem occurs, if the luciferin concentration is allowed to become limited. If this occurs, luminescence will decrease. Nevertheless, by keeping these limitations in mind and being aware of when they occur, using a luciferase reporter of promoter activity can be a useful tool in yeast. Corrections for hypoxia can be undertaken by using the P_{ACT1}-LUC reporter in an equivalent culture or experiment to that of the reporter of interest to indicate if cultures are becoming hypoxic as discussed previously (6).

2. Materials

2.1. Yeast Inoculum Preparation for Continuous Culture

1. *S. cerevisiae* strain CEN.PK113-7D (containing luciferase reporter stably transformed into the genome, if bioluminescence is to be monitored).
2. YPD: 1% yeast extract, 2% peptone, and 2% D-glucose (anhydrous).
3. 50 mL Flask.

2.2. Generation of the YRO in Continuous Culture

1. 3 L New Brunswick Scientific Bioflo 110 or 115 Bioreactor with water jacket and direct drive agitation equipped with two Rushton-type impellers, condenser, pH probe, and DO probe.
2. Pressurized air supply capable of at least 4 L/min.
3. Bioreactor medium: 10 g/L anhydrous glucose, 5 g/L ammonium sulfate, 0.5 g/L magnesium sulfate heptahydrate, 1 g/L yeast extract, 2 g/L potassium phosphate, 0.5 mL/L of 70% v/v sulfuric acid, 0.5 mL/L of antifoam A, 0.5 mL/L 250 mM calcium chloride, and 0.5 mL/L mineral solution A. (Mineral solution A consists of 40 g/L $FeSO_4 \cdot 7H_2O$, 20 g/L $ZnSO_4 \cdot 7H_2O$, 10 g/L $CuSO_4 \cdot 5H_2O$, 2 g/L $MnCl_2 \cdot 4H_2O$,and 20 mL/L 75% sulfuric acid.) (see Note 1).
4. Tubing: Silicone tubing i.d. 3/16 in., o.d. 9/32 in.; Norprene A-60-G tubing i.d. 1/16 in., o.d. 3/16 in. (see Note 2).
5. Reduction Couplers, sizes 1/16–1/8, 1/8–3/16, and 3/16–1/4 (see Note 2).
6. 2 N NaOH.
7. Media Bottles (10 L, 1 L, and 250 mL) with filter-vented cap and liquid exit port (see Note 3).
8. 30 mL Syringe and 21 g 1.5 in. needle.
9. 1 L Graduated Cylinder (suitable for autoclaving).

10. Chiller (circulating chilled water bath).
11. Waste collection container of choice (bucket, flask, or bottle) with at least a 4 L capacity.
12. Two (or more) 0.2 μM autoclavable air filters.

2.3. Luminescence Monitoring in Continuous Culture

1. Beetle luciferin (potassium salt).
2. 1 mL Syringe w/needle.
3. 60 mL syringe.
4. 16 gauge 1 in. needle.
5. Harvard Apparatus syringe pump.
6. Tubing: Clear plastic tubing i.d. 1/8 in., o.d. 1/4 in. (Nalgene); PTFE tubing i.d. 0.012 in., o.d. 0.03 in.; Silicone tubing i.d. 1/32 in., o.d. 3/32 in. (see Note 2).
7. Reduction Couplers, sizes 1/16–1/8, 1/8–3/16, and 3/16–1/4 (see Note 2).
8. Cole-Parmer Masterflex L/S Standard Drive 600 rpm Peristaltic Pump.
9. Black box (see Note 4).
10. Hamamatsu HC135-01 photomultiplier.
11. Two ring stands and clamps small enough to fit inside black box.
12. 50 mL plastic conical tube.
13. Aluminum foil.
14. Binder clip (1/2 in.).
15. Black cloth (2 × 10 ft, dimensions can vary).
16. Computer with data logger software (e.g., BioCommand by New Brunswick Scientific).
17. Computer with luminescence monitoring software (e.g., PMTMON by Tom Breeden, U. Virginia).

2.4. Luminescence Monitoring in Small Batch Culture

1. *S. cerevisiae* strain of choice (containing luciferase reporter stably transformed into the genome).
2. YPD: 1% yeast extract, 2% peptone, and 2% D-glucose (anhydrous).
3. Beetle luciferin (potassium salt).
4. 50 mL Flask.
5. Magnetic micro stirbar (~10 mm length).
6. Magnetic stirrer.
7. Styrofoam cup (see Note 5).
8. Black box (see Notes 4 and 5).

9. Hamamatsu HC135-01 photomultiplier.
10. Ring stand and clamp small enough to fit inside black box.
11. Computer with luminescence monitoring software (e.g., PMTMON by Tom Breeden, U. Virginia).

3. Methods

3.1. Yeast Inoculum Preparation for Continuous Culture and Bioluminescence Monitoring

1. In a 50 mL flask, prepare a 20 mL starter culture of yeast in YPD medium that will be used to inoculate the bioreactor. Inoculate 20 mL of YPD with a match-head-sized yeast colony or a scraping from an YPD plate (see Note 6).
2. Grow the starter culture for 20–30 h at 28–30°C with agitation.

3.2. Establishment of Respiratory Oscillations During Continuous Culture

Respiratory oscillations spontaneously arise in continuous cultures of certain strains of yeast when grown under a specific range of conditions. However, for this to occur, the culture must be sufficiently dense so that oscillations reinforce themselves. The quickest way to achieve this critical cell density is to inoculate the bioreactor with a starter culture of yeast and grow that bioreactor culture in batch overnight before beginning continuous culture the next day.

3.2.1. Bioreactor Setup

1. Prepare the Bioflo 110 (without baffles) for batch and continuous culture. Adjust two Ruston-type impellers on the agitator so that one is below the media level and one is at the air–media interface (when the vessel contains ~850 mL).
2. Autoclave the bioreactor and the necessary accessories (see Note 7).
3. Fill the bioreactor vessel with 850 mL sterile bioreactor media. A sterile 1 L graduated cylinder and sterile 1 L bottle with filter-vented cap can be used to measure and add the medium to the bioreactor.
4. Attach the loose end of tubing from the 250 mL bottle that has the filter-vented cap to a port of the bioreactor. Add 200 mL of sterile 2 N NaOH to the 250 mL bottle and load the Norprene tubing from the bottle's cap into the peristaltic pump that controls the culture's pH, but do not turn on the pump at this time (see Note 8).
5. Attach an autoclaved 0.2 μM air filter to the sparger inlet and connect the other end of the filter to a regulated pressurized air supply. Introduce filtered air into the bioreactor's media through the sparger at a flow rate of 0.9 L/min. Begin agitation at 550 rpm.

6. If the bioreactor has a water jacket, attach the water jacket and vapor condenser to the circulating water chiller set to operate at 4°C (see Note 9). Turn the bioreactor's temperature control to 30°C and let the media and condenser come to the desired temperatures.
7. Adjust the level of the stainless steel tube in the bioreactor that is to serve as the medium's outflow tube to the level of the media–air interface (see Note 10). Then connect a ~10 ft length of sterile silicone tubing to the media outflow port and load it into a peristaltic pump (turned off) and arrange the rest of the 10 ft tubing to deliver bioreactor waste to a collection container of choice.
8. Adjust (and maintain) the pH of the media to the desired pH (recommended pH 3.4–4) (see Note 11).
9. After the DO probe has polarized (see Note 12) and prior to inoculation, calibrate the DO probe.

3.2.2. Inoculation and Growth

1. Inoculate the bioreactor by injecting the 20 mL culture through the septum with a syringe and a 21 gauge needle. If luminescence from this culture is going to be monitored, see Note 13.
2. Grow the yeast in batch culture overnight. During this time, the DO of the culture gradually drops as the culture becomes denser and total respiration of the culture increases. Once the carbon sources have been consumed, the DO of the culture rises sharply (about 16–18 h after inoculation). Incubate the culture in this starved condition for 4–7 more hours before beginning continuous culture (however, see Note 14).
3. Begin continuous culture at a dilution rate of ~0.085/h (see Note 15). Set the outflow pump for 100% duty cycle to remove media as the level of the culture rises to the level of the removal tube within the bioreactor. Respiratory oscillations often begin about 12 h after the initiation of continuous culture.

3.3. Luminescence Monitoring in Continuous Culture

Luminescence of the continuous culture is constantly monitored by using a high speed peristaltic pump to move culture through a closed loop from the bioreactor, into a dark box, in front of a photomultiplier tube for measurement, and back to the bioreactor (see Fig. 2).

1. Connect a closed loop of autoclaved tubing to the bioreactor for luminescence monitoring. This loop includes a length of Norprene A-60-G tubing (3/16 in. o.d.) that passes through a high rpm peristaltic pump (e.g. Cole-Parmer Masterflex L/S) (turned off) and connects (by a coupler) to a ~10 ft length of

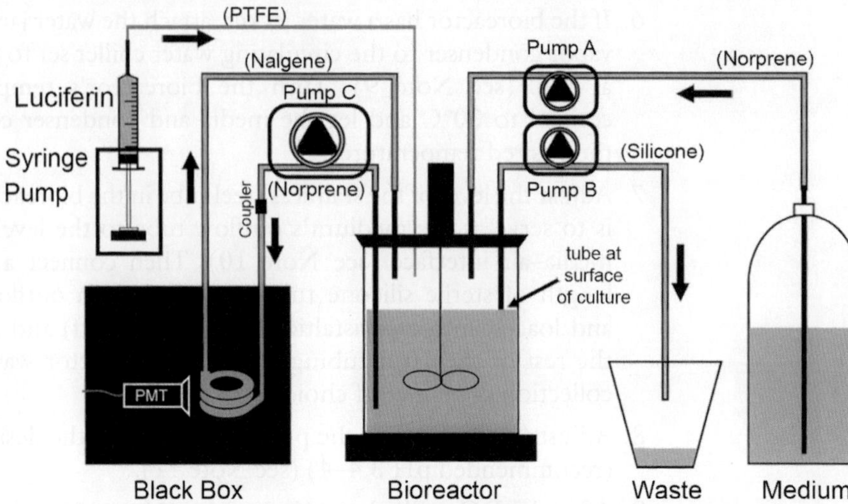

Fig. 2. A schematic diagram showing the setup for continuous monitoring of bioluminescence during continuous culture. Pump A is the peristaltic pump that supplies medium to the bioreactor. Pump B is the peristaltic pump that removes culture from the bioreactor. Pump C is the high speed peristaltic pump that moves culture from the bioreactor into the black box for luminescence monitoring and back to the bioreactor through the closed loop. The *large arrows* indicate the direction of flow through the different types of tubing indicated in *parentheses*. The coupler shown in the closed loop of pump C indicates the junction of Norprene tubing (needed to withstand the action of the high speed peristaltic pump C) and Nalgene tubing (needed because it is transparent). PMT is the photomultiplier tube.

transparent Nalgene tubing (1/4 in. o.d.) (see Notes 2 and 16). Connect the free end of the Norprene tubing of the loop to the media sampling port – this is where the circulating loop of culture leaves the bioreactor. Connect the other end of the loop (the free end of the transparent Nalgene tubing) to a port that returns the culture back to the vessel (see Note 16).

2. Pull a portion of the closed loop (comprising the majority of the transparent Nalgene tubing) through a light-tight port of the black box. Wrap the transparent Nalgene tubing of the loop around a 50 mL conical tube (or some other cylinder of approximate size) for several turns (see Note 17). Use a 1/2 in. binder clip to keep the tubing from unraveling from the conical tube (see Fig. 3). Within the black box, use a ring stand clamp to hold the cylinder and coiled tubing near a photomultiplier device so that light from yeast flowing through the transparent tubing can be detected by the photomultiplier (see Note 18). Close the black box and cover any light leaks with foil or black cloth.

3. Turn on the high rpm peristaltic pump to begin moving culture through the closed loop (see Note 13). The speed of the pump is not critical, but it should not be so slow that the culture is kept away from the bioreactor for more than

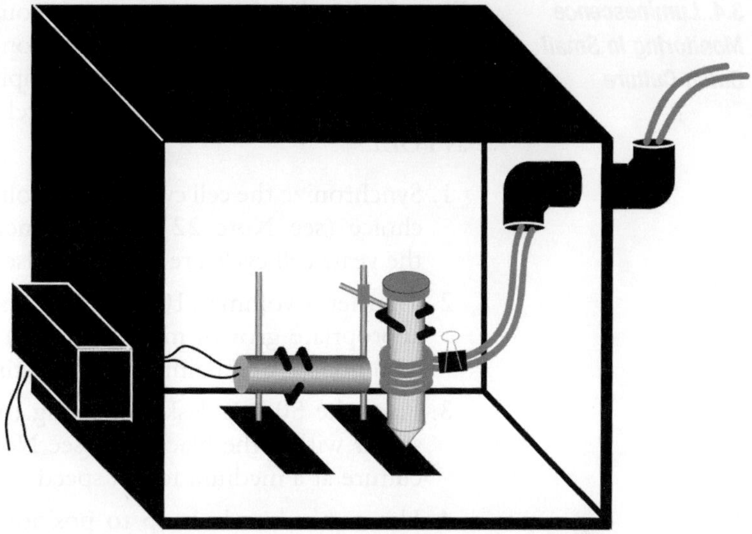

Fig. 3. A diagram showing the setup within the black box for continuous monitoring of bioluminescence during continuous culture. The front panel has been removed in this diagram to show the box's interior. The small box on the *left* and the pipe on the *right* of the black box are light-tight ports through the black box for wires and tubing, respectively. Within the box on the *left*, a ring stand and clamp hold a photomultiplier tube positioned to collect light. On the *right*, a ring stand and clamp support a 50 mL conical tube around which Nalgene tubing (from the bioreactor) is wrapped and held in place by a 1/2 in. binder clip.

a minute. A circuit time of about half a minute is preferable, which can be achieved with ~180 rpm.

4. Immediately after the closed loop is filled with culture (and culture from the closed loop can be seen returning to the bioreactor), lower the level of the outflow tube in the bioreactor to the new level of the culture–air interface. This is important in order for continuous culture to maintain the same dilution rate since some of the volume of the culture no longer resides in the reactor vessel.
5. Add 5 μM luciferin to the bioreactor's culture during a phase of the respiratory oscillation when dissolved oxygen is decreasing rapidly or near the trough (see Note 19). This can be done by injecting 425 μL of a 10 mM (2,000×) stock solution of luciferin into the bioreactor through the septum with a 1 mL syringe and needle.
6. Maintain a 5 μM concentration of luciferin in the bioreactor during continuous culture by adding luciferin to the media that feeds the culture or supplying a steady drip of luciferin from a syringe pump (see Notes 20 and 21).
7. Turn on the power to the photomultiplier device and begin recording bioluminescence.

3.4. Luminescence Monitoring in Small Batch Culture

For other applications, where continuous culture is not required, luciferase reporters can be used to monitor promoter activity in small batch cultures of yeast, for example, to monitor promoters for inducible genes such as *GAL1* or cell cycle-related genes such as *POL1*.

1. Synchronize the cell cycle of the bioluminescent yeast strain of choice (see Note 22). Various methods for synchronizing the yeast cell cycle are described elsewhere (6, 16).
2. Transfer a volume (10 mL) of the synchronized cells in the appropriate growth medium (YPD) with 50 μM luciferin to a 50 mL flask containing a micro-stirbar.
3. Place the 50 mL flask containing the culture on a magnetic stirrer within the black box (see Notes 4 and 5) and stir the culture at a medium to fast speed.
4. Use a stand and clamp to position a photomultiplier tube next to the stirred culture in the black box. Angle the photomultiplier tube so that it can capture the most light from the culture. Aluminum foil may be used to help direct more photons toward the photomultiplier.
5. Close the black box and begin recording (see Note 23).
6. Perform any necessary detrending of the luminescent signal (see Note 24).

4. Notes

1. Make 10 L of bioreactor medium in a 10 L bottle. Mark the level on the 10 L bottle for 10 L of ddH_2O at room temperature with tape and/or permanent marker before mixing the medium. Remove at least 200 mL of the water and then combine all components of the bioreactor medium ***except antifoam A and mineral solution A*** before autoclaving. Mix, stir, or shake as needed to dissolve all components. Add ddH_2O until the volume of media is close to the 10 L mark. Cover the mouth of the bottle with a loose fitting cap and/or aluminum foil and autoclave the medium for 45 min (sterilization time). Let it cool overnight. Add antifoam A and mineral solution A after the medium has cooled, and then bring the volume to the 10 L mark with sterile ddH_2O.
2. Tubing of different materials and sizes are needed for different tasks. The tubing that goes through peristaltic pumps needs to be both pliable and durable. The tubing that carries culture for luminescence monitoring needs to be flexible, sturdy, and

transparent. The small 3/16 in. (o.d.) Norprene tubing is ideal for peristaltic pumps because it is both pliable and durable. The larger 9/32 in. (o.d.) silicone tubing is also pliable enough for peristaltic pumps, but is not as durable as the Norprene tubing and should be inspected for wear between uses. When possible, the Norprene tubing should be used in peristaltic pumps. However, because the ports on the bioreactor and media bottles may not permit the smaller Norprene tubing to attach, it may be necessary to use tubing of a different size to make the connections and join the different sized tubing with plastic reduction couplers. This is also true for joining the types of tubing needed for monitoring luminescence during continuous culture. If different tubing sizes or types are used besides the ones recommended here, make sure that they possess the necessary characteristics for their purposes (see Fig. 2).

3. Bottles containing liquids that are to be added to the bioreactor need a filtered gas vent in their caps to reduce the risk of contaminating the continuous culture by air entering the bottle to replace displaced liquid. In addition, these bottles need a tube or pipe that penetrates the cap and extends to nearly the bottom of the bottle, through which the liquid in the bottle is removed and added to the bioreactor (usually by a peristaltic pump). These caps can be made using an appropriately sized two-hole rubber stopper with glass or metal tubes penetrating the holes with tight seals. If needed, the stopper can be held firmly in the bottle by an appropriately sized plastic cap that contains a wide hole drilled in the top to accommodate the glass or metal tubes coming through the stopper and large enough to allow the cap to screw down onto the bottle.
4. The light emitted from bioluminescent yeast is very dim compared with the amount of light in the environment. Even a room that is dark to the eye has enough stray photons from various sources to flood a sensitive light detecting photomultiplier with noise that can conceal the true bioluminescent signal. Therefore, luminescent measurements must be made in an enclosure that totally excludes light from the environment. These enclosures are often painted black inside and out (to absorb stray photons) and so are sometimes called "black boxes." A black box can be constructed from plywood and should include light-tight ports to permit tubing and wires to pass into and out of the box (see Fig. 3). Light-tight ports can be easily constructed out of black PVC elbows that are connected so that there are "corners" around which incident light cannot pass.
5. The motorized magnetic stirrer used to stir the yeast culture in the black box generates some heat. Depending on the

application, this heat may be useful to warm the culture to an optimum growth temperature for yeast. However, if this heat is undesirable for the application or too much heat is generated from the stirplate, excess heat can be dissipated from the culture by using a fan-cooled black box and/or elevating the 50 mL culture flask off of the magnetic stirrer by using an inverted Styrofoam cup cut to the desired height. (If fan cooling is used, a light-tight pathway for airflow must be constructed as mentioned in Note 4. For example, in Fig. 3, the small box on the left of the black box can serve as a light-tight path for airflow when a small fan is attached.)

6. This protocol will reproducibly generate respiratory oscillations for the MAT-a yeast strain CEN.PK113-7D (from Peter Kötter, U. Frankfurt, Germany), but other strains of CEN.PK may work as well. Other strains of yeast such as S288C (14) and IFO 0233 (15) also manifest robust YROs under certain conditions of continuous culture but may not be suited to the precise conditions described here. If bioluminescence will be monitored, then the appropriate strain containing the desired luciferase reporter should be used. If the luciferase reporter has been stably integrated into the genome of the yeast strain, continued selection with an antibiotic is not needed during the establishment of respiratory oscillations.

7. In addition to the autoclaved bioreactor, it is helpful to have the following items sterilized by autoclave and cooled before proceeding: one 0.2 μM air filter connected to ~3 in. of silicone tubing, one 250 mL bottle with a filter-vented cap and the outflow tube (see Note 3) connected to ~6 ft of Norprene A-60-G tubing (i.d. 1/16 and o.d. 3/16), one 1 L bottle with a filter-vented cap and the outflow tube connected to ~6 ft of Norprene A-60-G tubing, one separate filter-vented cap (that fits the 10 L bottle) with the outflow tube connected to ~6 ft of Norprene A-60-G tubing, one 1 L graduated cylinder, ~6 ft of silicone tubing (i.d. 3/16 and o.d. 9/32), and a ~10 ft length of the same silicone tubing. The exposed ends of all the tubing should be covered with aluminum foil before autoclaving. Also, the separate filter-vented cap that fits the 10 L bottle should be autoclaved in a covered beaker or completely wrapped in foil. It will be added to the 10 L bottle of medium later.

8. Attaching the bottle of NaOH at this time serves to keep the used tri-port inlet covered by sterile tubing. And it is important to install the tubing into a peristaltic pump (that is off) at this time to prevent back flow of the pressurized air from the bioreactor into the NaOH bottle.

9. The same water chiller can be used to cool the bioreactor and the vapor condenser, but the vapor condenser must be plumbed so that it can be continually cooled by the circulating water from the chiller. When the vapor condenser is kept cool (0–4°C), it helps to prevent the bioreactor from drying out as a result of continuously flowing air through the culture. The vent from the vapor condenser can be covered with an air filter to help minimize risk of culture contamination, but the filter sometimes becomes wet over time and air flow through it is reduced. An uncovered length (2–3 ft) of sterile silicone tubing from the condenser's vent works well to prevent culture contamination while permitting unrestricted air flow through the condenser. Also, for the condenser to work properly, all other avenues of gas flow from the bioreactor should be sealed. This includes unused tri-port inlets and other ports in the bioreactor's head plate. A small length of tubing with a knot tied in one end works well for sealing an unused port.
10. Since the volume of the bioreactor should be kept constant at ~850 mL during continuous culture, setting the level of the outflow tube (i.e., the tube to the waste) to the level of the stirred and aerated media at this time will establish the proper volume for the culture (see Fig. 2, "tube at surface of culture").
11. The pH of the media will often lag the readout from the probe so one should manually adjust the pH gradually until the desired pH is reached. However, accidentally overshooting the desired pH by less than one pH unit at this point does not noticeably affect the establishment of respiratory oscillations. At times during batch growth, the pH of the culture may rise above the desired pH, but this will not adversely affect the formation of respiratory oscillations once continuous culture begins.
12. Various DO probes require some length of time to polarize their electrodes before accurate oxygen concentrations can be made. It is recommended to allow 2–6 h (or a length of time specified by the manufacturer) after attaching the DO probe's wiring to the bioreactor before calibrating the DO probe.
13. The best time to begin moving culture through the closed loop for luminescence monitoring is prior to inoculating the bioreactor (or at least prior to the establishment of respiratory oscillations). The initial change of conditions that occurs when the high rpm pump begins moving culture through the closed loop can perturb oscillations that have already been established. The best way to avoid this perturbation is to have the culture moving through the closed loop from the beginning (during batch growth).

14. Batch growth (including the 4–7 h of starvation) has been found *not to be necessary* for the establishment of respiratory oscillations. One can begin continuous culture immediately after inoculation, but such a method may consume more media before oscillations begin (usually ~24 h after inoculation).
15. One needs to know the flow rate of the media supply pump in combination with the tubing used to know what pump speed results in a dilution rate of 0.085/h. For an 850 mL culture and media supplied through the pump by Norprene A-60-G tubing (i.d. 1/16 in. and o.d. 3/16 in.), a duty cycle of 34% will achieve a dilution rate ~0.085%. The speed of the outflow pump is not important as long as it removes culture at a faster rate than the supply pump adds medium to the culture. Setting the outflow pump to 100% is recommended.
16. The order of the loop in the direction of culture-flow should be as follows: media sampling port, Norprene tubing (through high rpm peristaltic pump), transparent Nalgene tubing, and inlet port (of choice). It is important that the high rpm peristaltic pump draws the culture from a port that has a stainless steel tube that extends below the surface of the mixed bioreactor culture. The lengths of the Norprene and Nalgene tubings can vary as needed to accommodate distances from bioreactor, pump, and black box. The connection between the Norprene tubing and the Nalgene tubing should be made just downstream of the high rpm peristaltic pump and should remain outside of the black box since a leak at this connection may be difficult to identify if it is within the black box.
17. Wrapping the transparent tubing around a cylinder provides an increased surface exposure of the culture to the light detecting photomultiplier tube. If the luminescence is sufficiently bright, fewer turns around the cylinder are required. The intensity of the luminescence signal can be increased by coating the cylinder with reflective aluminum foil before wrapping the tubing around it and can be further increased by wrapping the cylinder with a double layer of turns of the transparent Nalgene tubing from the loop.
18. If the black box is not completely light tight, background light can still interfere with the luminescent signal. Background light can be further reduced by encapsulating the entire photomultiplier and cylinder with aluminum foil and then covering both with a loose arrangement of black cloth. Also, room light can travel through the transparent Nalgene tubing carrying the culture into and out of the black box (by analogy with optic fibers); therefore, wrapping the exposed Nalgene tubing with foil and keeping the room lights

off (or dim) will help to reduce background light. If there is a small amount of unavoidable background room light leak detected by the photomultiplier, it is better to maintain the room light at a stable (dim) level than turning on (and off) the room lights to make adjustments to the apparatus.

19. A sudden delivery of luciferin to a bioluminescent strain of yeast in the respiro-fermentative phase can acutely drop the intracellular oxygen concentration, which can result in a phase shift of the oscillation. To avoid affecting the oscillation, charge the culture with luciferin during the respiratory phase of the oscillation when intracellular oxygen levels are already low.

20. Because the culture in the bioreactor is constantly being diluted during continuous culture, the concentration of luciferin will gradually decline if not constantly supplied at a concentration and rate that keeps up with the dilution rate of the culture. One easy way to do this (over a short term) is to add 5 μM luciferin to the medium that feeds the continuous culture; however, this method is not recommended for long-term experiments because luciferin degrades in the acidic medium over time. For long-term experiments, where luminescence needs to be measured for more than several hours, use a syringe pump to supply a steady drip of a concentrated stock of luciferin to the bioreactor. For example, a 120× stock of luciferin in water (i.e., 600 μM) supplied to the bioreactor at 1/120 of the culture's dilution rate (i.e., 0.6 mL/h) will maintain a constant 5 μM luciferin concentration in the culture without adversely affecting the dilution of the culture. 60 mL of luciferin at this concentration and pump speed can supply the bioreactor for more than 4 days. The stability of the luciferin in the syringe can be increased by shielding the luciferin from light and by chilling the syringe with several wraps of tubing carrying cold water from the bioreactor's condenser.

21. It can be difficult to regularly drip luciferin into the culture at slow pump speeds. If delivered to the bioreactor through one of its normal ports, luciferin can adhere to the inside of the vessel or headplate rather than dripping down into the culture. A steady drip into the culture can be achieved, however, if the luciferin is delivered to the culture through very thin rigid tubing (e.g., PTFE tubing i.d. 0.012 and o.d. 0.03). Use a 16 gauge needle to penetrate the septum of the bioreactor and while the needle is through the septum, thread a few inches of the autoclaved tubing through the needle so that the end of the tubing hangs freely in the reactor's vessel. Gently remove the needle from the septum leaving the tubing in place, held securely by the septum.

Attach the other end of the tubing to the syringe of the syringe pump that supplies luciferin to the vessel. The thin PTFE tubing can be connected to the syringe by constructing an adaptor from a cut p200 pipette tip and a short (~1 in.) piece of silicone tubing i.d. 1/32 and o.d. 3/32. This adaptor including the cut pipette tip should be autoclaved while attached to the PTFE tubing prior to use.

22. This protocol describes the steps needed to monitor the rhythms of cell cycle-related promoter activity. If the use of bioluminescence is not to observe cell cycle-related promoter activity, then cell cycle synchronization may not be required.
23. During batch culture, yeast will eventually begin respiring and consuming oxygen at a high rate. As a result, luminescence can decline due to limited oxygen. Luminescent reporters of promoter activity are not accurate once oxygen becomes limited. Oxygen limitation can be monitored with a parallel culture of luciferase driven by a strong constitutive promoter such as actin (*ACT1*).
24. The number of cells in a batch-grown culture increases over time. As a result, the total bioluminescence from the culture increases as well. To observe rhythmic promoter activity from a culture in which bioluminescence increases with cell density, it may be necessary to subtract from the luminescent signal the trend of luminescence that results from the increase in cell density. There are several methods to accomplish a trend correction. One procedure is to generate a polynomial trendline that best represents the growth of the culture and use this formula for baseline subtraction of the luminescence signal. Another method is to repeat the experiment using a parallel culture of the same strain that has not been synchronized; luminescence from this nonsynchronized culture can be used as a baseline for cell growth that can be subtracted from the luminescent trace from the synchronized culture.

References

1. Yamazaki, S., Numano, R., Abe, M., Hida, A., Takahashi, R., Ueda, M., Block, G. D., Sakaki, Y., Menaker, M., and Tei, H. (2000) Resetting central and peripheral circadian oscillators in transgenic rats, *Science* **288**, 682–685.
2. Izumo, M., Sato, T. R., Straume, M., and Johnson, C. H. (2006) Quantitative analyses of circadian gene expression in mammalian cell cultures, *PLoS Comput Biol* **2**, e136.
3. Brandes, C., Plautz, J. D., Stanewsky, R., Jamison, C. F., Straume, M., Wood, K. V., Kay, S. A., and Hall, J. C. (1996) Novel features of drosophila period Transcription revealed by real-time luciferase reporting, *Neuron* **16**, 687–692.
4. Millar, A. J., Short, S. R., Chua, N. H., and Kay, S. A. (1992) A novel circadian phenotype based on firefly luciferase expression in transgenic plants, *Plant Cell* **4,** 1075–1087.
5. Gooch, V. D., Mehra, A., Larrondo, L. F., Fox, J., Touroutoutoudis, M., Loros, J. J., and Dunlap, J. C. (2008) Fully codon-optimized luciferase uncovers novel temperature characteristics of the Neurospora clock, *Eukaryot Cell* 7, 28–37.

6. Robertson, J. B., Stowers, C. C., Boczko, E., and Johnson, C. H. (2008) Real-time luminescence monitoring of cell-cycle and respiratory oscillations in yeast, *Proc Natl Acad Sci U S A* **105**, 17988–17993.
7. Thompson, J. F., Hayes, L. S., and Lloyd, D.B. (1991) Modulation of firefly luciferase stability and impact on studies of gene regulation, *Gene* **103**, 171–177.
8. Mateus, C., and Avery, S. V. (2000) Destabilized green fluorescent protein for monitoring dynamic changes in yeast gene expression with flow cytometry, *Yeast* **16**, 1313–1323.
9. Brauer, M. J., Saldanha, A. J., Dolinski, K., and Botstein, D. (2005) Homeostatic adjustment and metabolic remodeling in glucose-limited yeast cultures, *Mol Biol Cell* **16**, 2503–2517.
10. Hoskisson, P. A., and Hobbs, G. (2005) Continuous culture–making a comeback?, *Microbiology* **151**, 3153–3159.
11. Zamamiri, A. Q., Birol, G., and Hjortso, M. A. (2001) Multiple stable states and hysteresis in continuous, oscillating cultures of budding yeast, *Biotechnol Bioeng* **75**, 305–312.
12. Murray, D. B., Engelen, F. A., Keulers, M., Kuriyama, H., and Lloyd, D. (1998) NO+, but not NO., inhibits respiratory oscillations in ethanol-grown chemostat cultures of Saccharomyces cerevisiae, *FEBS Lett* **431**, 297–299.
13. Tu, B. P., Kudlicki, A., Rowicka, M., and McKnight, S. L. (2005) Logic of the yeast metabolic cycle: temporal compartmentalization of cellular processes, *Science* **310**, 1152–1158.
14. Xu, Z., and Tsurugi, K. (2006) A potential mechanism of energy-metabolism oscillation in an aerobic chemostat culture of the yeast Saccharomyces cerevisiae, *FEBS J* **273**, 1696–1709.
15. Klevecz, R. R., Bolen, J., Forrest, G., and Murray, D. B. (2004) A genomewide oscillation in transcription gates DNA replication and cell cycle, *Proc Natl Acad Sci U S A* **101**, 1200–1205.
16. Futcher, B. (1999) Cell cycle synchronization, *Methods Cell Sci* **21**, 79–86.

Chapter 5

Linearizer Gene Circuits with Negative Feedback Regulation

Dmitry Nevozhay, Rhys M. Adams, and Gábor Balázsi

Abstract

Gene functional studies consist of phenotyping cells with altered gene expression. Improving the precision of current gene expression control techniques would enable more detailed studies of gene function. Here, we provide protocols for building synthetic gene constructs for tuning the expression of a gene in all the cells of a population precisely and uniformly, achieving expression levels proportional to the extracellular inducer concentration.

Key words: Gene expression systems, TetR, Linearizer, Dose–response

1. Introduction

Gene expression is a crucial step connecting genotype to phenotype through the production of proteins that determine most observable phenotypes in populations of living cells. Therefore, developing methods of gene expression control is critical for understanding gene function. For example, gene deletion (1) and overexpression (2) techniques have contributed immensely to our understanding of the genotype–phenotype connection. However, these methods are aimed solely at altering average protein or mRNA levels in the cell population in a drastic manner, and therefore allow only limited, qualitative control of gene activity. The relatively new method of RNA interference (3) suffers from similar problems, including massive off-target effects (4).

Gene expression systems (small synthetic gene networks that consist of a regulator controlling the expression of a gene of interest) permit more precise, reversible, quasi-quantitative control of protein levels in a cell population. A small molecule inducer or a co-repressor added to the growth medium affects the regulator's activity and thereby indirectly controls target gene

Attila Becskei (ed.), *Yeast Genetic Networks: Methods and Protocols*, Methods in Molecular Biology, vol. 734, DOI 10.1007/978-1-61779-086-7_5, © Springer Science+Business Media, LLC 2011

expression between two extremes that define the "dynamic range." Based on natural scenarios of inducible/repressible regulatory control (5), several types of gene expression systems are conceivable, such as the T-Rex (6), Tet-On, and Tet-Off (7) systems. The transcriptional regulator components of these gene expression systems (TetR, rtTA, and tTA, respectively) utilize a common DNA-binding domain (8) and therefore bind to the same DNA sequence motif (*tet*O2). These and similar gene expression systems are widely used to control gene expression in various cell types and organisms (9, 10). While TetR-based systems offer the convenience of continuous and reversible gene expression control, their use is hampered by nonlinear (sigmoidal) dose–responses (11) and uncontrolled fluctuations around mean expression levels that prevent truly precise control of gene activity. For these reasons, improving the performance of gene expression systems is highly important for functional genetics.

Autoregulation (when a gene product regulates its own synthesis) is a frequent theme in gene regulatory networks (12), indicating that feedback might alter the properties of gene expression and provide selective advantage in certain situations. Accordingly, ***negative autoregulation*** (self-repression) has been shown to reduce gene expression noise (13–15), to speed the response times of transcription (16), or to be the basis of robust genetic oscillators (17, 18).

In addition to these functions, we recently showed that negative feedback can linearize the dose–response of TetR repressor-based synthetic gene circuits in yeast (19), complementing recent findings of response alignment in yeast signaling cascades due to negative feedback (20). This linear dose–response prior to saturation and low gene expression noise (21, 22) together indicate that TetR-based gene circuits with negative feedback could be used to precisely and uniformly control the gene expression of all cells within a population, which would be highly useful in future functional genomics studies. For example, if a library of yeast strains – each carrying a linearly regulatable gene from the genome – were established in the wild-type (2) or the corresponding knockout mutant (23), then the phenotypic effect of precisely tuned yeast protein levels could be investigated at the genomic scale, in a massively parallel fashion, in various environments. Similar synthetic linearizer constructs could also be useful in employing budding yeast to study the function of genes from closely related, clinically important ***Hemiascomycota*** (24) such as ***Candida albicans***, or even more distant fungi.

While the TetR-based feedback system produces a linear dose–response with low noise over a wide range of inducer concentrations even at the lowest measurable yEGFP expression levels (19), it has several limitations that may require the construction of linearizers from different components, depending on specific experimental needs. In particular, our system is based on a pair of modified GAL1 promoters, and therefore would be strongly repressed in glucose-containing media. For this reason, other TetR-repressible promoters (based on the CMV or ADH1 promoters, for example) may be constructed to decouple gene expression control from the growth medium.

In addition, based on computational models of the tetR-based linearizer gene circuit we suggest that similar linearizers can be feasibly built using other repressor/inducer pairs, provided that (1) the inducer–repressor and repressor–DNA dissociation constants are very low; and (2) the repressor and gene of interest (reporter) are expressed from identical promoters. Furthermore, to obtain a linear dose–response, no significant additional feedback should be present in the system. Additional feedback could appear if the controlled gene product (1) has a strong effect on cell growth; or (2) if the regulator is involved in additional endogenous feedback loops. Since low repressor–DNA and repressor–inducer dissociation constants, as well as slow inducer uptake/outflux are required to obtain a linear dose–response, we suggest that future systems utilize TetR/ATc or other conjugate repressor/inducer pairs with these properties. Therefore, we describe below in detail the steps that we have taken in building a gene expression linearizer based on the TetR repressor, the yEGFP reporter, and the modified GAL10 promoter, which we hope will be a useful guide for building future linearizers using other promoter–repressor–inducer–reporter combinations, including GFP-tagged versions of endogenous proteins (25).

2. Materials

2.1. Strains

1. *Saccharomyces cerevisiae* haploid YPH500 strain (α, *ura3-52, lys2-801, ade2-101, trp1Δ63, his3Δ200, leu2Δ1*) (Stratagene, La Jolla, CA).
2. *Escherichia coli* XL-10 Gold strain (Stratagene, La Jolla, CA).

2.2. Plasmids and Oligonucleotides

1. Plasmid pRS4D1 (21, 26) containing the modified bidirectional promoter $P_{GAL1\text{-}10}$, which controls expression of both *yEGFP* and *tetR* genes. The promoter $P_{GAL1\text{-}10}$ was modified by inserting two *tetO2* sites downstream of the *GAL1* TATA-box and is referred to as $P_{GAL1\text{-}D12}$ here. The cassette is flanked by the *ADH1* and *CYC1* transcription terminators. The genes *ampR* and *TRP1* are used for selection in *E. coli* and as an auxotrophic marker in *S. cerevisiae*, respectively.
2. Plasmid pRS403 (Stratagene, La Jolla, CA). This plasmid uses *HIS3* as an auxotrophic marker in *S. cerevisiae*.

3. Oligonucleotides used for PCRs and sequencing:

Backbone-r	5′-CGCGTTGGCCGATTCATTAATGC-3′
Before2 TRP-r	5′-CACATATATTACGATGCTGTTCTATTAAA TGCTTCC-3′
Gal1-D12-r	5′-GAAGTAATATCTCTATCACTGATAGGGAGA TCTCTATC-3′
GALSeqA-f	5′-CAAACCTCTGGCGAAGAATTG-3′
GALSeqB-f	5′-GCGGCCGCCCTTTAGTGAGGG-3′
GALSeqC-f	5′-ACCCCGGATCCTATTAAAATG-3′
GALSeqD-r	5′-GATCTTAGCTAGCCGCGGTAC-3′
GALSeqE-r	5′-TGAATAATTCTTCACCTTTAG-3′
GALSeqG-r	5′-ATTCAACCCTCACTAAAGGGC-3′
Insert-f	5′-AATTGGAGCGACCTCATGCTATACCTG-3′
PvuII-f	5′-ACGCCAGCTGAATTGGAGCGACCTCATG-3′
PvuII-r	5′-TAATGCAGCTGGATCTTCGAGCGTCC-3′
tetR-BamHI-f	5′-GCGCGGATCCTATTAAAATGTCTAGATTAGA TAAAAG-3′
tetRcut-f	5′-AATTAAGAGCTCTTAAGACCCACTTTCAC-3′
tetRcut-r	5′-GCCCGACTAGTGAGAATGCATTATATGCAC TCAGCGCT-3′
tetR-XhoI-r	5′-GCGCCTCGAGTTAAGACCCACTTTCACA TTTAAG-3′

2.3. Enzymes, Kits, Media, and Transformation Components

1. Molecular biology grade (MBG) water.
2. Enzymes: *Afl*II, *Age*I, *Ahd*I, *Bam*HI, *Pvu*II, *Sac*I, *Spe*I, *Xho*I, T4 DNA ligase, DNA Polymerase I Large (Klenow) Fragment, Phusion™ Hot Start High-Fidelity DNA Polymerase (New England Biolabs, Beverly, MA), and Paq5000™ DNA Polymerase, PfuUltra™ II Fusion HS DNA Polymerase (Stratagene, La Jolla, CA) and their respective buffers.
3. 1 mM (for reactions with the Klenow fragment) and 10 mM (for polymerase chain reactions, PCR) deoxynucleotide triphosphates (dNTP) solution.
4. 0.5 M Ethylenediamine tetraacetic acid.
5. Glycerol 80% v/v in water, autoclaved.
6. Phosphate-buffered saline (PBS), sterile.
7. Single-stranded carrier DNA from Salmon testes (Sigma–Aldrich, St. Louis, MO). Prepare stock solutions in a TE

buffer (1 mM EDTA, 10 mM Tris–HCl, pH 8.0) at a concentration of 2 mg/ml, and store at −20°C.

8. 100 mM and 1 M lithium acetate (Sigma–Aldrich, St. Louis, MO), filter sterilized.
9. Polyethylene glycol 50% w/v (PEG MW 3350).
10. QIAquick Gel Extraction kit, QIAquick PCR purification kit, QIAprep Spin Miniprep kit, DNeasy Blood & Tissue kit (QIAGEN, Germantown, MD).
11. Lennox L Broth (LB Broth, Research Products International, Mt. Prospect, IL), or similar medium: 10 g of tryptone, 5 g of sodium chloride, 5 g of yeast extract, and 1.5 g of Tris–HCl per 1 l of water, autoclaved and ampicillin at 50 mg/ml concentration for *E. coli* propagation and selection.
12. Components for gel electrophoresis: agarose 0.8%, Tris–acetate–EDTA (TAE) electrophoresis buffer (ISC BioExpress, Kaysville, UT), bromphenol blue gel loading buffer 4× (Amresco, Solon, OH), 2-log DNA ladder 50 μg/ml, and ethidium bromide 10 mg/ml.
13. Anhydrotetracycline (ATc, ACROS Organics, Geel, Belgium). Prepare a stock solution of 5 mg/ml by diluting 25 mg of ATc in 5 ml of ethanol and store at −20°C for up to 6 months.
14. Glucose 20% w/v and galactose 20% w/v.
15. YPD medium and plates. Dissolve 5 g of yeast extract, 10 g of peptone, and 19 mg of adenine in 450 ml of water, autoclave for 20 min, and add 50 ml of glucose 20%. For plates, add 7.5 g of agar to medium but before autoclaving.
16. SC medium and plates. Dissolve 3.35 g of yeast nitrogen base without aminoacids, 0.7 g of drop-out supplement mix without histidine, tryptophan, leucine, uracil, and 19 mg of adenine in 450 ml of water. Add 38 mg of uracil, 190 mg of leucine, and 38 mg of tryptophan for SC-his and omit tryptophan for SC-his-tryp. Autoclave for 20 min and add 50 ml of sugar (either glucose 20% or galactose 20%). For plates, add 7.5 g of agar to medium but before autoclaving.

2.4. Equipment

1. Flow cytometer.
2. Equipment for gel electrophoresis.
3. Thermocycler for PCR.
4. Thermomixer.
5. Centrifuge.
6. Spectrophotometer.
7. Shaking incubator at 30°C for *S. cerevisiae*.

8. Shaking incubator at 37°C for *E. coli.*
9. Miscellaneous laboratory plastic disposables.

2.5. Stochastic Simulation and Mathematical Modeling

A software environment such as Matlab (The Mathworks, Inc) with the Sim-Biology, Symbolic Math, and Statistics toolboxes is used for stochastic simulations, calculations, and statistics. Alternatively, noncommercial, freely available software such as Octave (27), Dizzy (28), iBioSim,[1] or R[2] could be used for the same purpose.

3. Methods

The negative feedback circuit described below is based on the tetracycline repressor protein (TetR) which was originally identified in prokaryotes (29, 30), but is widely used nowadays for conditional gene expression regulation in bacteria (13, 31, 32), yeast (26, 33, 34), insects (35), and mammalian cells (9). This protein binds with high affinity to specific DNA sequences called *tetO* sites, usually introduced in the promoter region to make it repressible. TetR binding can be abolished by the addition of tetracycline or its analogs (9).

The following protocols can be used to recreate the plasmids and respective yeast strains carrying integrated negative feedback circuits that are also available from our laboratory by request (19). In addition, the reporter yEGFP gene in these protocols can be replaced with another gene of interest for which precise linear regulation of expression is needed.

3.1. Construction of Regulatory and Reporter Plasmids for Building the Synthetic Gene Construct with Negative Feedback

In this section, we describe the assembly of the two plasmids (reporter plasmid with *yEGFP* gene and regulatory plasmid with *tetR* gene) used to build synthetic gene constructs in yeast (Fig. 1) (19). Both plasmids contain the ampicillin resistance gene *ampR* for selection in *E. coli.*

3.1.1. Removal of tetR Gene Expressed from the P_{GAL10} Promoter in the pRS4D1 Plasmid

The *tetR* gene downstream of the P_{GAL10} promoter in the parental pRS4D1 plasmid (Fig. 2a) (21, 26) should be deleted so that *tetR* can be expressed solely from the $P_{GAL1\text{-}D12}$ promoter in the final regulatory plasmid. Two protocols are described below in which *tetR* is either completely deleted (steps 1–6) (Fig. 2b) or replaced

[1] http://www.async.ece.utah.edu/iBioSim/

[2] http://www.r-project.org/

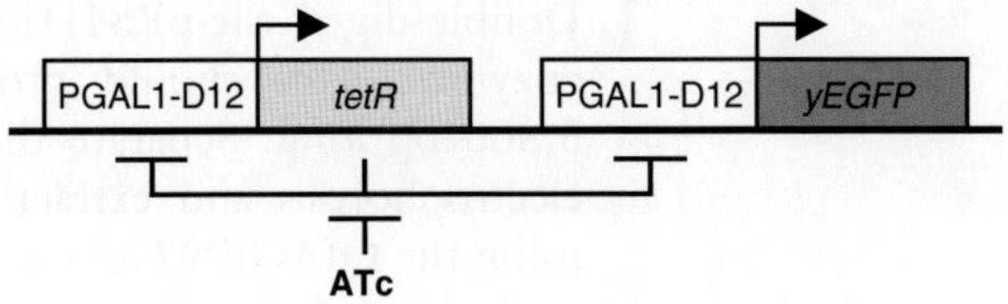

Fig. 1. Diagram of the gene regulatory cascade with negative feedback, consisting of the *yEGFP* reporter and the *tetR* repressor that also regulates its own expression.

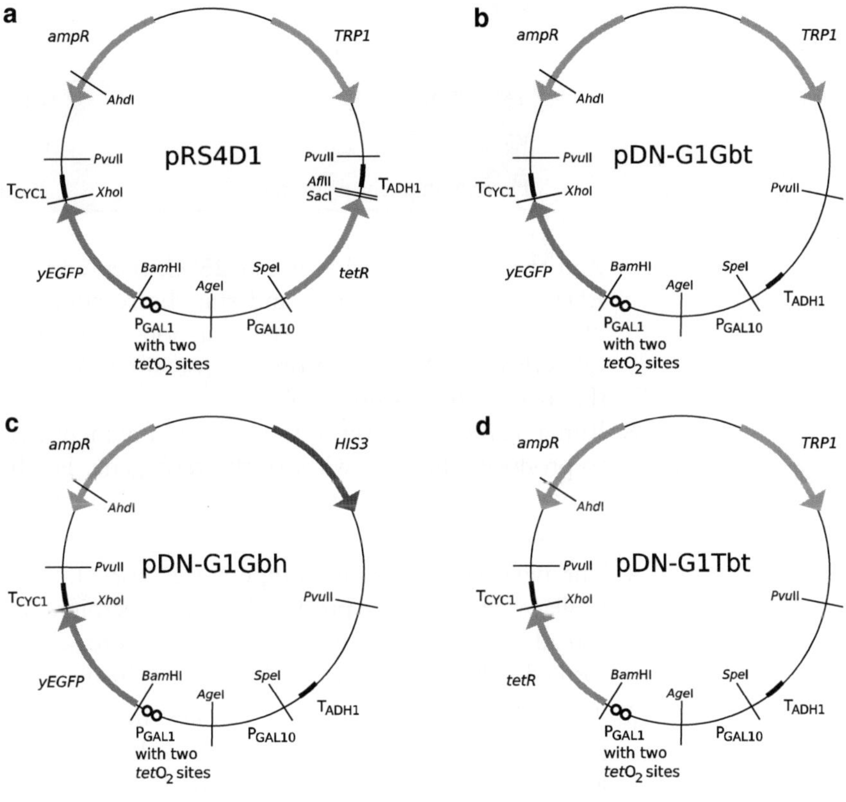

Fig. 2. Map of the plasmids used to build the gene expression linearizer with negative feedback (19). (**a**) The original pRS4D1 plasmid (21, 26) that was used for subsequent cloning procedures. (**b**) The intermediate pDN-G1Gbt plasmid created in Subheading 3.1.1, from which the *tetR* gene is deleted downstream of the P_{GAL10} promoter. (**c**) The pDN-G1Gbh reporter plasmid created in Subheading 3.1.2, in which the *yEGFP* gene is expressed from the $P_{GAL1\text{-}D12}$ promoter. (**d**) The regulatory pDN-G1Tbt plasmid created in Subheading 3.1.3, with the *tetR* gene expressed from the $P_{GAL1\text{-}D12}$ promoter.

with a nonfunctional gene fragment lacking the ATG start codon (steps 7–11). The completion of either protocol will result in a final plasmid product without a functional *tetR* gene expressed from P_{GAL10}.

1. Double-digest the pRS4D1 plasmid using the *Afl*II and *Spe*I enzymes. This should produce two fragments 600 and 5,800 bp long. Separate these fragments using agarose gel electrophoresis and extract the large (5,800 bp) fragment using the QIAGEN Gel Extraction kit (or similar), according to the manufacturer's protocol.
2. Convert the sticky ends of the extracted fragment from the previous step to blunt ends in the reaction with the Klenow fragment, as below:

Water	5.6 μl
Buffer 2 (New England Biolabs) 10×	3 μl
dNTP 1 mM	1 μl
Klenow 2 U	0.4 μl
Plasmid fragment from step 1	20 μl

   Run the reaction for 15 min at 25°C, then add 0.612 μl of EDTA 0.5 M to the tube and run the reaction for another 20 min at 75°C. Purify the reaction product using the QIAGEN PCR Purification kit (or similar), according to the manufacturer's protocol.
3. Run a ligation reaction of the blunted fragment from step 2, to produce a plasmid without the *tetR* gene. For better efficiency, run the ligation reaction overnight at room temperature (see Note 1).
4. The next day, transform competent *E. coli* XL-10 cells (or a similar strain) with the product from the overnight ligation and spread bacteria onto a plate with ampicillin to select for transformed cells. Grow these cells in the 37°C incubator overnight.
5. The next day, pick several colonies from the plate, and propagate them overnight in LB medium with ampicillin added. Purify the plasmids using the QIAGEN Miniprep kit (or similar), according to the manufacturer's protocol.
6. Send the plasmid samples to sequencing to confirm that the product has the proper DNA sequence. We suggest the Insert-f and GalSeqE-r primers for sequencing.
7. Double-digest the pRS4D1 plasmid using the *Sac*I and *Spe*I restriction enzymes, which should produce two fragments 600 and 5,800 bp long. Separate these fragments using agarose gel electrophoresis and extract the large (5,800 bp) fragment using the QIAGEN Gel Extraction kit (or similar), according to the manufacturer's protocol.
8. Amplify the nonfunctional *tetR* fragment by PCR (30× cycles: 10 s at 98°C; 30 s at 68°C; 30 s at 72°C; Phusion HS

polymerase) using the pair of primers tetRcut-f and tetRcut-r, and the original pRS4D1 plasmid as the template. Separate the final 270-bp PCR product using agarose gel electrophoresis and extract it using the QIAGEN Gel Extraction kit (or similar), according to the manufacturer's protocol.

9. Double-digest the product from the previous PCR step using the *Sac*I and *Spe*I restriction enzymes, and purify it using the QIAGEN PCR Purification kit (or similar), according to the manufacturer's protocol.
10. Use both the 5,800-bp fragment from step 1 and the 270-bp PCR product from step 3 for sticky-end ligation to produce a plasmid with a nonfunctional *tetR* gene fragment lacking the ATG start codon downstream of the P_{GAL10} promoter.
11. Follow steps 4–6.

3.1.2. Construction of the Reporter Plasmid with the HIS3 Yeast Selective Marker

In this section, a protocol for producing the reporter plasmid will be provided by transferring a cassette containing the $P_{GAL1\text{-}D12}$ promoter, the *yEGFP* gene, and the flanking terminators into the pRS404 vector with the *HIS3* gene as the auxotrophic marker for *S. cerevisiae*. The final product from this step will be the reporter plasmid (Fig. 2c).

1. Amplify the cassette containing the $P_{GAL1\text{-}D12}$ promoter, the *yEGFP* gene, and the flanking terminators by PCR (30 × cycles: 30 s at 98°C; 30 s at 54°C; 75 s at 72°C; Phusion HS polymerase) using the pair of primers PvuII-f, PvuII-r and the plasmid product from Subheading 3.1.1 as a template. Purify the final PCR product using the agarose gel electrophoresis and QIAGEN Gel Extraction kit (or similar), according to the manufacturer's protocol.
2. Digest both the PCR product from step 1 and plasmid pRS403 with the *Pvu*II restriction enzyme. Separate the reaction products using agarose gel electrophoresis. Extract the digested PCR product and the large (4,000 bp) backbone fragment of the digested pRS403 plasmid and purify them using the QIAGEN Gel Extraction kit (or similar), according to the manufacturer's protocol.
3. Perform a blunt-end ligation reaction using both the purified PCR product and the plasmid backbone fragment (4,000 bp) from step 2 to produce a plasmid with the cassette inserted into the pRS403 backbone. For better efficiency, run the ligation reaction overnight at room temperature.
4. Follow the *E. coli* transformation and DNA preparation procedures as described in steps 4 and 5 of Subheading 3.1.1.
5. Sequence the plasmid samples together with the primers GalSeqA-f, GalSeqB-f, GalSeqC-f, GalSeqD-r, and GalSeqG-r to

confirm proper insertion of the cassette into the plasmid (see Note 2).

3.1.3. Construction of the Regulatory Plasmid with the TRP1 Yeast Selective Marker

In this part, the regulatory plasmid will be built by replacing the *yEGFP* gene downstream from the $P_{GAL1\text{-}D12}$ promoter with the *tetR* gene. The final product of this protocol will be the regulatory plasmid with negative feedback (Fig. 2d).

1. Amplify the functional *tetR* gene by PCR (30× cycles: 30 s at 98°C; 30 s at 65°C; 50 s at 72°C; Phusion HS polymerase) using the pair of primers tetR-BamHI-f, tetR-XhoI-r and the pRS4D1 plasmid as a template. Purify the final 600-bp PCR product using the QIAGEN PCR Purification kit (or similar), according to the manufacturer's protocol.
2. Double-digest the PCR product from step 1 and the final plasmid from either Protocol A or B with *Bam*HI and *Xho*I restriction enzymes. Separate the products of these reactions using agarose gel electrophoresis. Extract the digested PCR product and the large (5,300 bp) backbone fragment of the digested plasmid, and purify them using the QIAGEN Gel Extraction kit (or similar), according to the manufacturer's protocol.
3. Join the purified PCR product and the plasmid backbone fragment from step 2 by sticky-end ligation, to produce the plasmid with the *tetR* gene downstream of the $P_{GAL1\text{-}D12}$ promoter.
4. Follow the *E. coli* transformation and DNA preparation (steps 4 and 5 of Subheading 3.1.1).
5. Send the plasmid samples to sequencing together with the GalSeqB-f, GalSeqC-f, GalSeqD-r, and Backbone-r primers to confirm proper insertion of the *tetR* gene into the plasmid.

3.2. Integration of Synthetic Gene Constructs into the Yeast Genome

In this section, we describe a two-step integration process of both plasmids into the *S. cerevisiae* genome, starting with the reporter plasmid followed by the regulatory plasmid. The final product will be a yeast strain with the synthetic gene construct where both genes are present in single copies. Both protocols are based on the modified lithium acetate procedure (36, 37). Please note that yeast cells should be grown in darkness.

Transformation procedures were designed such that the first reporter plasmid is linearized by *Age*I enzyme and integrated into the yeast genome in the native $P_{GAL1\text{-}10}$ promoter locus. Subsequently, the regulatory plasmid is linearized by the *Ahd*I enzyme and integrated into the *ampR* gene of the previously integrated reporter plasmid. As a result, both parts (reporter and regulatory) are placed nearby on the same chromosome, eliminating variability of gene expression due to different integration sites.

3.2.1. Integration of the Reporter Plasmid with the HIS3 Yeast Selective Marker

1. Pick a single colony of the haploid YPH500 strain from the plate and inoculate 5 ml of YPD medium. Incubate overnight in a shaking incubator at 30°C, 300 rpm.
2. Set up a linearization reaction for the reporter plasmid obtained in Subheading 3.1.2 with the *Age*I restriction enzyme (see Note 3).
3. The next morning, prepare a flask with 10 ml of fresh YPD medium and add 5 ml from the overnight culture (step 1) to the flask. Incubate it in a shaking incubator at 30°C at 300 rpm for 4–5 h.
4. During the incubation, purify the digested reporter plasmid from step 2 using the QIAGEN PCR Purification kit (or similar), according to the manufacturer's protocol and elute the plasmid in 50 μl of water.
5. Harvest the cells obtained from step 3 and centrifuge them at 3,000 × *g* for 5 min.
6. Discard the supernatant and resuspend the cells in 10 ml of water.
7. Centrifuge the cells again at 3,000 × *g* for 5 min.
8. While centrifuging the cells, boil carrier DNA from salmon testes for 5 min and keep it on ice afterward until transformation (see Note 4).
9. Discard the supernatant from step 7 and resuspend the cells in 1 ml of lithium acetate 100 mM. Transfer the suspension to a 1.5-ml tube.
10. Centrifuge the cells in a minicentrifuge at 6,000 × *g* for 1 min.
11. Discard the supernatant and resuspend the cells in 0.2 ml of 100 mM lithium acetate. Transfer 50 μl of cell suspension to a separate 1.5-ml tube (see Note 5).
12. Centrifuge the transferred cells in a separate 1.5-ml tube at 6,000 × *g* for 1 min. Discard the supernatant.
13. Add the following to the tube with the cell pellet from step 12 (in the order listed):

    240 μl of polyethylene glycol MW 3350 50% w/v

    36 μl of lithium acetate 1 M

    25 μl carrier DNA from salmon testes 2 mg/ml from step 8

    50 μl purified digested plasmid from step 4 (0.1–10 μg).
14. Vortex the tube until the cells are completely resuspended. Use a pipette if needed.
15. Incubate the tube at 30°C for 30 min.
16. Incubate the tube at 42°C for another 20–25 min (heat shock).

17. Centrifuge the suspension in a minicentrifuge at 6,000 × *g* for 1 min and carefully remove the supernatant by pipetting.
18. Resuspend the cells in water.
19. Spread the cells on selective SC-his plates.
20. Incubate at 30°C for 2 days.
21. Select appropriate clones using flow cytometry (Subheading 3.5.1) and PCR (Subheading 3.5.2). Make backup stocks (Subheading 3.5.4).

3.2.2. Integration of the Regulatory Plasmid with the TRP1 Yeast Selective Marker

1. Pick a single colony of the yeast strain with the integrated reporter plasmid obtained in Subheading 3.2.1 and inoculate 5 ml of SC-his medium supplemented with 2% glucose. Incubate overnight in a shaking incubator at 30°C, 300 rpm.
2. Linearize the regulatory plasmid obtained in Subheading 3.1.3 using the *Ahd*I restriction enzyme (see Note 3).
3. The next morning prepare a flask with 10 ml of fresh SC-his medium supplemented with 2% glucose and add 5 ml of the overnight culture from step 1 to the flask. Incubate it in a shaking incubator at 30°C, 300 rpm for 4–5 h.
4. During the incubation, purify the digested reporter plasmid from step 2 using the QIAGEN PCR Purification kit (or similar), according to the manufacturer's protocol and elute the plasmid in 50 μl of water.
5. Follow steps 5–18 from Subheading 3.2.1, using the regulatory (instead of the reporter) plasmid in this procedure.
6. Spread cells on selective SC-his-tryp plates (see Note 6).
7. Incubate at 30°C for 2 days.
8. Select appropriate clones using PCR (Subheading 3.5.3) and make backup stocks (Subheading 3.5.4).

3.3. Fluorescence Measurements and Data Processing

This section describes the quantitative assessment of dose–response linearity in the synthetic TetR-based gene circuit with negative feedback. Reporter (*yEGFP*) expression over the cell population will be measured by flow cytometry at increasing ATc concentrations, followed by metrics of linearity.

1. Pick a single colony of the yeast strain with integrated reporter and regulatory plasmids (Subheading 3.2.2) from the plate and inoculate 1 ml of SC-his-tryp medium supplemented with 2% glucose. Incubate overnight in a shaking incubator at 30°C, 300 rpm.
2. The next morning centrifuge cells at 3,000 × *g* for 5 min, discard supernatant, and resuspend them in SC-his-tryp medium supplemented with 2% galactose (see Note 7).

3. Prepare a set of tubes with SC-his-tryp medium supplemented with 2% galactose and inoculate with cells prepared in step 2 so that the final OD_{600} of the culture is 0.01. Add ATc in increasing concentrations (0–500 ng/ml, see Note 8).
4. Grow the cells overnight (16 h) in a shaking incubator at 30°C, 300 rpm.
5. The next day take 100–200 μl of every overnight culture, centrifuge at 3,000 × *g* for 5 min, and resuspend in 400 μl of PBS. Run the samples on a flow cytometer until 50,000–100,000 cells are collected.
6. Preprocess the data to transform the original log-binned fluorescence intensity values to linear scale and to filter out contributions from cellular debris. A narrow forward and side scatter gate must be used for data analysis to minimize external variability due to cell size and shape. Calculate the yEGFP fluorescence mean by averaging the fluorescence values of at least ~10,000 cells after preprocessing. Calculate the noise (defined as the coefficient of variation, CV), dividing the standard deviation by the mean for each sample.
7. Plot the gene expression mean and noise (CV) for increasing ATc concentrations. Linearity prior to saturation can be assessed both using linear regression (a standard technique that yields the R^2 value) and the L1-norm. The L1-norm measures the distance of the dose–response from an ideal, linear "target function" as the area between a nonlinear fit to the dose–response data and the linear target function. The area enclosed by these functions can be determined by numerical integration using the trapeze method (the same procedure can also be applied to measure the difference of two dose–responses). The range of inducer concentrations over which linearity is assessed depends on the experimenter's objective: the linearity of dose–response should be measured in the induction regime where linearity is desired. However, in most cases the objective is to have a linear dose–response from no induction up to saturation (or extending as close to these extremes as possible). Therefore, we recommend measuring linearity up to 90% saturation, because both linearity metrics (linear regression and the L1-norm) were robust up to this induction level in our linearizer constructs (19).

3.4. Mathematical and Computational Modeling

1. Determine a system of chemical reactions that constitute a model. For example, for TetR negative autoregulation the following statements provide a skeleton for modeling:
   (a) TetR represses itself and the downstream reporter gene.
   (b) TetR and reporter proteins degrade at roughly constant rates given by cell growth.

(c) TetR binds nearly irreversibly to ATc and is inactivated when bound.

(d) In order for ATc to bind to TetR, it must diffuse into the cell from outside the cell.

These observations can be converted into the Dizzy code in Box 1, or written as the set of ODEs

$$\begin{aligned} \dot{w} &= bxy - dw, \\ \dot{x} &= a_x F(x) - bxy - dx, \\ \dot{y} &= C - bxy - fy, \\ \dot{z} &= a_z F(x) - dz, \end{aligned} \tag{1}$$

where w, x, y, and z are concentrations of TetR species bound to ATc, TetR species unbound to ATc, intracellular ATc, and reporter proteins, respectively. The parameters a_x and a_z represent maximal gene expression of TetR and the gene of interest, respectively, b represents

Box 1
Dizzy code for the TetR gene expression system with negative feedback

```
//Rates

ax=100;                 // Maximal TetR production rate
az=100;                 // Maximal GFP production rate
b=4;                    //ATc TetR Binding rate
d=ln(2)/2;              //TetR and GFP degradation/dilution rate
f=ln(2)/(45/60);        //ATc diffusion rate
kd=0.1;                 //TetR promoter dissociation constant
n=4;                    //Hill coefficient

//Species

w=0;                    //Bound/Inactive TetR
x=0.5;                  //Unbound/active TetR
y=0;                    //Intracellular ATc
z=0.5;                  //Reporter GFP
T=[w+x];                //Total TetR
C=5;                    //Extracellular ATc

//Reactions

influx,        C->C+y,   f;
outflux,       y->,      f;
xproduction,   ->x,      [ax*kd^n/(kd^n+x^n)];
xdegrade,      x->,      d;
xbind,         x+y->w,   b;
wdegrade,      w->,      d;
zprod,         ->z,      [az*kd^n/(kd^n+x^n)];
zdegrade,      z->,      d;
```

ATc binding to TetR, C is ATc influx into the cell and is directly related to extracellular ATc concentrations, d is the rate of degradation/dilution, f is the combined rate of ATc dilution and diffusion out of the cell, and $F(x)$ is a function representing repression and can be approximated with the Hill function

$$F(x) = \frac{k_\mathrm{d}^n}{k_\mathrm{d}^n + x^n}, \quad (2)$$

where k_d is the dissociation constant, and n is the Hill coefficient.

2. Estimate model parameters to describe a specific system. ATc/TetR binding rates (b) and ATc diffusion (f) rates can be estimated from literature (38, 39). For slowly degrading proteins such as yEGFP, the growth rate of a cell can be used to estimate d (~ln(2)/μ, where μ is the growth rate). The production rates ($aF(x)$) and ATc influx (C) change depending on promoter strength, number of *tet*O2 sites, and ATc degradation and may have to be obtained from fitting. Experimental GFP observations collected when cells are close to a stationary distribution can be used to fit the gene expression and ATc influx rates at steady state using nonlinear fitting methods such as the Nelder–Mead algorithm implemented in fminsearch in Matlab.
3. Estimate the dose–response characteristic, $z(C)$. Assuming high free ATc retention within cells, low TetR degradation, and identical upstream and downstream promoter dynamics, Eq. 1 can be rewritten at steady state as

$$\begin{aligned} y &\approx \frac{C}{bx}, \\ a_x F(x) &\approx bx\frac{C}{bx} \Rightarrow x = F^{-1}\left(\frac{C}{a_x}\right), \quad (3) \\ z &= \frac{a_z}{d}F(x) \approx \frac{a_z C}{a_x d}, \end{aligned}$$

which implies a linear dose–response to ATc whose slope is dependent on the ratio of maximal promoter strengths for TetR and the downstream gene, ATc influx into the cell, and downstream protein degradation. The assumptions required for linear response will break down at saturation, when ATc is no longer able to sequester TetR and at low levels of induction when basal *tetR* expression is significant (19).
4. Include additional reactions/molecular details if needed. Perform stochastic simulations based on the Gillespie algorithm (40) in the software Dizzy (28) or iBioSim.[3] Adjust

[3]http://www.async.ece.utah.edu/iBioSim/

parameters to match the measured values of the coefficient of variation (CV, noise).

3.5. Supplementary Protocols

3.5.1. Checking the Number of Integrations by Flow Cytometry

1. Pick several individual colonies of the yeast strain containing reporter plasmid from the transformation plate and inoculate 1 ml of SC-his medium supplemented with 2% galactose.
2. Grow cells overnight (16 h) in a shaking incubator at 30°C, 300 rpm.
3. The next day take 100–200 μl of every overnight culture, centrifuge at 3,000 × *g* for 5 min, and resuspend in 400 μl of PBS. Read cells on a flow cytometer until 50,000–100,000 cells are collected.
4. Preprocess data to transform the original log-binned fluorescence intensity values to a linear scale and filter out contributions from cellular debris. Calculate the mean by averaging the fluorescence values of cells after preprocessing.
5. Compare the mean fluorescence values for the samples. Cell cultures containing multiple integrations of the reporter plasmid will have higher fluorescence level compared to the rest of the clones.

3.5.2. Checking the Number of Integrations of the Reporter Plasmid by PCR

1. Pick several individual colonies of the yeast strain with the integrated reporter plasmid obtained in Subheading 3.2.1 and inoculate 1 ml of SC-his medium supplemented with 2% glucose. Incubate overnight in a shaking incubator at 30°C at 300 rpm.
2. The next day extract genomic DNA from the overnight cultures using the QIAGEN DNeasy Blood & Tissue kit (or similar), according to the manufacturer's protocol.
3. Run a set of PCRs (35× cycles: 20 s at 95°C; 20 s at 54°C; 50 s at 72°C; Paq5000™ DNA Polymerase) using the pair of primers GalSeqA-f and Gal1-D12-r and genomic DNA preparations from step 2 as templates (see Note 9).
4. Separate PCR products using gel electrophoresis. Only PCR done with genomic preparations obtained from the clones with multiple integrations of the reporter plasmid will result in an 850 bp product (see Note 10).

3.5.3. Checking the Number of Integrations of the Regulatory Plasmid by PCR

1. Pick several individual colonies of the yeast strain with integrated reporter and regulatory plasmids obtained in Subheading 3.2.2 and inoculate 1 ml of SC-his-tryp medium supplemented with 2% glucose. Incubate overnight in a shaking incubator at 30°C, 300 rpm.
2. The next day extract genomic DNA from the overnight cultures using the QIAGEN DNeasy Blood & Tissue kit (or similar), according to the manufacturer's protocol.

3. Run a set of PCR (35× cycles: 20 s at 98°C; 20 s at 60°C; 120 s at 72°C; *PfuUltra*™ II Fusion HS DNA Polymerase) using the pair of primers tetR-BamHI-f, before2trp-r and genomic preparations from step 2 as templates (see Note 9).
4. Separate the PCR products using gel electrophoresis. Only PCR done with genomic preparations obtained from the clones with multiple integrations of the regulatory plasmid will result in a 3,200 bp product (see Note 11).

3.5.4. Preparing Stock of the Selected Yeast Clones

1. Pick a single colony from the transformation plate with the yeast strain of interest and inoculate 1 ml of respective selective SC medium containing 2% glucose (see Note 12). Incubate overnight in a shaking incubator at 30°C, 300 rpm.
2. Pour 812 μl of overnight culture into a plastic vial and add 188 μl of 80% glycerol, then mix.
3. Store vials with frozen stocks at −80°C (see Notes 13).

4. Notes

1. Please note that blunt end ligation destroys the *Afl*II site in the original pRS4D1 plasmid, keeping the *Spe*I site intact.
2. Due to the fact that blunt end ligation is used in this case, it is possible to obtain two orientations of the cassette in the resulting plasmid. We chose an orientation that is similar to the orientation of the cassette in the source plasmid with respect to the mutual position of the *yEGFP* and *ampR* genes (Fig. 2c).
3. Approximately 0.1–10 μg of plasmid DNA is used for one transformation.
4. We usually use a thermocycler for this purpose, which can boil DNA and automatically chill it afterward to keep it ready for further transformation steps.
5. Depending on the density of the cell suspension the amount of transfer can be modified. Around 2–3 mm^3 of cell pellet per transformation should remain after the next centrifugation.
6. Please note that we are using SC-his-tryp plates to maintain selection for the already integrated reporter plasmid (*HIS3* auxotrophic marker, Subheading 3.2.1) and for the regulatory plasmid (*TRP1* auxotrophic marker).
7. It is important to wash the cells, because glucose from the overnight incubation medium can repress expression from the $P_{GAL1\text{-}D12}$ promoter.
8. Please note that ATc is extremely light sensitive. Therefore, the cells have to be grown in complete darkness. Avoid

exposing the cultures to light during the tube setup. ATc must be protected from light using metal foil while working with it on the bench. In addition, the inducer should be added to the tubes at the end, immediately before each individual tube was transferred to the shaking incubator.

9. We used high concentrations of primers (50 μM) for all genomic diagnostic PCRs as contrary to the usual 5–10 μM primer concentration used for preparatory PCR.
10. The plasmids used for integration (Fig. 1c) should be used as positive controls, and due to circularity, should normally give the same PCR product as the yeast genome samples with multiple integrations (approximately 850 bp). It is also reasonable to include diagnostic PCR genome samples from yeast clones which have suspected multiple integrations based on the results of flow cytometry screening (Subheading 3.5.1). Results of both flow cytometry screening and diagnostic PCR (Subheadings 3.5.1 and 3.5.2, respectively) complement each other and should be assessed together to chose clones with single integrations.
11. As for previous protocols, the plasmid used for integration (Fig. 1d) should be used as positive control and should normally give the same PCR product as yeast genome samples with multiple integrations (approximately 3,200 bp). However, due to the length of the anticipated product, the lack of the band on the gel should be treated as an argument in favor of single integration, but not as definite proof of it. The quality of genomic DNA preparation and reaction parameters might affect the efficiency of diagnostic PCR and even multiple integrants can lack the band sometimes. Therefore, we recommend running a parallel control PCR for products with similar length (3,200 bp) using the same genomic sample preparations.
12. Use SC-his medium for the yeast strain obtained in Subheading 3.2.1 and SC-his-tryp medium for the yeast strain obtained in Subheading 3.2.2.
13. Stocks may be stored indefinitely.

Acknowledgments

We thank J. J. Collins for some of the constructs, yeast strains, and discussions. We also thank K. F. Murphy, K. Josić, R. Agarwal, T. Żal, A. Żal, G. Chodaczek, M. Stamatakis, W. Blake, T. F. Cooper, and B. Dutta for valuable comments and discussions. This work was supported by M. D. Anderson Cancer Center start-up funds.

References

1. Hughes, T. R., Marton, M. J., Jones, A. R., Roberts, C. J., Stoughton, R., Armour, C. D., Bennett, H. A., Coffey, E., Dai, H., He, Y. D., Kidd, M. J., King, A. M., Meyer, M. R., Slade, D., Lum, P. Y., Stepaniants, S. B., Shoemaker, D. D., Gachotte, D., Chakraburtty, K., Simon, J., Bard, M., and Friend, S. H. (2000) Functional discovery via a compendium of expression profiles. *Cell* **102**, 109–26.
2. Sopko, R., Huang, D., Preston, N., Chua, G., Papp, B., Kafadar, K., Snyder, M., Oliver, S. G., Cyert, M., Hughes, T. R., Boone, C., and Andrews, B. (2006) Mapping pathways and phenotypes by systematic gene overexpression. *Mol Cell* **21**, 319–30.
3. Elbashir, S. M., Harborth, J., Lendeckel, W., Yalcin, A., Weber, K., and Tuschl, T. (2001) Duplexes of 21-nucleotide RNAs mediate RNA interference in cultured mammalian cells. *Nature* **411**, 494–8.
4. Birmingham, A., Anderson, E. M., Reynolds, A., Ilsley-Tyree, D., Leake, D., Fedorov, Y., Baskerville, S., Maksimova, E., Robinson, K., Karpilow, J., Marshall, W. S., and Khvorova, A. (2006) 3′ UTR seed matches, but not overall identity, are associated with RNAi off-targets. *Nat Methods* **3**, 199–204.
5. Wall, M. E., Hlavacek, W. S., and Savageau, M. A. (2004) Design of gene circuits: lessons from bacteria. *Nat Rev Genet* **5**, 34–42.
6. Yao, F., Svensjo, T., Winkler, T., Lu, M., Eriksson, C., and Eriksson, E. (1998) Tetracycline repressor, tetR, rather than the tetR-mammalian cell transcription factor fusion derivatives, regulates inducible gene expression in mammalian cells. *Hum Gene Ther* **9**, 1939–50.
7. Urlinger, S., Baron, U., Thellmann, M., Hasan, M. T., Bujard, H., and Hillen, W. (2000) Exploring the sequence space for tetracycline-dependent transcriptional activators: novel mutations yield expanded range and sensitivity. *Proc Natl Acad Sci U S A* **97**, 7963–8.
8. Hillen, W., and Berens, C. (1994) Mechanisms underlying expression of Tn10 encoded tetracycline resistance. *Annu Rev Microbiol* **48**, 345–69.
9. Berens, C., and Hillen, W. (2003) Gene regulation by tetracyclines. Constraints of resistance regulation in bacteria shape TetR for application in eukaryotes. *Eur J Biochem* **270**, 3109–21.
10. Berens, C., and Hillen, W. (2004) Gene regulation by tetracyclines. *Genet Eng (N Y)* **26**, 255–77.
11. Kramer, B. P., Weber, W., and Fussenegger, M. (2003) Artificial regulatory networks and cascades for discrete multilevel transgene control in mammalian cells. *Biotechnol Bioeng* **83**, 810–20.
12. Thieffry, D., Huerta, A. M., Perez-Rueda, E., and Collado-Vides, J. (1998) From specific gene regulation to genomic networks: a global analysis of transcriptional regulation in Escherichia coli. *Bioessays* **20**, 433–40.
13. Becskei, A., and Serrano, L. (2000) Engineering stability in gene networks by autoregulation. *Nature* **405**, 590–3.
14. Austin, D. W., Allen, M. S., McCollum, J. M., Dar, R. D., Wilgus, J. R., Sayler, G. S., Samatova, N. F., Cox, C. D., and Simpson, M. L. (2006) Gene network shaping of inherent noise spectra. *Nature* **439**, 608–11.
15. Ramsey, S. A., Smith, J. J., Orrell, D., Marelli, M., Petersen, T. W., de Atauri, P., Bolouri, H., and Aitchison, J. D. (2006) Dual feedback loops in the GAL regulon suppress cellular heterogeneity in yeast. *Nat Genet* **38**, 1082–7.
16. Rosenfeld, N., Elowitz, M. B., and Alon, U. (2002) Negative autoregulation speeds the response times of transcription networks. *J Mol Biol* **323**, 785–93.
17. Stricker, J., Cookson, S., Bennett, M. R., Mather, W. H., Tsimring, L. S., and Hasty, J. (2008) A fast, robust and tunable synthetic gene oscillator. *Nature* **456**, 516–9.
18. Elowitz, M. B., and Leibler, S. (2000) A synthetic oscillatory network of transcriptional regulators. *Nature* **403**, 335–8.
19. Nevozhay, D., Adams, R. M., Murphy, K. F., Josic, K., and Balázsi, G. (2009) Negative autoregulation linearizes the dose-response and suppresses the heterogeneity of gene expression. *Proc Natl Acad Sci U S A* **106**, 5123–8.
20. Yu, R. C., Pesce, C. G., Colman-Lerner, A., Lok, L., Pincus, D., Serra, E., Holl, M., Benjamin, K., Gordon, A., and Brent, R. (2008) Negative feedback that improves information transmission in yeast signalling. *Nature* **456**, 755–61.
21. Blake, W. J., Kærn, M., Cantor, C. R., and Collins, J. J. (2003) Noise in eukaryotic gene expression. *Nature* **422**, 633–7.
22. Newman, J. R., Ghaemmaghami, S., Ihmels, J., Breslow, D. K., Noble, M., DeRisi, J. L., and Weissman, J. S. (2006) Single-cell proteomic analysis of S. cerevisiae reveals the architecture of biological noise. *Nature* **441**, 840–6.
23. Giaever, G., Chu, A. M., Ni, L., Connelly, C., Riles, L., Veronneau, S., Dow, S., Lucau-Danila, A., Anderson, K., Andre, B., Arkin, A. P., Astromoff, A., El-Bakkoury, M., Bangham, R., Benito, R., Brachat, S., Campanaro,

S., Curtiss, M., Davis, K., Deutschbauer, A., Entian, K. D., Flaherty, P., Foury, F., Garfinkel, D. J., Gerstein, M., Gotte, D., Guldener, U., Hegemann, J. H., Hempel, S., Herman, Z., Jaramillo, D. F., Kelly, D. E., Kelly, S. L., Kotter, P., LaBonte, D., Lamb, D. C., Lan, N., Liang, H., Liao, H., Liu, L., Luo, C., Lussier, M., Mao, R., Menard, P., Ooi, S. L., Revuelta, J. L., Roberts, C. J., Rose, M., Ross-Macdonald, P., Scherens, B., Schimmack, G., Shafer, B., Shoemaker, D. D., Sookhai-Mahadeo, S., Storms, R. K., Strathern, J. N., Valle, G., Voet, M., Volckaert, G., Wang, C. Y., Ward, T. R., Wilhelmy, J., Winzeler, E. A., Yang, Y., Yen, G., Youngman, E., Yu, K., Bussey, H., Boeke, J. D., Snyder, M., Philippsen, P., Davis, R. W., and Johnston, M. (2002) Functional profiling of the Saccharomyces cerevisiae genome. *Nature* **418**, 387–91.

24. Wapinski, I., Pfeffer, A., Friedman, N., and Regev, A. (2007) Natural history and evolutionary principles of gene duplication in fungi. *Nature* **449**, 54–61.
25. Huh, W. K., Falvo, J. V., Gerke, L. C., Carroll, A. S., Howson, R. W., Weissman, J. S., and O'Shea, E. K. (2003) Global analysis of protein localization in budding yeast. *Nature* **425**, 686–91.
26. Blake, W. J., Balázsi, G., Kohanski, M. A., Isaacs, F. J., Murphy, K. F., Kuang, Y., Cantor, C. R., Walt, D. R., and Collins, J. J. (2006) Phenotypic consequences of promoter-mediated transcriptional noise. *Mol Cell* **24**, 853–65.
27. Eaton, J. W., Bateman, D., and Hauberg, S. (2008) *GNU Octave Manual*, Network Theory Ltd.
28. Ramsey, S., Orrell, D., and Bolouri, H. (2005) Dizzy: stochastic simulation of large-scale genetic regulatory networks (supplementary material). *J Bioinform Comput Biol* **3**, 437–54.
29. Hillen, W., Klock, G., Kaffenberger, I., Wray, L. V., and Reznikoff, W. S. (1982) Purification of the TET repressor and TET operator from the transposon Tn10 and characterization of their interaction. *J Biol Chem* **257**, 6605–13.
30. Yang, H. L., Zubay, G., and Levy, S. B. (1976) Synthesis of an R plasmid protein associated with tetracycline resistance is negatively regulated. *Proc Natl Acad Sci U S A* **73**, 1509–12.
31. Dublanche, Y., Michalodimitrakis, K., Kummerer, N., Foglierini, M., and Serrano, L. (2006) Noise in transcription negative feedback loops: simulation and experimental analysis. *Mol Syst Biol* **2**, 41.
32. Hooshangi, S., Thiberge, S., and Weiss, R. (2005) Ultrasensitivity and noise propagation in a synthetic transcriptional cascade. *Proc Natl Acad Sci U S A* **102**, 3581–6.
33. Becskei, A., Kaufmann, B. B., and van Oudenaarden, A. (2005) Contributions of low molecule number and chromosomal positioning to stochastic gene expression. *Nat Genet* **37**, 937–44.
34. Murphy, K. F., Balázsi, G., and Collins, J. J. (2007) Combinatorial promoter design for engineering noisy gene expression. *Proc Natl Acad Sci U S A* **104**, 12726–31.
35. Bello, B., Resendez-Perez, D., and Gehring, W. J. (1998) Spatial and temporal targeting of gene expression in Drosophila by means of a tetracycline-dependent transactivator system. *Development* **125**, 2193–202.
36. Gietz, R. D., Schiestl, R. H., Willems, A. R., and Woods, R. A. (1995) Studies on the transformation of intact yeast cells by the LiAc/SS-DNA/PEG procedure. *Yeast* **11**, 355–60.
37. Amberg, D. C., Burke, D. J., and Strathern, J. N. (2005) Methods in Yeast Genetics. Cold Spring Harbor Laboratory Press.
38. Sigler, A., Schubert, P., Hillen, W., and Niederweis, M. (2000) Permeation of tetracyclines through membranes of liposomes and Escherichia coli. *Eur J Biochem* **267**, 527–34.
39. Schubert, P., Pfleiderer, K., and Hillen, W. (2004) Tet repressor residues indirectly recognizing anhydrotetracycline. *Eur J Biochem* **271**, 2144–52.
40. Gillespie, D. T., Lampoudi, S., and Petzold, L. R. (2007) Effect of reactant size on discrete stochastic chemical kinetics. *J Chem Phys* **126**, 034302.

Chapter 6

Measuring In Vivo Signaling Kinetics in a Mitogen-Activated Kinase Pathway Using Dynamic Input Stimulation

Megan N. McClean, Pascal Hersen, and Sharad Ramanathan

Abstract

Determining the in vivo kinetics of a signaling pathway is a challenging task. We can measure a property we termed pathway bandwidth to put *in vivo* bounds on the kinetics of the mitogen-activated protein kinase (MAPk) signaling cascade in *Saccharomyces cerevisiae* that responds to hyperosmotic stress [the High Osmolarity Glycerol (HOG) pathway]. Our method requires stimulating cells with square waves of oscillatory hyperosmotic input (1 M sorbitol) over a range of frequencies and measuring the activity of the HOG pathway in response to this oscillatory input. The input frequency at which the pathway's steady-state activity drops precipitously because the stimulus is changing too rapidly for the pathway to respond faithfully is defined as the pathway bandwidth. In this chapter, we provide details of the techniques required to measure pathway bandwidth in the HOG pathway. These methods are generally useful and can be applied to signaling pathways in *S. cerevisiae* and other organisms whenever a rapid reporter of pathway activity is available.

Key words: MAP kinase, Microfluidics, HOG pathway, Bandwidth, Frequency-response, Kinetics

1. Introduction

MAP kinase cascades are ubiquitous highly conserved phosphorylation cascades found in signaling pathways throughout the eukaryotic kingdom (1–3). The MAP kinases regulate diverse cellular processes, including differentiation, apoptosis, and proliferation (2). These cascades generally consist of three highly conserved kinases: a MAP kinase kinase kinase (MAPKKK), a MAP kinase kinase (MAPKK), and a MAP kinase (MAPK). When a cell is exposed to an external stimuli components of the appropriate MAP kinase pathway, including upstream kinases and the MAPK cascade, are sequentially activated by phosphorylation. The phosphorylated and activated MAPK triggers appropriate

Attila Becskei (ed.), *Yeast Genetic Networks: Methods and Protocols*, Methods in Molecular Biology, vol. 734,
DOI 10.1007/978-1-61779-086-7_6, © Springer Science+Business Media, LLC 2011

transcriptional and regulatory responses within the cell that lead to altered gene expression and protein activity (4, 5).

In *Saccharomyces cerevisiae*, the MAP kinase cascade that responds to increased external osmolarity is called the High Osmolarity Glycerol, or HOG pathway. The HOG pathway has two branches through which it receives input. One branch works through the Sho1 membrane protein and the MAPKKK Ste11. The other branch utilizes a phosphorelay system (involving the proteins Sln1, Ypd1, and Ssk1) and two semiredundant MAPKKKs (Ssk2 and Ssk22). When the HOG pathway is stimulated by increased osmolarity, the MAP kinase of the pathway, Hog1, is phosphorylated and localizes to the nucleus where it interacts with various transcription factors and begins the cell's transcriptional response to osmotic stress. The localization of Hog1 tagged with GFP (Hog1-GFP) can therefore be used as a reporter of the activity of the HOG pathway.

The HOG pathway is a well-studied system and much effort has been placed into measuring its kinetics and modeling the pathway's dynamics (6, 7). However, much of this work has been done in vitro and in silico. Here, we report a method for measuring the kinetics of all reactions in the pathway in vivo by measuring a property called pathway bandwidth (8). Pathway bandwidth puts a lower bound on the in vivo reaction rates in a cellular signaling pathway; no reaction can be slower than the pathway bandwidth.

For a signaling pathway responding to oscillatory input, the pathway bandwidth is defined as a critical frequency of input f_c above which the pathway can no longer respond faithfully to the input signal but either averages over the incoming signal or barely responds. We developed a theory and experimental technique for measuring the bandwidth of the HOG pathway, the pathway in *S. cerevisiae* which responds to hyperosmotic stress (8). To measure the bandwidth of the HOG pathway, we built a novel microfluidic device which allowed us to expose yeast cells to oscillating osmotic conditions (between 0 and 1 M sorbitol) while confined to a growth chamber. We then followed the activity of the pathway (by monitoring Hog1-GFP localization) real-time using a fluorescence microscope and measured the amplitude of this localization as the response of the pathway. The critical frequency of input f_c above which the Hog1-GFP localization was significantly reduced at steady state was taken to be the pathway bandwidth and found to be ~$4.6 \times 10^{-3}\ s^{-1}$. Furthermore, we were able to differentiate between the two input branches to the pathway and found that the Sho1 input branch is slower than the Ssk1 input branch with a bandwidth of ~$2.6 \times 10^{-3}\ s^{-1}$.

In this chapter, we describe the methods used to measure signaling pathway bandwidth. These methods can be adapted for use with a variety of signaling pathways in yeast and other organisms, and are therefore generally applicable to the study of a wide range of signaling questions.

2. Materials

2.1. Yeast Strains and Culture

1. Yeast strain ySR255 (S288C Mat a leu2Δ0 lys2Δ0 Hog1-GFP::HIS3 HTB2-mCHERRY-URA3). Htb2 was tagged with mCherry in the Hog1-GFP strain from the yeast GFP collection (9) using primers oSR421 (tactagggctgttaccaaatactcctcctctactcaagccGGTGACGGTGCTGGTTTA) and oSR422 (aaaagaaaacatgactaaatcacaatacctagtgagtgacTCGATGAATTCGAGCTCG) and plasmid pSR101 (mCherry-pTEF-caURA3-AMP in the pRS406 backbone (10)). Standard molecular biology and yeast transformation techniques were employed (11).
2. Synthetic complete yeast media (SC): 6.7 g YNB w/o AA (MP Biomedicals LLC, Pasadena, CA), 2 g CSM amino acid supplement (MP Biomedical LLC, Pasadena, CA), 20 g glucose dissolved in 1 l of water, autoclaved at 121°C at 15lb/sq of pressure.
3. 1 M Sorbitol synthetic complete media: 91.1 g of D-sorbitol is dissolved in synthetic complete yeast media, and this solution is brought up to 500 ml and sterile filtered using a 0.2-μm Supor machV membrane in the 0.75-mm filter unit from Nalgene.
4. 14 ml Polypropylene round bottom tubes.

2.2. Photolithography

1. 5,080 dpi photolithography transparency mask (Pageworks, Cambridge, MA).
2. 4″ Silicon Wafers (P(100) 0–100 Ω cm SSP 500 μm Test 100 crystal orientation, one-side polished) (University Wafers, South Boston, MA).
3. SU8 2050 (Microchem Corp., Newton, MA).
4. Propylene glycol methyl ether acetate (PGMEA) (Sigma–Aldrich, Saint Louis, MO).
5. AB-M Mask Aligner (ABM Inc., San Jose, CA).
6. Headway Spin Coater Model PWM32 (Headway Research, Garland, TX).
7. Veeco Profilometer, Model Detak 6M (Veeco Instruments, Plainview, NY).
8. 3-Aminopropyl-triethoxysilane, 99% (Sigma–Aldrich, Saint Louis, MO).
9. Hot Plate, Model EchoTherm HP30 (Torrey Pines Scientific, San Diego, CA).

2.3. Microfluidics

1. Dow Corning SYLGARD 184 Elastomer kit containing Sylgard 184 base, Sylgard 184 curing agent (polydimethylsiloxane, PDMS) (Ellsworth Adhesive Systems, Germantown, WI).

2. Cover Glass 23 × 60 mm No. 1 (VWR International Inc., Pittsburgh, PA).
3. 3M Scotch tape, Matte Finish, Magic Tape (3M, St. Paul, MN).
4. Plasma-Preen Cleaner/Etcher (Terra Universal, Fullerton, CA).
5. VWR Gravity Convection Oven Model 1300U (VWR International Inc., Pittsburgh, PA).
6. Harris Uni-Core 1.5 mm PDMS punches (Ted Pella Inc., Redding, CA).
7. 16 G 1½ and 21 G 1½ PrecisionGlide needles (Becton–Dickinson, Franklin Lakes, NJ).

2.4. Cell Loading and Adhesion

1. 1× Phosphate-buffered saline (calcium and magnesium-free, pH 7.4) (Mediatech Inc., Herndon, VA).
2. D-Sorbitol.
3. ConA loading solution: 2 mg/ml concanavalin A (conA) (supplied as a lyophilized, white powder. Essentially salt-free and carbohydrate-free, MP Biomedical LLC, Pasadena, CA), 5 mM $MnSO_4$, 5 mM $CaCl_2$, dissolved in 1× phosphate-buffered saline.

2.5. Microscopy

1. Microscope slide holder (1-mm thick aluminum, cut 1.25 × 3″ with 0.9″ × 0.9 inner square hole).
2. Lab Labeling Tape (VWR International, Pittsburgh, PA).
3. Zeiss 200M Fluorescence Microscope (Carl Zeiss MicroImaging, Inc., Thornwood, NY).
4. Orca-II-ER Camera (Hamamatsu, Bridgewater, NJ).
5. 100×/1.45 NA plan α fluor objective (Carl Zeiss MicroImaging, Inc., Thornwood, NY).
6. Metamorph (Molecular Devices, Sunnyvale, CA).

2.6. Fluid Control

1. RS-232 8-Channel 1-Amp N-Channel FET Controller Board (controlanything.com, Osceola, MO).
2. Fluidic switch, LFAA1201418H Model (The Lee Company, Westbrook, CT).
3. Intramedic tubing, Becton–Dickinson 0.86 mm inner diameter, 1.27 mm.
4. Outer diameter (Becton–Dickinson, Franklin Lakes, NJ).
5. Visual Basic software (Microsoft Visual Basic Express Edition 2008) (Microsoft, Redmond, WA).

6. Computer with available serial port.
7. DCTX-1216 12 V dc 1.2 A Wall Transformer (Allelectronics. Com, Van Nuys, CA).
8. 100 ft PVS for LIF Soft Tubing 0.042″ inner diameter (Lee Company, Westbrook, CT).
9. VWR Talon regular clamp holder (VWR International Inc., Pittsburgh, PA).
10. Screw cap tubes (15 and 50 ml Axygen Scientific, Union City, CA).
11. High vacuum grease silicone lubricant (Dow Corning, Midland, MI).
12. O-Ring stand (VWR International Inc., Pittsburgh, PA).
13. Tube clamps (VWR International Inc., Pittsburgh, PA).
14. Small binder clips for fluid control (VWR International Inc., Pittsburgh, PA).

2.7. Image and Data Analysis

1. ImageJ (http://rsbweb.nih.gov/ij/index.html).
2. Matlab (The Mathworks Inc., Natick, MA).

3. Methods

To measure signaling pathway response in vivo over different input frequencies, we use a microfluidic device that allows for rapid periodic changes in media while cells are continuously monitored under an inverted fluorescence microscope.

Rapid changes in media are difficult to achieve in conventional microfluidic devices. Our device has stimulating (1 M sorbitol) and nonstimulating media entering through the two inlets of a Y-shaped flow cell, as shown in Fig. 1. The flow of the media to the flow cell is gravity-driven and the flow velocity within the flow cell is proportional to the pressure drop ΔP between the inlet and outlet (12). The dimensions of the flow cell are shown in Fig. 4, which is a diagram of the mask used to make the flow cell. At these length scales and with an average flow rate of 7,500 $\mu m^3/s$ the Reynolds number Re of the fluids in the flow chamber stays $Re < 2{,}300$ and therefore flow in the flow cell is laminar (12, 13). The only mixing occurs by diffusion which scales as $\sqrt{(Dx/u)}$ with D representing the diffusion constant of the media, u the speed of the laminar flow, and x the distance from the point of union of the two fluids, measured along the direction of the flow. Near the point where the two fluids meet, mixing is minimal. The pressure difference between the two fluids is changed by using a computer-controlled

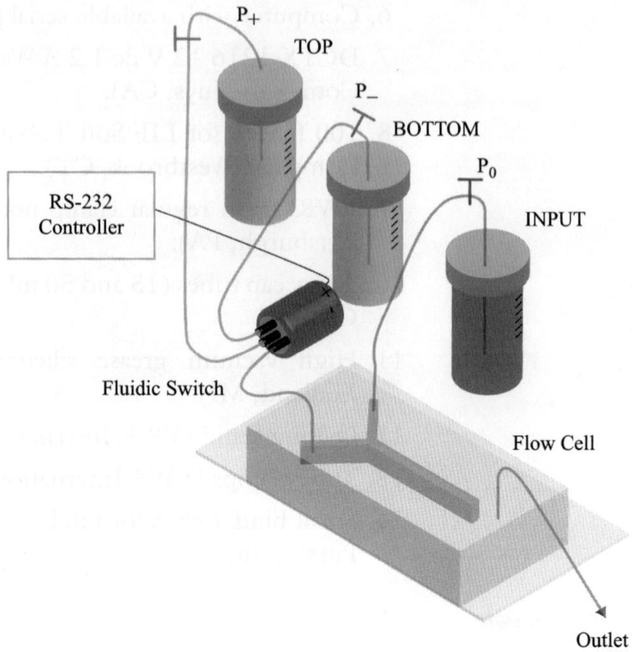

Fig. 1. One of the input arms of the Y-shaped flow cell is fed by a stimulating media (1 M sorbitol synthetic complete yeast media, *dark gray*) contained in the reservoir INPUT at a hydrostatic pressure head P_0. The other arm is fed by nonstimulating media (synthetic complete yeast media, *light gray*) from one of the two reservoirs, TOP at a hydrostatic pressure head P_+ or BOTTOM at P_-. The choice between the two reservoirs, TOP or BOTTOM, is made by a fluidic switch which is controlled using the RS-232 controller. When reservoir BOTTOM is chosen, the fluid from reservoir INPUT fills most of the chamber, while when TOP is chosen, the fluid from TOP fills the chamber. Periodic changes in which reservoir (TOP or BOTTOM) feeds nonstimulating media to the flow cell allow a change in the environment of the cells at a tunable period *T*.

fluidic switch to change the reservoir being used. By changing the relative pressure between the stimulating and nonstimulating media we can sweep the separation line across the width of the flow cell. This allows us to rapidly switch the conditions to which the cells in the flow cell are exposed. The media can be changed as frequently as twice per second without perturbing cell adhesion. Cell adhesion is achieved by functionalizing the glass coverslip with conA as described below. conA is a lectin which binds specifically to mannosyl and glucosyl residues in the yeast cell wall (14). Appropriate alignment is achieved by observing the interface between the stimulating and nonstimulating media in real time by using phase contrast microscopy as detailed below. Due to the difference in refractive index between the two fluids the interface is clearly visible (Fig. 2).

During the course of the experiment cells are stimulated with programmed input waves. The switching is controlled by a RS-232 relay controller which controls power to the fluidic switch.

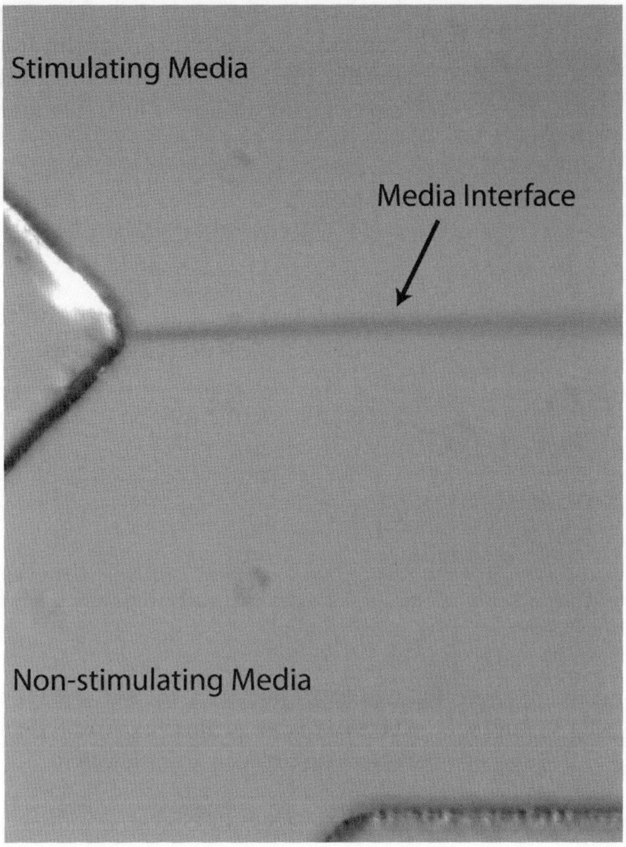

Fig. 2. Picture of the interface between stimulating (1 M sorbitol) and nonstimulating media in the flow cell under phase microscopy at ×40 magnification. The inlets where media enters the flow cell are on the *left* of the picture.

The controller interfaces with a custom-written Visual Basic program. The frequency and duty cycle of the input wave are adjustable. Image acquisition is controlled by using a multidimensional acquisition program. In our particular experiments, we used MetaMorph's multidimensional acquisition feature to acquire differential interference contrast (DIC), mCherry, and GFP images at fixed intervals. Autofocusing is achieved using Metamorph's built-in autofocusing software in the mCherry channel. Emission from GFP is visualized at 528 nm (38 nm bandwidth) upon excitation at 490 nm (20 nm bandwidth) and emission of mCherry is visualized at 617 nm (73 nm bandwidth) upon excitation at 555 nm (28 nm bandwidth). Images are processed using a custom-written ImageJ program to identify the nucleus of each cell using thresholding in the mCherry channel and then measuring the Hog1-GFP signal that is colocalized in that region. The ImageJ macro returns intensity and location data for each feature analyzed. A custom-written Matlab program is then used to analyze the ImageJ data, identify cells throughout the time course, and measure Hog1-GFP nuclear localization for each cell. The response of the pathway is then computed as the average of the amplitudes across the time course. Finally, the bandwidth of the pathway is computed using Matlab's curve fitting utility to fit the amplitude data across different input frequencies f_i to the equation:

$$X = \sqrt{\frac{G}{1 + (f_i \tau_c)^2}} + \delta, \tag{1}$$

where $f_c = 1/\tau_c$ is taken as the critical input frequency, or pathway bandwidth. G represents the gain of the system. The term δ takes into account a constant offset that is the result of changes in autofluorescence due to changes in cell size as water enters and leaves the cell.

It is also consistent to fit across input periods T to the equation:

$$X = A \frac{\left(1 - e^{-k_{on}T/2}\right)\left(1 - e^{-k_{off}T/2}\right)}{\left(1 + e^{-\left(k_{on}+k_{off}\right)T/2}\right)} + \delta, \tag{2}$$

where the larger of k_{on} or k_{off} is taken as the pathway bandwidth f_c. Here, A represents the gain of the system and δ is again a constant offset due to cell size change.

3.1. Timeline

The timeline for running a flow cell experiment is given here. Several of the steps need to be performed days and hours before the microscopy experiment is started:

Prior to day 1: Prepare the microfluidic mask. This mask can be reused many times to make multiple flow cells.

Day 1: Pour uncured PDMS into flow cell molds. Cure in convection oven overnight.

Day 2: Cut out the plasma bond PDMS flow cell. Cure overnight. Inoculate yeast into 4 ml of synthetic complete media to grow to saturation overnight.

Day 3: Run the flow cell experiment.

Five hours prior: Reinoculate yeast from the saturated culture approximately 5 h before you would like to begin microscopy.

One hour prior: Prepare the setup shown in Fig. 1 approximately 1 h before you would like to begin microscopy.

Immediately before: Check the line, load cells into the flow cell.

During: Set up a time course acquisition using the appropriate multidimensional acquisition software.

After: Thoroughly clean the setup.

Following days: Image processing and data analysis

3.2. Fabrication of the Microfluidic Mask (see Note 1)

1. Sonicate silicon wafers for 5–10 min in an acetone bath. Rinse with isopropyl alcohol. Dry wafers with a nitrogen gun. Bake wafer at 200°C for at least 10 min to remove moisture. Cool wafer with nitrogen gun until it is cool to the touch.
2. Spin coat the wafer with desired thickness of SU8 2050. For our masks we coated the wafers to approximate thickness of 100 μm using a Headway Spin Coater Model PWM32. The spin program is as follows:

   Step 1: Speed 500 rpm/s, Ramp 100 rpm/s, 10 s

   Step 2: Speed 1,000 rpm/s, Ramp 300 rpm/s, 30 s

   Step 3: Speed 0 rpm/s, ramp 500 rpm/s, 0 s
3. Prebake coated wafer at 65°C for 10 min.
4. Using a 28-gauge needle and syringes filled with PGMEA, carefully remove the SU8 edge on the wafer while the wafer is spinning at 1,000 rpm on the Headway Spin Coater.
5. Place the wafer at 95°C for 50 min, allow wafer to cool to 65°C by changing hot plate temperature to 65°C and waiting for temperature to adjust (approximately 15 min on Torrey Pines HP30 hot plates).
6. Cover the wafer with a 360-nm long-pass filter and the transparency mask and expose for 1 min at 25 mW power on the AB-M Mask Aligner (see Note 2).
7. Put wafers at 65°C (1 min), allow hot plate temperature to ramp to 95°C (approximately 7 min), keep wafers at 95°C for 10 min. Allow temperature to ramp back down to 45°C (approximately 20 min). Move wafer to the bench and allow cooling at room temperature for 5–10 min.
8. Put wafer in PGMA with sporadic stirring for 10 min or until unexposed SU8 is removed. Rinse with isopropyl alcohol when it is clear that the unexposed SU8 has been removed. Dry wafer with the nitrogen gun (see Note 3).
9. Measure mask features using a contact profilometer.
10. Place mask in an appropriately sized petri dish.
11. Incubate the mask in a fume hood in an enclosed vacuum chamber at 6 mmHg pressure for 3 h with two to three drops of 3-aminopropyl-triethoxysilane in a separate disposable aluminum tray (see Note 4).

3.3. Flow Cell Preparation (see Note 5)

1. Flow cell masks patterned with SU8 are constructed on 4″ silicon wafers using standard photolithography techniques for microfluidics (12). Once the microfluidic mask has been made (see above) it can be reused many times to make multiple flow cells. The design of the mask is shown in Fig. 3.
2. Place the mask in a petri dish if you have not done so already.

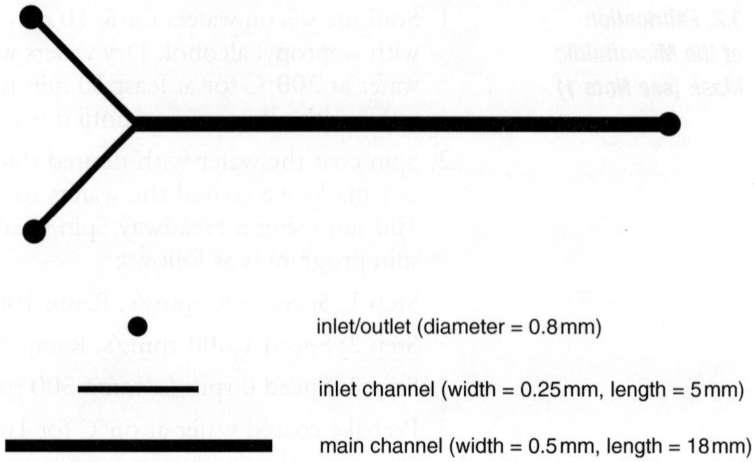

Fig. 3. Diagram of the mask pattern used to make the PDMS flow cell. The height of the channels is 100 μm. The height is set by how thickly the SU8 is applied as explained in Subheading 3.2.2.

3. Prepare PDMS by mixing curing agent and polymer in a 1:9 ratio by weight (see Note 6).
4. Degas the PDMS in a vacuum desiccator. The amount of pressure is not important, but will affect the amount of time required to degas the PDMS. Be cautious that your mixing container is not overfull, or the PDMS will bubble over during degassing.
5. Pour the PDMS carefully into the petri dish so that new air bubbles are not created. If bubbles form, remove them carefully with a 21 G 1/2 gauge needle without scratching the mask.
6. Cure the PDMS at 65°C overnight.
7. Using a razor blade, cut out a 15 × 40 mm rectangle of PDMS surrounding the flow cell design. Cut gently without pressure so as not to break the mask. When the PDMS separates from the underlying mask remove the rectangle of PDMS from the mold. The mold can now be reused by simply refilling the hole made in the cured PDMS (see Note 7).
8. Punch holes for the inlets and outlets using a PDMS puncher.
9. Clean the PDMS block with scotch tape. With the feature side face up on the bench, apply tape being careful not to touch the feature-side of the block with your gloves. Repeat this three times (see Note 8).
10. Plasma clean the PDMS block and a 23 × 60 mm coverslip for 30 s at 100 W plasma power at 30 mTorr base pressure and 200 mTorr process pressure.

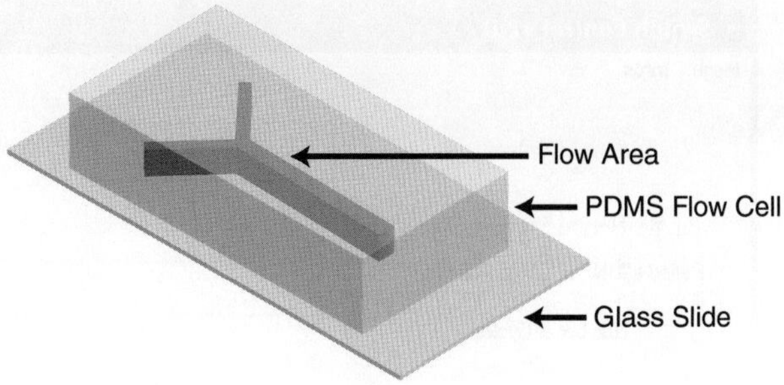

Fig. 4. The microfluidic flow cell after plasma cleaning and bonding to the cover glass.

11. Apply the PDMS to the coverslip feature-side down to create the finished microfluidic flow cell shown in Fig. 4.
12. Cure the flow cell in the convection oven at 65°C for several hours or overnight (see Note 9).

3.4. Preparation of Yeast Samples

1. Grow yeast cells to saturation overnight in standard yeast synthetic complete (SC) yeast media (11) at 30°C with shaking (OD600 ~2).
2. Reinoculate cells into fresh media several hours before the experiment is started. Reinoculate at a low density in SC media in a conical flask such that cells are in early exponential growth phase (OD600 ~0.05 for haploid yeast) just before being loaded into the flow cell (approximately 100 μl in 25 ml of SC media for our strains and media).

3.5. Preparation of the Switch and Liquid Handling

1. The experimental setup with the liquid containers for feeding the flow cell and switching media in the flow cell is shown in Fig. 1. Punch a hole in the plastic top of two 50 ml screw cap tubes and one 15 ml screw cap tube with a 16 G 1/2 gauge needle.
2. Label one 50 ml tube TOP and fill it with 45 ml of synthetic complete yeast media (SC). Label one 50 ml tube INPUT and fill it with 45 ml of 1 M sorbitol synthetic complete media. Fill the 15 ml tube with 13 ml of SC and label it BOTTOM.
3. Suspend the tubes on the ring stand using the clamp holders. The TOP tube should be the highest, followed by the INPUT, and then the BOTTOM tube.
4. Connect the RS-232 controller to your computer through the desired COM serial port. Attach a power source to the RS-232 controller but do not plug it in yet.
5. Open the Switch Control software. The interface for the software is shown in Fig. 5 (see Note 10).

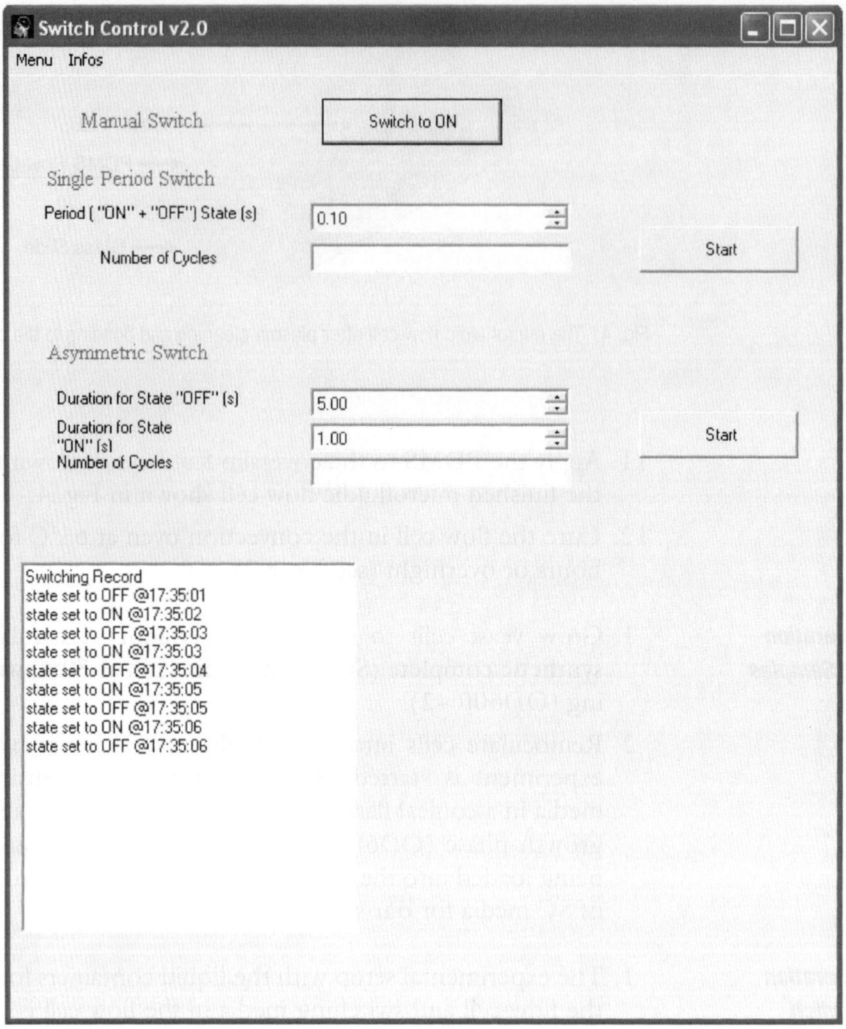

Fig. 5. The Visual Basic software user interface for controlling the RS-232 relay controller. The user is able to define both symmetric and asymmetric input waves. The switching times are recorded and can be saved from "Menu." The "OFF" state refers to when the flow cell is filled with media from the TOP reservoir and the "ON" state refers to when the flow cell is filled with media from the INPUT reservoir (1 M sorbitol).

6. Connect the switch to the RS-232 microcontroller relay R1. Connect a power source to the switch.
7. Plug in power sources for the switch. Check that the COM port is communicating with the RS-232 controller by switching the switch on and off several times using the software.
8. Attach 60 cm of soft tubing with inner diameter 0.042″ to the outlet of the Lee switch (see Note 11). With the other end of the tubing in a conical flask full of sterile water clean and remove bubbles from the switch by running MiliQ water through both outlets using 10 s "ON," 10 s "OFF" switching for about 5–10 min.

9. Insert 60 cm of the intramedic tubing into each media tube. Allow media to flow to the bottom of the tubing before clamping the tubing with a binder clip (see Note 12).
10. Attach 5 cm of the soft tubing to the inlets of the Lee switch.
11. Insert the end of the tubing from the TOP tube into the soft tubing attached to the upper inlet on the switch. Insert the end of the tubing from the BOTTOM tube into the soft tubing attached to the lower inlet on the switch. Insert 2″ of intramedic tubing into the soft-tubing attached to the outlet of the Lee switch.
12. To ensure that there are no air bubbles trapped in the switch use the software to turn the switch to the "ON" state. Unclamp the BOTTOM tube and allow media to flow until there are no bubbles. Repeat for the TOP tube with the switch in the "OFF" state (see Note 13).
13. Clean a brand new flow cell by injecting the cell with 70% ethanol followed by sterile water by syringe injection. Make sure to fill the flow cell with water completely, so that the inlets and outlets are covered with fluid to prevent air bubbles.
14. Attach the flow cell to the metal slide holder with lab tape.
15. Tape the Lee company switch to the microscope stage.
16. With the flow cell on the microscope stage, insert the intramedic tubing from the fluidic switch outlet into the flow cell inlet closest to you (see Fig. 1).
17. Unclamp the intramedic tubing from the INPUT tube and insert it into the other flow cell inlet.
18. Allow the flow cell to fill with media.
19. Cut 8″ of intramedic tubing and insert it into the flow cell outlet. Allow this tubing to fill with media. Put the end of the tubing into the waste collection container and fill the waste container with 50 ml of sterile water. Make sure that the end of the outlet tube is completely submerged (see Note 14).
20. Unclamp all tubing if you have not done so already.
21. By changing the relative heights of the TOP, BOTTOM, and INPUT tubes set the line under the microscope using phase contrast microscopy so that it matches the diagram in Fig. 6 in the "ON" and "OFF" states. The "ON" state means that the flow cell is filled with the INPUT media and "OFF" means that the flow cell is filled with media from the TOP tube (see Note 15).
22. Once the line is set you are ready to load cells into the flow cell. Clamp all tubing to avoid leaks and remove the flow cell in its holder from the microscope.

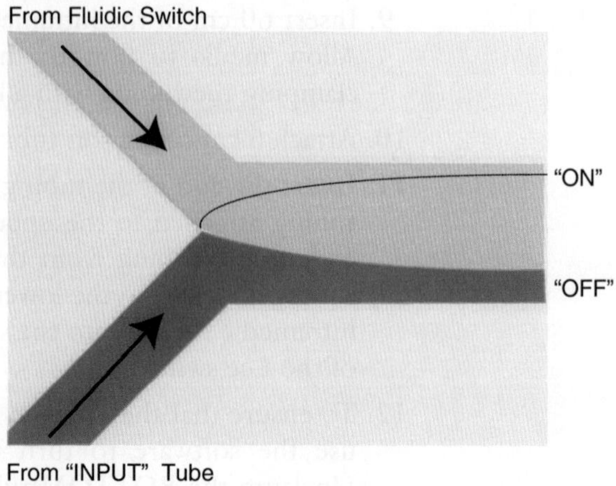

Fig. 6. Top view of flow cell. The figure shows the appropriate orientations of the media interface for the "ON" and "OFF" states. The input is "ON" when the flow cell is filled with media from the INPUT reservoir (*dark gray*) and "OFF" when the flow cell is filled with media from the TOP reservoir (*light gray*).

3.6. Preparation of the conA Loading Solution

1. Gather the appropriate supplies: 45 ml of sterile H_2O, 2.5 ml of 1 M $CaCl_2$ in H_2O, 2.5 ml of 1 M $MnSO_4$ in H_2O, and conA (MP Biomedicals cat. no. 150710/CAS #11028710/ EC #2342582 supplied as a lyophilized white powder. Essentially salt-free and carbohydrate-free, molecular weight not available).
2. Mix the water, $CaCl_2$, and $MnSO_4$. The pH of this solution should be between 6 and 7.
3. Dissolve conA to a concentration of 20 mg/ml in this solution.
4. Dilute this conA solution in sterile water to make a 2 mg/ml solution for use in the flow cell (see Note 16).

3.7. Loading Cells into the Microfluidic Chip

1. Clean flow cell with 70% ethanol by injection. Use a syringe that has intramedic tubing slipped over a 16 G 1/2 gauge needle to syringe inject the ethanol. Then inject MiliQ water. Use 1–2 ml of ethanol and water.
2. Load the 2 mg/ml conA solution into the flow cell using a syringe. Allow conA to incubate in the flow cell at room temperature for 15 min (see Note 17).
3. While conA is incubating spin down cells (3000 *g*) growing in a 14-ml falcon tube (~4 ml of cells in log-phase growth.)
4. Resuspend cells in 4 ml sterile water, and spin down again. Discard the supernatant. Resuspend cells in remaining supernatant by gently shaking the tube. If needed add a few hundred microliters of water to the cells.
5. When the conA is incubated, load the cells.

6. Allow cells to sit in the flow cell for 5 min at room temperature on the bench.
7. Check that cells have stuck at 40× magnification. Stuck cells will not move when the flow cell is gently tapped.
8. Put the flow cell back on the microscope and replace the appropriate tubing. Recheck the media interface.
9. Unclamp all tubing and check for leaks. Make sure that the switch is in the "OFF" position.

3.8. Microscopy Time Course (see Note 18)

1. With the switch in the "OFF" position allow cells to equilibrate for 20 min.
2. Switch the microscope objective to 100×.
3. Set the parameters for the time course on the microscope software. Set the software so that the stage position, time, and channel of each image are recorded in a log file.
4. Set stage positions to acquire in the middle of the flow channel. This is important as the middle of the channel is not exposed to diffusion region near the media interface and therefore sees the appropriate input signal.
5. Once cells have equilibrated start the multidimensional acquisition and then start fluid switching.
6. Acquire a DIC (10 ms), mCherry (50 ms), and GFP (200 ms) exposure at each time point. Take pictures every n seconds where $n =$ (the period of the switching)/10 (see Note 19).
7. Periodically check the experiment for leaks.
8. Periodically check that the "ON" and "OFF" states of the line are correct by pausing the multidimensional acquisition and monitoring switching by eye using phase microscopy. This will require movement of the stage to monitor different regions of the flow cell, thus the stage position of interest must be marked in the microscopy software.

3.9. Cleaning the Setup

1. Once the experiment is over, clamp all tubing. Remove all tubing and clean with 70% ethanol by injection from a syringe with a 16 G 1/2 gauge needle.
2. Repeat cleaning of the switch using sterile water as described previously.

3.10. Image Processing (see Note 20)

1. We used a custom-written ImageJ macro to threshold on the Htb2-mCherry images and then measure the colocalization of Hog1-GFP with these nuclear regions. However, any program which allows you to collect intensity data for Hog-GFP in the cell's nucleus will work.

2. Run the ImageJ macro to extract values for Hog1-GFP colocalization with the Htb2-mCherry tagged nucleus.
3. Allow the ImageJ macro to run (see Note 21).
4. This macro returns an excel file which contains the intensity data for both the GFP and mCherry channels.
5. The Matlab program takes the intensity data, a background file (found by measuring the average background in each image slide using ImageJ), and the log file recorded by the multidimensional acquisition routine.
6. Use the Matlab program to reconstruct time traces for the nuclear intensity for each cell in the experiment.
7. Calculate the amplitude of the cellular response to the input by finding the mean amplitude of each cell (over the first few periods in the experiment so that photobleaching does not have a significant effect) then computing the mean and standard deviation of these means.

3.11. Bandwidth Measurement

1. Amplitude distributions are collected for Hog1 localization over a range of input frequencies.
2. The bandwidth is found using Matlab's Curve Fitting Toolbox and fitting the amplitude data to Eq. 1 or 2. The variable f_c is taken to be the pathway bandwidth.

4. Notes

1. All fabrication steps for creation of the microfluidic mask were performed in a nanofabrication facility with an appropriate clean room. Similar facilities are available at many universities and research institutes.
2. Failure to filter wavelengths below 350 nm will result in overexposure of the top portion of the SU8 resist film leaving negative sidewall profiles or T-topping. If you see these features, then check whether you have used the correct bandpass filter. Alternatively, reduce exposure time.
3. Problems in the fabrication process often become apparent during development. If the developed mask pattern does not remain in contact with the wafer or there is excessive cracking this indicates an under cross-linking condition and can be corrected by increasing the exposure time or increasing the postexposure bake.
4. Silanization of the mask is an optional step. However, it allows for easy removal of the polymerized PDMS used for creation

of the flow cell and is highly recommended to increase the useful lifetime of the mask.

5. Always wear nitrile gloves when handling PDMS. Oil from your skin can compromise the polymerization of the PDMS.
6. This is most easily done in a disposable plastic container (such as a drinking cup) because leftover PDMS will eventually ruin most containers. Aim for about 10 ml of PDMS the first time you use the mask. During subsequent flow cell construction you will need less PDMS to fill the mold because not all cured PDMS is removed after each round of construction. Mix the PDMS *very* thoroughly with a plastic disposable fork.
7. Cracking masks is inevitable. It is recommended to have two to three identical masks on hand so that breaking a mask does not cause experimental delays.
8. If desired, the flow cell can be stored in this state (with tape covering the attachment side) for several days before plasma cleaning and bonding to the coverslip.
9. This curing step can help remove small bubbles than might have formed between the PDMS and the coverslip. Do not cure for longer than overnight, as PDMS will shrink with excessive drying affecting the channels and seal with the coverslip.
10. Visual Basic code for controlling the switch is available upon request.
11. Vacuum grease can be used to ease insertion of the switch outlet into the tubing. We have found that we do not need additional connectors for connecting tubing to the switch or to the microfluidic chamber. Vacuum grease can also be used to seal minor leaks.
12. This may require using a syringe with a 21 G 1/2 needle to start flow. Be cautious when clamping the tubing with the binder clip. Care must be taken to prevent leaks as they are messy and expensive around microscopy equipment.
13. This step is crucial. Even small bubbles will aggregate in the switch becoming very large over the course of the experiment until they eventually release and perturb fluid flow or cell adhesion upon reaching the interior of the flow cell.
14. If the outlet tube is not submerged in an waste container containing a large amount of liquid the pressure in the flow cell with change drastically over the course of the experiment altering the alignment of the interface between the stimulus and nonstimulus.
15. It is often easier to set the line using a lower magnification than used to observe cells over the course of the experiment. Try 40× magnification and phase illumination for setting the line.

16. This conA solution may be stored at −20°C indefinitely and thawed immediately before use. It can be refrozen several times before its efficacy is reduced. Storage in aliquots of about 500 μl is recommended.
17. When loading flow cells always make sure that the flow cell outlets and inlets are covered with large droplets of MiliQ water to prevent drying and air bubble formation.
18. We used Metamorph imaging software to control the microscopy time course. We used the built-in multidimensional acquisition utility to acquire images at programmed stage positions, time points, and wavelengths. However, the instructions are easily modified to work with any microscope control software.
19. Pictures were never acquired more rapidly than once every 10 s. To maintain focus throughout the experiment we used Metamorph's autofocus routine to autofocus on the mCherry labeled nucleus once every few timepoints.
20. Matlab and ImageJ codes are available upon request.
21. The thresholding values for the mCherry channel might need to be manually adjusted so that they find the cell nucleus correctly. The values depend on the background of your camera.

References

1. Pearson, G., Robinson, F., Beers Gibson, T., Xu, B. E., Karandikar, M., Berman, K., and Cobb, M. H. (2001) Mitogen-activated protein (MAP) kinase pathways: regulation and physiological functions, *Endocrine reviews* **22**, 153–183.
2. Raman, M., Chen, W., and Cobb, M. H. (2007) Differential regulation and properties of MAPKs, *Oncogene* **26**, 3100–3112.
3. Robinson, M. J., and Cobb, M. H. (1997) Mitogen-activated protein kinase pathways, *Current opinion in cell biology* **9**, 180–186.
4. Banuett, F. (1998) Signalling in the yeasts: an informational cascade with links to the filamentous fungi, *Microbiol Mol Biol Rev* **62**, 249–274.
5. Gustin, M. C., Albertyn, J., Alexander, M., and Davenport, K. (1998) MAP kinase pathways in the yeast Saccharomyces cerevisiae, *Microbiol Mol Biol Rev* **62**, 1264–1300.
6. Klipp, E., Nordlander, B., Kruger, R., Gennemark, P., and Hohmann, S. (2005) Integrative model of the response of yeast to osmotic shock, *Nature biotechnology* **23**, 975–982.
7. Janiak-Spens, F., Cook, P. F., and West, A. H. (2005) Kinetic analysis of YPD1-dependent phosphotransfer reactions in the yeast osmoregulatory phosphorelay system, *Biochemistry* **44**, 377–386.
8. Hersen, P., McClean, M. N., Mahadevan, L., and Ramanathan, S. (2008) Signal processing by the HOG MAP kinase pathway, *Proceedings of the National Academy of Sciences of the United States of America* **105**, 7165–7170.
9. Huh, W. K., Falvo, J. V., Gerke, L. C., Carroll, A. S., Howson, R. W., Weissman, J. S., and O'Shea, E. K. (2003) Global analysis of protein localization in budding yeast, *Nature* **425**, 686–691.
10. Sikorski, R. S., and Hieter, P. (1989) A system of shuttle vectors and yeast host strains designed for efficient manipulation of DNA in *Saccharomyces cerevisiae*, *Genetics* **122**, 19–27.
11. Burke, D. D., D and T. Stearns. (2000) Methods in Yeast Genetics: A Cold Spring Harbor Laboratory Course Manual, in *Methods in Yeast Genetics: A Cold Spring Harbor Laboratory Course Manual* 2000 ed., Cold Spring Harbor Laboratory Woodbury, NY.
12. Beebe, D. J., Mensing, G. A., and Walker, G. M. (2002) Physics and applications of

microfluidics in biology, *Annual review of biomedical engineering* **4**, 261–286.
13. Weigl, B. H., Bardell, R. L., and Cabrera, C. R. (2003) Lab-on-a-chip for drug development, *Advanced drug delivery reviews* **55**, 349–377.
14. Agarwal, B. a. G., IJ. (1968) Protein carbohydrate interaction vii: Physical and chemical studies of concanavalin a, the hemagglutinin of the jack bean., *Arch Biochem Biophys* **124**,11.

Part II

Mathematical Modelling of Network Behavior

Chapter 7

Stochastic Analysis of Gene Expression

Xiu-Deng Zheng and Yi Tao

Abstract

In this chapter, stochasticity in gene expression is investigated using Ω-expansion technique. Two theoretical models are considered here, one concern the stochastic fluctuations in a single-gene network with negative feedback regulation, and the other the additivity of noise propagation in a protein cascade. All of these theoretical analyses may provide a basic framework for understanding stochastic gene expression.

Key word: Gene expression network, Feedback regulation, Genetic cascade, Noise, One-step process, Omega-expansion, Monte Carlo algorithm

1. Introduction

Stochastic fluctuations in genetic works are inevitable as chemical reactions are probabilistic and many genes, RNAs, and proteins are present in low numbers per cell (1). To identify the source of noise in gene expression, a single fluorescent reporter gene, i.e., the green fluorescent gene (*gfp*), is used to incorporate into the chromosome of *Bacillus subtilis.* Then, by varying independently the rates of transcription and translation of the reporter gene, the resulting changes in the phenotypic noise characteristics can be quantitatively measured (2). The results provide the first direct evidence of the biochemical origin of phenotypic noise, demonstrating that the level of phenotypic variation in an isogenic population can be regulated by genetic parameters. This result is consistent with a long-standing hypothesis that protein fluctuations depend on the number of proteins made per mRNA transcript (3–9). Similarly, two types of *gfp* are inserted into *Escherichia coli* chromosome, and their correlation is used to infer the source of the fluctuations (10). Besides, in the study, the stochastic gene expression in *Saccharomyces cerevisiae* (11), it is found that the stochasticity arising from transcription contributes

Attila Becskei (ed.), *Yeast Genetic Networks: Methods and Protocols*, Methods in Molecular Biology, vol. 734, DOI 10.1007/978-1-61779-086-7_7, © Springer Science+Business Media, LLC 2011

significantly to the level of heterogeneity within a eukaryotic clonal population, in contrast to observations prokaryotes (2) and that such noise can be modulated at the translation level.

The results suggested that eukaryotes differ from prokaryotes because promoter fluctuations and transcriptional reinitiation produce a nonmonotonous transcription noise (see also ref. 9).

In the past decade, the stochastic models of gene expression have received considerable interest (1, 12–14). Elf and Ehrenberg presented a general method that allows rapid characterization of the stochastic properties of intracellular networks, i.e., fast evaluation of fluctuations in biochemical networks with the linear noise approximation (12). Paulsson reviewed the theoretical models of stochastic gene expression (1). Some experimental studies have revealed some important properties in stochastic gene expression (13, 14). For example, *E. coli* and *B. subtilis* are used to confirm the translation bursting hypothesis on gene expression noise and to distinguish the intrinsic and extrinsic noise (13); *S. cerevisiae* is used to confirm the validity of the translational bursting hypothesis in eukaryotes and to investigate the effect of transcriptional induction on the fluctuations in gene expression (13). Furthermore, some artificial networks, for example, gene-regulatory cascade, gene network with positive or negative feedback, are set up to study the stochastic effects on situations including the cell cycle, circadian rhythms, and aging(13, 14).

In this chapter, we consider only two models using the Ω-expansion technique, one concerns the stochastic fluctuations in a single-gene network with negative feedback regulation, and the other the additivity of noise propagation in a protein cascade.

2. Stochastic Fluctuations in a Single-Gene Network with Negative Feedback Regulation

It is a commonly held idea that negative feedback provides a noise-reduction mechanism (13). A single-gene negative-feedback system in *E. coli* was engineered by Becskei and Serrano (15). They compared the variability generated by this regulatory network with that generated in the absence of feedback control. Their result shows a decrease in gene-expression variability because of the negative feedback, i.e., negative autoregulation provides a noise-reduction mechanism.

In this section, stochastic fluctuations in a single-gene network with negative feedback are investigated (see also refs. 1, 9, 16). The analysis of this theoretical model will show how to measure the noise in gene expression and why the noise can be reduced by the negative feedback regulation.

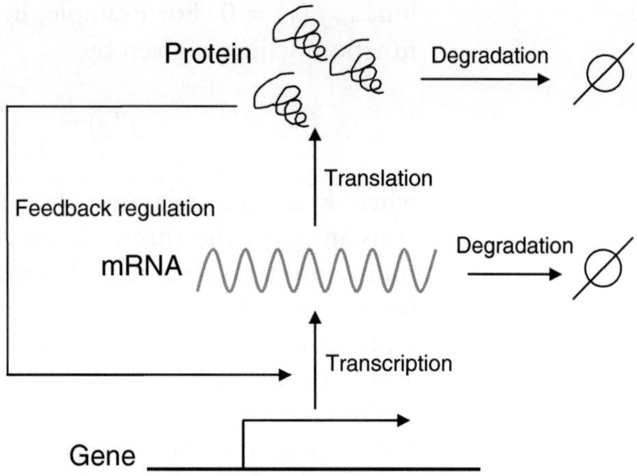

Fig. 1. A single-gene network with feedback regulation.

2.1. Network Model

Consider a single-gene network with feedback regulation, i.e., the gene transcription is regulated by its protein product (see Fig. 1), which is also called transcription factor (TFT). To model gene expression, we need to make some biochemical assumptions, which are (1) the transcription initiation is assumed to be a pseudofirst order reaction, i.e., the initial reversible binding of an RNA polymerase (RNAP) to the promoter region and subsequent formation of an open complex achieve rapid equilibrium; (2) the TFTs tend to act by binding the promoter region and shielding it from RNAP, and the reactions for the binding of TFTs to the promoter region are considered to be in equilibrium and simply change the fraction of RNAP bound as a closed complex, thereby changing the transcription rate; (3) similar to the transcription initiation, the translation initiation of a single-mRNA molecule is assumed to proceed with a pseudofirst-order rate k_P; (4) we take each transcription and translation initiation reaction to be independent; and (5) we assume that mRNA and protein molecules degrade with rates γ_R and γ_P, respectively, where the decay rate γ gives a half-life of $\ln 2/\gamma$ (6).

Let $x(t)$ and $y(t)$ represent the concentrations of mRNA and protein at time t, respectively. According to the above assumptions, the deterministic macroscopic rates equation can be given by

$$\begin{aligned}\frac{dx}{dt} &= -\gamma_R x + f(y),\\ \frac{dy}{dt} &= k_P x - \gamma_P y,\end{aligned} \tag{1}$$

where $f(y)$ is the transcription rate, which is defined as the function of protein concentration. Here we consider only the situation with negative feedback regulation, i.e., $df(y)/dy<0$ and

$\lim_{y\to\infty} f(y) = 0$. For example, in general, we take $f(y)$ as a Hill function, which is given by

$$f(y) = \frac{k_{\text{max}}}{1 + (y/k_{\text{d}})^{\beta}}, \tag{2}$$

where k_{max} is the maximum of the transcription rate, k_{d} is a binding constant (i.e., the threshold protein concentration at which the transcription rate is at half of its maximum value), and the parameter β is the Hill coefficient and determines the steepness of the repression curve (17, 18). For example, the *cl* repressor protein acts on the promoter P_{R} and P_{RM} of phage λ with a k_{d} about 50 and 1,000 nm, respectively (6, 19). Typical biological values β range from 1 (hyperbolic control) to over 30 (sharp switching) (6).

For the deterministic stability of Eq. 1, we have the following results. Let $(x*, y*)$ be the equilibrium of Eq. 1, i.e., $(x*, y*)$ is the solution of equation

$$\begin{aligned} -\gamma_{\text{R}} x + f(y) &= 0, \\ k_{\text{P}} x - \gamma_{\text{P}} y &= 0. \end{aligned} \tag{3}$$

It is easy to see that the equilibrium $(x*, y*)$ must be unique since $\mathrm{d}f(y)/\mathrm{d}y < 0$ and $\lim_{y\to\infty} f(y) = 0$. Notice that the Jacobian matrix of Eq. 1 about $(x*, y*)$ is

$$\begin{pmatrix} -\gamma_{\text{R}} & f'(y*) \\ k_{\text{P}} & -\gamma_{\text{P}} \end{pmatrix} \tag{4}$$

with eigenvalues

$$\lambda_{1,2} = \frac{-(\gamma_{\text{R}} + \gamma_{\text{P}}) \pm \sqrt{(\gamma_{\text{R}} + \gamma_{\text{P}})^2 - 4(\gamma_{\text{R}}\gamma_{\text{P}} - k_{\text{P}} f'(y*))}}{2}, \tag{5}$$

where $f'(y) = \mathrm{d}f(y)/\mathrm{d}y$. Thus, $(x*, y*)$ is globally asymptotically stable since the real parts of both eigenvalues λ_1 and λ_2 are negative.

2.2. Noise and Steady-State Statistics

Now let us consider the stochastic fluctuations (noise) in the dynamics Eq. 1. Let Ω be the size of the system (normally, this parameter is defined as the volume) (20). The numbers of mRNA and protein molecules can be expressed as $n_{\text{R}} = \Omega x$ and $n_{\text{P}} = \Omega y$. For the stochastic fluctuations in the copy numbers of mRNA and protein, the probabilities of having n_{R} mRNAs and n_{P} proteins can be described by a birth-and-death Markov process with events

$$\begin{aligned} n_{\text{R}} &\xrightarrow{\Omega f(n_{\text{P}}/\Omega)} n_{\text{R}} + 1, \\ n_{\text{R}} &\xrightarrow{\gamma_{\text{R}} n_{\text{R}}} n_{\text{R}} - 1, \\ n_{\text{P}} &\xrightarrow{k_{\text{P}} n_{\text{R}}} n_{\text{P}} + 1, \\ n_{\text{P}} &\xrightarrow{\gamma_{\text{P}} n_{\text{P}}} n_{\text{P}} - 1. \end{aligned}$$

Clearly, this is also called the one-step process in physics (20). According to Paulsson (9), we use a logarithmic gain

$$H_{\mathrm{R,R}} = \frac{\partial \ln(\gamma_{\mathrm{R}} n_{\mathrm{R}} / \Omega f(n_{\mathrm{P}}/\Omega))}{\partial \ln n_{\mathrm{R}}} \tag{6}$$

to measure how the balance between production and elimination of mRNA is affected by itself. The scale-free parameter is closely related to the elasticity of metabolic control analysis and the apparent kinetic orders of biochemical systems theory (9). For example, if $H_{\mathrm{ij}} = 2$, then a 1% increase in component n_{j} will cause n_{i} to decrease by increasing its death-to-birth ratio by approximately 2%. The elasticity is then a normalized measure for the strengths of the kinetic nonlinearities. Similarly, we have also that

$$\begin{aligned} H_{\mathrm{R,P}} &= \frac{\partial \ln(\gamma_{\mathrm{R}} n_{\mathrm{R}} / \Omega f(n_{\mathrm{P}}/\Omega))}{\partial \ln n_{\mathrm{P}}}, \\ H_{\mathrm{P,R}} &= \frac{\partial \ln(\gamma_{\mathrm{P}} n_{\mathrm{P}} / k_{\mathrm{P}} n_{\mathrm{R}})}{\partial \ln n_{\mathrm{R}}}, \\ H_{\mathrm{P,P}} &= \frac{\partial \ln(\gamma_{\mathrm{P}} n_{\mathrm{P}} / k_{\mathrm{P}} n_{\mathrm{R}})}{\partial \ln n_{\mathrm{P}}}. \end{aligned} \tag{7}$$

Omega-expansion. The technique of the Omega-expansion of the master equation provides the answer to the question how the deterministic macroscopic equation emerges from stochastic description in terms of a master equation (19). This method will develop in the form of a power series expansion in a parameter Ω. The parameter is used to govern the size of the fluctuations and the size of the jumps in the master equation. So if this parameter is large enough, the jumps are relatively small. In general, Ω is considered to be the size of the reaction system.

Let $\Phi(n_{\mathrm{R}}, n_{\mathrm{P}}; t)$ denote the joint probability that the numbers of mRNA and protein molecules equal exactly n_{R} and n_{P} at time t, respectively. From the properties of the one-step process, the master equation of $\Phi(n_{\mathrm{R}}, n_{\mathrm{P}}; t)$ is

$$\begin{aligned} \partial_t \Phi(n_{\mathrm{R}}, n_{\mathrm{P}}; t) =& \left(E_{\mathrm{R}}^{+1} - 1\right)\gamma_{\mathrm{R}} n_{\mathrm{R}} \Phi(n_{\mathrm{R}}, n_{\mathrm{P}}; t) \\ &+ \left(E_{\mathrm{R}}^{-1} - 1\right)\Omega f\left(\frac{n_{\mathrm{P}}}{\Omega}\right)\Phi(n_{\mathrm{R}}, n_{\mathrm{P}}; t) \\ &+ \left(E_{\mathrm{P}}^{+1} - 1\right)\gamma_{\mathrm{P}} n_{\mathrm{P}} \Phi(n_{\mathrm{R}}, n_{\mathrm{P}}; t) \\ &+ \left(E_{\mathrm{P}}^{-1} - 1\right)k_{\mathrm{P}} n_{\mathrm{R}} \Phi(n_{\mathrm{R}}, n_{\mathrm{P}}; t), \end{aligned} \tag{8}$$

where the symbol E represents the a step operator, which is defined as $E^{\pm 1} g(n) = g(n \pm 1)$ (20). Since Eq. 8 cannot be solved explicitly, we need to develop an approximate solution for large Ω. van Kampen pointed out that the joint probability distribution $\Phi(n_{\mathrm{R}}, n_{\mathrm{P}}; t)$ can be anticipated to have a sharp maximum around the macroscopic values $n_{\mathrm{R}}(t) = \Omega x(t)$ and $n_{\mathrm{P}}(t) = \Omega y(t)$

determined by Eq. 1 with the width of order $n_R^{1/2}, n_P^{1/2} \sim \Omega^{1/2}$ (20). To utilize this insight, let

$$\begin{aligned} n_R(t) &= \Omega x(t) + \Omega^{1/2}\xi_R(t), \\ n_P(t) &= \Omega y(t) + \Omega^{1/2}\xi_P(t), \end{aligned} \tag{9}$$

where $x(t)$ and $y(t)$ are the stationary solution of Eq. 1, and $\xi_R(t)$ and $\xi_P(t)$ are two new stochastic variables associated with number fluctuations. Equation 9 here is a time-dependent transformation from the variables $n_R(t)$ and $n_P(t)$ to the new variables $\xi_R(t)$ and $\xi_P(t)$. The joint probability distribution $\Phi(n_R, n_P; t)$ is now rewritten as a function of ξ_R and ξ_P, i.e.,

$$\Phi(n_R, n_P; t) = \phi(\xi_R, \xi_P; t). \tag{10}$$

From van Kampen (20), since $n_R \to n_R \pm 1 \Leftrightarrow \xi_R \to \xi_R \pm \Omega^{-1/2}$ and $n_P \to n_P \pm 1 \Leftrightarrow \xi_P \to \xi_P \pm \Omega^{-1/2}$, the step operators in the master equation (Eq. 7.8) can be expressed as

$$\begin{aligned} E_R^{\pm 1} &= 1 \pm \Omega^{-1/2}\frac{\partial}{\partial \xi_R} + \frac{1}{2}\Omega^{-1}\frac{\partial^2}{\partial \xi_R^2} \pm \cdots, \\ E_P^{\pm 1} &= 1 \pm \Omega^{-1/2}\frac{\partial}{\partial \xi_P} + \frac{1}{2}\Omega^{-1}\frac{\partial^2}{\partial \xi_P^2} \pm \cdots. \end{aligned} \tag{11}$$

The time derivative in the master equation (Eq. 7.8) is taken with constants n_R and n_P, i.e., $\mathrm{d}\xi_R/\mathrm{d}t = -\Omega^{1/2}(\mathrm{d}x/\mathrm{d}t)$ and $\mathrm{d}\xi_P/\mathrm{d}t = -\Omega^{1/2}(\mathrm{d}y/\mathrm{d}t)$. Hence, the master equation expressed in the new variables takes the form

$$\begin{aligned} \partial_t \Phi(n_R, n_P; t) &= \frac{\partial \phi(\xi_R, \xi_P; t)}{\partial t} - \Omega^{1/2}\frac{\mathrm{d}x}{\mathrm{d}t}\frac{\partial \phi(\xi_R, \xi_P; t)}{\partial \xi_R} \\ &\quad - \Omega^{1/2}\frac{\mathrm{d}y}{\mathrm{d}t}\frac{\partial \phi(\xi_R, \xi_P; t)}{\partial \xi_P} \\ \Leftrightarrow \frac{\partial \phi}{\partial t} &- \Omega^{1/2}\frac{\mathrm{d}x}{\mathrm{d}t}\frac{\partial \phi}{\partial \xi_R} - \Omega^{1/2}\frac{\mathrm{d}y}{\mathrm{d}t}\frac{\partial \phi}{\partial \xi_P} \\ &= \gamma_R\left(\Omega^{-1/2}\frac{\partial}{\partial \xi_R} + \frac{1}{2}\Omega^{-1}\frac{\partial^2}{\partial \xi_R^2} + \cdots\right)\left(\Omega x(t) + \Omega^{1/2}\xi_R\right)\phi \\ &\quad + \Omega f\left(y(t) + \Omega^{-1/2}\xi_P\right)\left(-\Omega^{-1/2}\frac{\partial}{\partial \xi_R} + \frac{1}{2}\Omega^{-1}\frac{\partial^2}{\partial \xi_R^2} + \cdots\right)\phi \\ &\quad + \gamma_P\left(\Omega^{-1/2}\frac{\partial}{\partial \xi_P} + \frac{1}{2}\Omega^{-1}\frac{\partial^2}{\partial \xi_P^2} + \cdots\right)\left(\Omega y(t) + \Omega^{1/2}\xi_P\right)\phi \\ &\quad + k_P\left(\Omega x(t) + \Omega^{1/2}\xi_R\right)\left(-\Omega^{-1/2}\frac{\partial}{\partial \xi_P} + \frac{1}{2}\Omega^{-1}\frac{\partial^2}{\partial \xi_P^2} + \cdots\right)\phi, \end{aligned} \tag{12}$$

where for the term $f(y(t) + \Omega^{-1/2}\xi_P)$, we take its Taylor expansion about $\xi_P = 0$.

$$f\left(y(t) + \Omega^{-1/2}\xi_{\mathrm{P}}\right) = f(y(t)) + \Omega^{-1/2}f'(y(t))\xi_{\mathrm{P}} + \frac{1}{2}\Omega^{-1}f''(y(t))\xi_{\mathrm{P}}^2 + \cdots. \tag{13}$$

Now, we are in a position to collect the several powers of Ω. Collecting terms in Eq. 12 of order Ω^0,

$$\frac{\partial\phi}{\partial t} = -\frac{\partial}{\partial\xi_{\mathrm{R}}}(-\gamma_{\mathrm{R}}\xi_{\mathrm{R}} + f'(y)\xi_{\mathrm{P}})\phi + \frac{1}{2}(\gamma_{\mathrm{R}}x + f(y))\frac{\partial^2\phi}{\partial\xi_{\mathrm{R}}^2} - \frac{\partial}{\partial\xi_{\mathrm{P}}}(k_{\mathrm{P}}\xi_{\mathrm{R}} - \gamma_{\mathrm{P}}\xi_{\mathrm{P}})\phi + \frac{1}{2}(k_{\mathrm{P}}x + \gamma_{\mathrm{P}}y)\frac{\partial^2\phi}{\partial\xi_{\mathrm{P}}^2} \tag{14}$$

with boundary conditions $\lim_{\xi_{\mathrm{R}}\to\pm\infty}\phi = \lim_{\xi_{\mathrm{P}}\to\pm\infty}\phi = 0$ and $\lim_{\xi_{\mathrm{R}}\to\pm\infty}\partial\phi/\partial\xi_{\mathrm{R}} = \lim_{\xi_{\mathrm{P}}\to\pm\infty}\partial\phi/\partial\xi_{\mathrm{P}} = 0$. This equation is a linear Fokker–Planck equation, whose coefficients depend on time t through $x(t)$ and $y(t)$, and it is called the linear noise approximation (20) and also called the fluctuation-dissipation theorem in physics (1, 9).

As pointed out by Elf and Ehrenberg (12), the Ω-expansion means that the master equation is Taylor expanded near macroscopic system trajectories or stationary solutions in powers of $1/\sqrt{\Omega}$. When the master equation is approximated near a macroscopically stable stationary solution, terms of the first order in $1/\sqrt{\Omega}$ give the macroscopic rate equation, and terms of the second order in $1/\sqrt{\Omega}$ give the linear noise approximation. The basic idea behind the Ω-expansion is that the relative fluctuations around the constant average concentrations will tend to decrease with the inverse of the square root of Ω. However, the linear noise approximation is true when the fluctuations are sufficiently small comparing to the corresponding means, and it can also give very good estimates of fluctuations in molecular numbers when they are larger than the corresponding means (see also ref. 1, 9, 20).

Steady-state statistics. When the system state is near the equilibrium of Eq. 1, $(x*, y*)$, Eq. 14 can be approximated as

$$\frac{\partial\phi}{\partial t} = -\frac{\partial}{\partial\xi_{\mathrm{R}}}(a_{11}\xi_{\mathrm{R}} + a_{12}\xi_{\mathrm{P}})\phi + D_{\mathrm{R}}\frac{\partial^2\phi}{\partial\xi_{\mathrm{R}}^2} - \frac{\partial}{\partial\xi_{\mathrm{P}}}(a_{21}\xi_{\mathrm{R}} + a_{22}\xi_{\mathrm{P}})\phi + D_{\mathrm{P}}\frac{\partial^2\phi}{\partial\xi_{\mathrm{P}}^2}, \tag{15}$$

where

$$D_{\mathrm{R}} = \gamma_{\mathrm{R}}x*, \quad D_{\mathrm{P}} = \gamma_{\mathrm{P}}y*, \tag{16}$$

and

$$a_{11} = -\gamma_{\mathrm{R}}, \quad a_{12} = f'(y*), \quad a_{21} = k_{\mathrm{P}}, \quad a_{22} = -\gamma_{\mathrm{P}}. \tag{17}$$

Clearly, the matrix $(a_{ij})_{2\times 2}$ is exactly the Jacobian matrix of Eq. 1 about the equilibrium $(x*, y*)$. However, Eq. 15 characterizes the stochastic properties of the system when the system state is near the equilibrium $(x*, y*)$, i.e., the steady-state statistics.

Notice that the expectations of ξ_R and ξ_P are $\langle\xi_R\rangle = \int_{-\infty}^{\infty}\int_{-\infty}^{\infty}\xi_R\phi\,d\xi_R\,d\xi_P$ and $\langle\xi_P\rangle = \int_{-\infty}^{\infty}\int_{-\infty}^{\infty}\xi_P\phi\,d\xi_R\,d\xi_P$, respectively. Then, using the boundary conditions of Eq. 14, we have that

$$\begin{aligned}\frac{d\langle\xi_R\rangle}{dt} &= a_{11}\langle\xi_R\rangle + a_{12}\langle\xi_P\rangle,\\ \frac{d\langle\xi_P\rangle}{dt} &= a_{21}\langle\xi_R\rangle + a_{22}\langle\xi_P\rangle.\end{aligned} \tag{18}$$

Similarly, we have also that

$$\begin{aligned}\frac{d\langle\xi_R^2\rangle}{dt} &= 2a_{11}\langle\xi_R^2\rangle + 2a_{12}\langle\xi_R\xi_P\rangle + 2D_R,\\ \frac{d\langle\xi_R\xi_P\rangle}{dt} &= (a_{11}+a_{22})\langle\xi_R\xi_P\rangle + a_{21}\langle\xi_R^2\rangle + a_{12}\langle\xi_P^2\rangle,\\ \frac{d\langle\xi_P^2\rangle}{dt} &= 2a_{21}\langle\xi_R\xi_P\rangle + 2a_{22}\langle\xi_P^2\rangle + 2D_P.\end{aligned} \tag{19}$$

It is easy to see that dynamical properties of Eq. 18 is same to that of Eq. 1, i.e., the point $(0,0)$ (i.e., $\langle\xi_R\rangle = 0$ and $\langle\xi_P\rangle = 0$) is a globally asymptotically stable equilibrium of Eq. 18. This implies that for large time t, the expected numbers of mRNA and protein molecules equal exactly $\langle n_R\rangle = \Omega x*$ and $\langle n_P\rangle = \Omega y^*$, respectively. It is also easy to see that the equilibrium of Eq. 7.19 (i.e., the solution of equations $d\langle\xi_R^2\rangle/dt = 0$, $d\langle\xi_P^2\rangle/dt = 0$, and $d\langle\xi_R\xi_P\rangle/dt = 0$) are given by

$$\begin{aligned}\langle\xi_R^2\rangle &= \frac{-D_1a_{22}^2 - D_2a_{12}^2 - D_1(a_{11}a_{22} - a_{12}a_{21})}{(a_{11}+a_{22})(a_{11}a_{22} - a_{12}a_{21})}\\ \langle\xi_R\xi_P\rangle &= \frac{D_1a_{21}a_{22} + D_2a_{11}a_{12}}{(a_{11}+a_{22})(a_{11}a_{22} - a_{12}a_{21})}\\ \langle\xi_P^2\rangle &= \frac{-D_2a_{11}^2 - D_1a_{21}^2 - D_2(a_{11}a_{22} - a_{12}a_{21})}{(a_{11}+a_{22})(a_{11}a_{22} - a_{12}a_{21})}\end{aligned} \tag{20}$$

and it must be asymptotically stable for Eq. 19, i.e., the real parts of eigenvalue of the Jacobian matrix of Eq. 19 $\begin{bmatrix} a_{11} & a_{12} & 0\\ a_{21} & a_{11}+a_{22} & a_{12}\\ 0 & a_{21} & a_{22}\end{bmatrix}$ must be negative. This implies that for large time t, the variances of the numbers of mRNA and protein molecules, denoted, respectively, by σ_R^2 and σ_P^2, and the covariance of the numbers of mRNA and protein molecules, denoted by $\mathrm{cov}_{R,P}$, can be given by

$$\begin{aligned}
\sigma_R^2 &= \langle n_R^2 \rangle - \langle n_R \rangle^2 = \Omega \langle \xi_R^2 \rangle \\
&= \frac{\gamma_R \gamma_P (\gamma_R + \gamma_P) \langle n_R \rangle + \gamma_P f'(y^*)^2 \langle n_P \rangle - \gamma_R f'(y^*) k_P \langle n_R \rangle}{(\gamma_R + \gamma_P)(\gamma_R \gamma_P - f'(y^*) k_P)}, \\
\sigma_P^2 &= \langle n_P^2 \rangle - \langle n_P \rangle^2 = \Omega \langle \xi_P^2 \rangle \\
&= \frac{\gamma_R \gamma_P (\gamma_R + \gamma_P) \langle n_P \rangle + \gamma_R k_P^2 \langle n_R \rangle - \gamma_P f'(y^*) k_P \langle n_P \rangle}{(\gamma_R + \gamma_P)(\gamma_R \gamma_P - f'(y^*) k_P)}, \\
\mathrm{cov}_{R,P} &= \langle (n_R - \langle n_R \rangle)(n_P - \langle n_P \rangle) \rangle = \Omega \langle \xi_R \xi_P \rangle \\
&= \frac{\gamma_R \gamma_P (k_P \langle n_R \rangle + f'(y^*) k_P \langle n_P \rangle)}{(\gamma_R + \gamma_P)(\gamma_R \gamma_P - f'(y^*) k_P)},
\end{aligned} \tag{21}$$

respectively.

Measuring noise. For the noise measure, Paulsson (1, 9) suggested that the noise in the number of mRNA (or protein) molecules should be measured by the normalized variance, i.e., the variance σ_R^2 (or σ_P^2) is normalized by the squared average $\langle n_R \rangle^2$ (or $\langle n_P \rangle^2$) since the variance is a second-order moment. According to this definition and Eqs. 6 and 7, we have that

$$\begin{aligned}
\frac{\sigma_R^2}{\langle n_R \rangle^2} &= \frac{1}{\langle n_R \rangle H_{R,R}} - \frac{H_{R,P}}{H_{R,R}} \frac{\mathrm{cov}_{R,P}}{\langle n_R \rangle \langle n_P \rangle}, \\
\frac{\sigma_P^2}{\langle n_P \rangle^2} &= \frac{1}{\langle n_P \rangle H_{P,P}} - \frac{H_{P,R}}{H_{P,P}} \frac{\mathrm{cov}_{R,P}}{\langle n_R \rangle \langle n_P \rangle},
\end{aligned} \tag{22}$$

i.e., from a statistical viewpoint, for both mRNA and protein, the total noise can be decomposed into two basic components, one concerns the contribution of average number of molecules (system size), and other the contribution of interactions between mRNA and protein (see also ref. 16). According to Paulsson's interpretation (9), for the terms $1/\langle n_R \rangle H_{R,R}$ and $1/\langle n_P \rangle H_{P,P}$, $H_{R,R}$ and $H_{P,P}$ are interpreted as the statistical bias to return to the average rather than deviate further, and for the terms $-\frac{H_{R,P}}{H_{R,R}} \frac{\mathrm{cov}_{R,P}}{\langle n_R \rangle \langle n_P \rangle}$ and $-\frac{H_{P,R}}{H_{P,P}} \frac{\mathrm{cov}_{R,P}}{\langle n_R \rangle \langle n_P \rangle}$, the factors $\frac{H_{R,P}}{H_{R,R}}$ and $\frac{H_{P,R}}{H_{P,P}}$ are called the normalized susceptibility factors. Similarly, the normalized covariance can be given by

$$\begin{aligned}
\frac{\mathrm{cov}_{R,P}}{\langle n_R \rangle \langle n_P \rangle} = &- \frac{H_{R,R} H_{P,P}}{H_{R,R} H_{P,P} - H_{R,P} H_{P,R}} \\
&\left[\frac{1}{\langle n_P \rangle H_{P,P}} \frac{H_{R,P}}{H_{R,R}} \frac{H_{R,R}/\tau_R}{H_{R,R}/\tau_R + H_{P,P}/\tau_P} \right. \\
&\left. + \frac{1}{\langle n_R \rangle H_{R,R}} \frac{H_{P,R}}{H_{P,P}} \frac{H_{P,P}/\tau_P}{H_{R,R}/\tau_R + H_{P,P}/\tau_P} \right],
\end{aligned} \tag{23}$$

where $\tau_R = 1/\gamma_R$ and $\tau_P = 1/\gamma_P$ measure the expected life-times of mRNA and protein molecules, respectively. In the above equation, the factor $\frac{H_{R,R}H_{P,P}}{H_{R,R}H_{P,P}-H_{R,P}H_{P,R}}$ measures the relative importance of $H_{R,R}H_{P,P}$, where it must be bigger than one since the feedback is negative. Notice that the terms $\frac{H_{R,R}/\tau_R}{H_{R,R}/\tau_R+H_{P,P}/\tau_P}$ and $\frac{H_{P,P}/\tau_P}{H_{R,R}/\tau_R+H_{P,P}/\tau_P}$ are the time averages for mRNA and protein, respectively (9). Hence, the term $\frac{1}{\langle n_P\rangle H_{P,P}}\frac{H_{R,P}}{H_{R,R}}\frac{H_{R,R}/\tau_R}{H_{R,R}/\tau_R+H_{P,P}/\tau_P}$ represents how the fluctuations in the number of protein molecules affect the interactions between mRNA and protein, and, similarly, the term $\frac{1}{\langle n_R\rangle H_{R,R}}\frac{H_{P,R}}{H_{P,P}}\frac{H_{P,P}/\tau_P}{H_{R,R}/\tau_R+H_{P,P}/\tau_P}$ represents how the fluctuations in the number of mRNA molecules affect the interactions between mRNA and protein. Notice also that the interactions between mRNA and protein in statistics mainly reflect how the fluctuations in the reaction rates are affected by the fluctuations in the numbers of mRNA and protein molecules. Thus, for the mRNA–protein system, we define that (1) the intrinsic noises of mRNA and protein are measured by $1/\langle n_R\rangle H_{R,R}$ and $1/\langle n_P\rangle H_{P,P}$, respectively, and (2) the extrinsic noises of mRNA and protein are measured by $-\frac{H_{R,P}}{H_{R,R}}\frac{\mathrm{cov}_{R,P}}{\langle n_R\rangle\langle n_P\rangle}$ and $-\frac{H_{P,R}}{H_{P,P}}\frac{\mathrm{cov}_{R,P}}{\langle n_R\rangle\langle n_P\rangle}$, respectively (see also refs. 9, 16). Clearly, the extrinsic noise represents only how the total noise deviates from the intrinsic noise.

Effects of negative feedback on the protein noise. Notice that $H_{R,R} = H_{P,P} = 1$, $H_{P,R} = -1$, and $H_{R,P} = -\langle b\rangle\tau_P f'(y^*)$, where $\langle b\rangle$ is the expected number of protein molecules produced per mRNA transcript (called also the burst size) (5, 9, 12), which is defined as $\langle b\rangle = k_P/\gamma_R$. Thus, the protein noise can be expressed as

$$\frac{\sigma_P^2}{\langle n_P\rangle^2} = \frac{1}{\langle n_P\rangle} + \frac{1}{\langle b\rangle\tau_P f'(y*)}\left(\frac{\sigma_R^2}{\langle n_R\rangle^2} - \frac{1}{\langle n_R\rangle}\right). \tag{24}$$

This also implies that

$$\frac{\sigma_P^2}{\langle n_P\rangle^2} = \frac{\gamma_P}{\langle n_P\rangle}\left[\frac{\gamma_P}{\langle b\rangle} - f'(y*)\right]^{-1}\left[\frac{1}{\langle b\rangle} + \frac{\gamma_R - f'(y*)}{\gamma_R + \gamma_P}\right], \tag{25}$$

and the protein extrinsic noise (1, 9) is

$$\frac{\sigma_P^2}{\langle n_P\rangle^2} - \frac{1}{\langle n_P\rangle} = \frac{\gamma_R(\gamma_P + f'(y*))}{\langle n_P\rangle(\gamma_R + \gamma_P)}\left[\frac{\gamma_P}{\langle b\rangle} - f'(y*)\right]^{-1}, \tag{26}$$

where $f'(y*) = -\gamma_P\beta/\langle b\rangle[1-(\gamma_P y^*/k_{\max}, \langle b\rangle)]$, which represents the strength of the negative feedback at the equilibrium $(x*, y*)$. Obviously, the contribution of the protein extrinsic noise to the total noise is negative, if $\gamma_P + f'(y*) < 0$. This result shows clearly that the negative feedback can be used to reduce the protein noise (see also refs. 1, 9, 12, 16, 21).

2.3. Monte Carlo Simulation

The Monte Carlo simulation is one of the most important methods for the stochastic analysis of gene expression (1). Here, we first give a short description for the Monte Carlo algorithm, which is mainly from Gillespie (22, 23). This algorithm can be easily achieved by using MATLAB or MATHEMATIC.

Let us consider a well-stirred molecules of N reactants $\{S_1, S_2, \ldots, S_N\}$, which interact through M reactions $\{R_1, R_2, \ldots, R_M\}$. The system is confined in a constant volume Ω. Let $x_i(t)$ denote the number of molecules of species S_i in the system at time t for $i = 1, 2, \ldots, N$, i.e., the system state can be expressed as

$$\mathbf{x}(t) = (x_1(t), x_2(t), \ldots, x_N(t)).$$

Our main goal is to estimate $\mathbf{x}(t)$ with initial $\mathbf{x}(t_0) = \mathbf{x}_0$ using Monte Carlo simulation. The changes of system state are the consequences of chemical reactions $\{R_1, R_2, \ldots, R_M\}$. For reaction channel R_i, $(i = 1, 2, \ldots, M)$, $\mu_i = (\mu_{i1}, \mu_{i2}, \ldots, \mu_{iN})$ is the state-change vector where μ_{ij} is the change of S_j molecular population because of one R_i reaction, i.e., the system state will immediately jumps from $\mathbf{x}$ to $\mathbf{x} + \mu_i$ when one R_i reaction occurs, and $\alpha_i(\mathbf{x})$ is the propensity function defined as that

$$\alpha_i(\mathbf{x})\mathrm{d}t \triangleq \text{the probability that one } R_i \text{ reaction will occur in the next time interval } [t, t + \mathrm{d}t).$$

According to the above definitions, the stochastic simulation algorithm (SSA) for constructing an exact numerical realization of the process $\mathbf{x}(t)$ is given in the following.

Step 0. Initialize the time $t = t_0$ and the system's state $\mathbf{x}(t_0) = \mathbf{x}_0$, and input values of μ_i for $i = 1, 2, \ldots, M$.

Step 1. Calculate all $\alpha_i(\mathbf{x})$ and the sum $\alpha_0(\mathbf{x}) = \sum_{i=1}^{M} \alpha_i(\mathbf{x})$.

Step 2. Generate two random numbers r_1 and r_2 using the uniform distribution in the unit interval, and take $\tau = (1/\alpha_0(x)) \ln(1/r_1)$ and i with

$$\sum_{k=1}^{i-1} \alpha_k(\mathbf{x}) < r_2\alpha_0(\mathbf{x}) < \sum_{k=1}^{i} \alpha_k(\mathbf{x}).$$

Step 3. Put $t = t + \tau$ and $\mathbf{x} = \mathbf{x} + \mu_i$.

Step 4. Record $(\mathbf{x}, t)$ as desired, and return to step 1, or else end this simulation.

In our model, for example, $f(y)$ is taken as the Hill-type function (see Eq. 2). Figures 2 and 3 show how the Hill coefficient β and dissociation constant k_d act on the protein expression, respectively. In Fig. 2, to identify how the protein noise is affected by the Hill coefficient β, the expectation $\langle n_\mathrm{p} \rangle$ is to be fixed for different values of β. In Fig. 2a, two simulation trajectories of protein molecular number corresponding to $\beta = 2$ and $\beta = 4$, respectively, are plotted, where the histograms on the right show the probability distribution of the protein number. In Fig. 2b, it is easy to see that the protein noise will decrease with the increase of β. Figure 3 shows the effects of the dissociation constant k_d on the expectation $\langle n_P \rangle$ and noise $\sigma_P^2/\langle n_P \rangle^2$, where the Hill coefficient β is taken as $\beta = 4$.

3. Stochasticity in Multiple-Gene Network

In this section, we consider only an example for the stochasticity in multiple-gene network, i.e., additivity of noise propagation in a protein cascade. Regulatory cascades are ubiquitous in biological

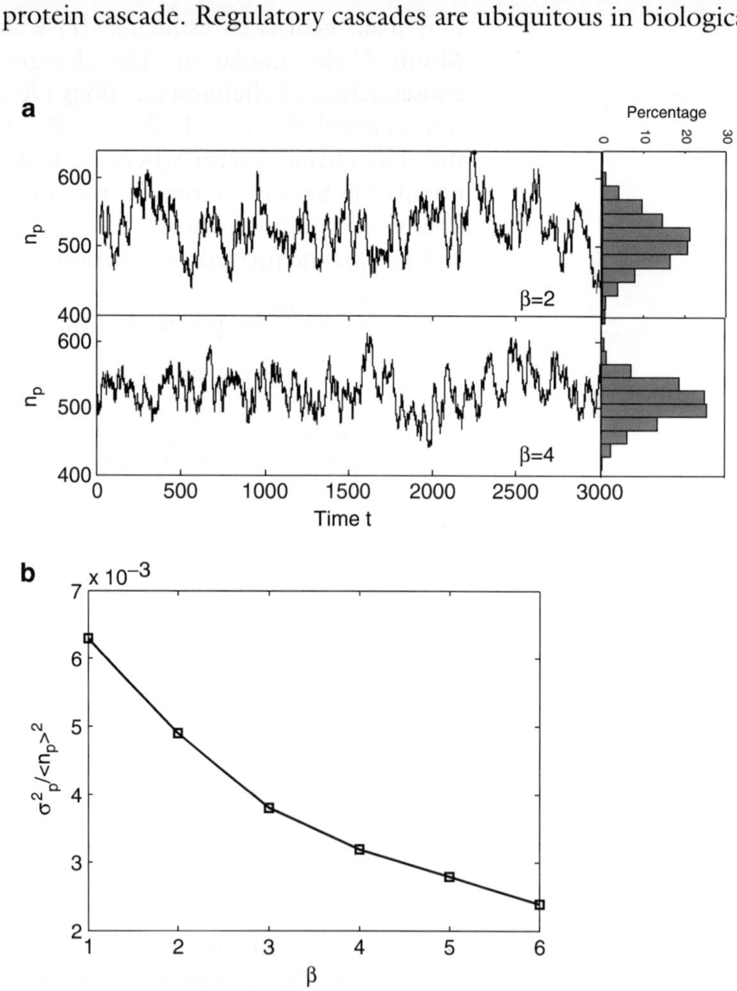

Fig. 2. Monte Carlo simulation – effect of Hill coefficient β on the protein noise: (**a**) Two simulation trajectories of protein molecular number corresponding to $\beta = 2$ and $\beta = 4$, respectively, the histogram on the right gives the probability distribution of protein copy number; (**b**) The total protein noise versus different values of Hill coefficient β. The parameters γ_R, γ_P, $\langle b \rangle$, $k_{\max}$, and k_d are taken as: $\ln 2/\gamma_R = 2$ min, $\ln 2/\gamma_P = 60$ min, $\langle b \rangle = 4$, $k_{\max} = 3$ and $k_d = 360/\ln 2$ which guarantees $\langle n_p \rangle$ to be fixed for different values of β.

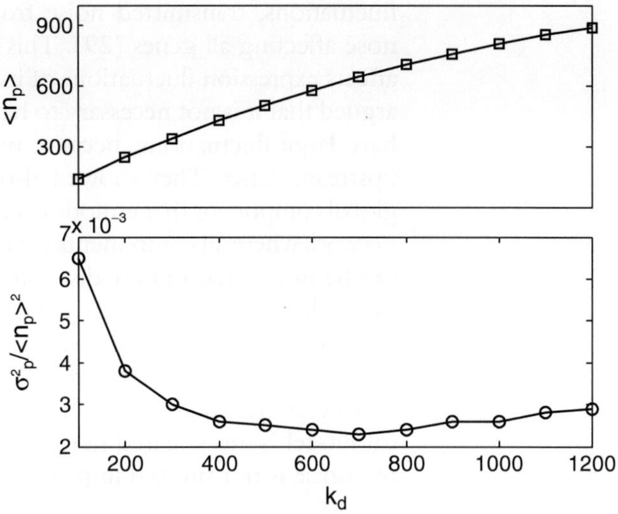

Fig. 3. The effects of the dissociation constant k_{d} on the expectation $\langle n_P \rangle$ and noise $\sigma_P^2/\langle n_P \rangle^2$, where the Hill coefficient β is taken as $\beta = 4$, and the other parameters are taken the same in Fig. 2.

systems (24–28). Recently, Hooshangi et al. presented the construction of three synthetic transcriptional cascades comparing one, two, and three repression stages and analyzed their dynamic and steady-state behaviors (28). Their main goal is to show how the depth of the cascade affects ultrasensitivity, temporal response, and phenotypic variations among cell population. A previous theoretical model developed by Thattai and van Oudenaarden (24) was used to approximate the mechanisms of noise attenuation in ultrasensitivity signaling cascades. Thattai and van Oudenaarden argued that cascades operating near saturation have output signal fluctuations that are bounded in magnitude, even as the number of noisy cascade stages becomes large and that cascades with ultrasensitive transfer functions can be made to simultaneously implement thresholding and noise reduction (24).

To quantify how noise propagates through gene networks, Pedraza and van Oudenaarden (29) presented a synthetic network consisted of four genes, of which three were monitored in single *E. coli* cells by cyan, yellow, and red fluorescent proteins (CFP, YFP, and RFP) and measured expression correlations between genes in single cells. In this synthetic network, the first gene, *lacI*, is constitutively transcribed and codes for the lactose repressor, which downregulates the transcription of the second gene, *tetR*, that is bicistronically transcribed with *cfp* (i.e., *tetR* and *cfp* are transcribed at the same time). The gene product of *tetR*, the tetracycline repressor, in turn downregulates the transcription the third gene, reported by YFP. The fourth gene, *rfp*, is under the control of the lambda repressor promoter. Pedraza and van Oudenaarden found that noise in a gene was determined by its intrinsic fluctuations, transmitted noise from upstream genes, and global nose affecting all genes (29). This means that the noise in a gene affects expression fluctuations of its downstream genes. Thus, they argued that it is not necessary to have low numbers of molecules to have large fluctuations, because noise could be transmitted from upstream genes. They showed also that the noise has a correlated global component that is modulated by the network. Thus, even in a network where all components have low-intrinsic noise, fluctuations can be substantial and the distribution of expression levels depends on the interactions between genes.

For the stochastic genetic cascade, one of the most important theoretical problems is to reveal the relationship between noise and signal transduction in regulated protein cascades. A theoretical model is investigated here, and the main goal is to show how the noise is transmitted in protein cascade.

3.1. Theory

Consider a synthetic transcriptional cascade consisted of N genes, where (a) the mRNA species and protein species corresponding to gene i are denoted by X_i and Y_i, respectively, for $i = 1, 2, \ldots, N$; (b) we here assume that the transcriptional level of mRNA X_i is only regulated by protein Y_{i-1} for $i = 2, 3, \ldots, N$; and (c) the transcription rate of mRNA species X_1 is assumed to be a constant (see Fig. 4) (see also ref. 24).

Let $x_i(t)$ and $y_i(t)$ denote the concentrations of mRNA X_i and protein Y_i, respectively, at time t for $i = 1, 2, \ldots, N$. The macroscopic rate equation of mRNA and protein can be given by

$$\begin{aligned} \frac{dx_1}{dt} &= -\gamma_1 x_1 + f_1, \\ \frac{dy_1}{dt} &= K_1 x_1 - \tilde{\gamma}_1 y_1, \\ \frac{dx_i}{dt} &= -\gamma_i x_i + f_i(y_{i-1}) \quad \text{for } i = 2, 3, \ldots, N, \\ \frac{dy_i}{dt} &= K_i x_i - \tilde{\gamma}_i y_i \qquad\quad \text{for } i = 2, 3, \ldots, N, \end{aligned} \tag{27}$$

where γ_i and $\tilde{\gamma}_i$ are the decay rates of mRNA X_i and protein Y_i, respectively, K_i the translation rate of protein Y_i

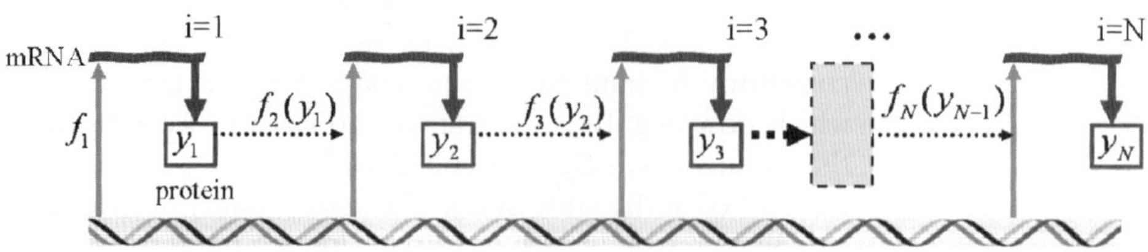

Fig. 4. Modeling genetic cascade, where the transcriptional rate of mRNA X_1 is assumed to be constant, and the transcriptional rate of mRNA X_i are regulated only by protein Y_{i-1} for $i = 2, 3, \ldots, N$. The concentration of protein Y_i is denoted by y_i for $i = 1, 2, \ldots, N$.

$(i = 1, 2, \ldots, N)$, f_1 is the transcription rate of protein X_1, which is assumed to be a constant, and $f_i(y_{i-1})$ are the transcription rates of mRNA X_i that depends on the concentration of protein Y_{i-1} for $i = 2, 3, \ldots, N$. Similar to Eq. 2, the function $f_i(y_{i-1})$ can be taken as the Hill-type function

$$f_i(y_{i-1}) = \frac{k_{\max}^{(i)}}{1 + \left(y_{i-1}/k_{\mathrm{d}}^{(i)}\right)^{\beta_i}} \quad (28)$$

for $i = 2, 3, \ldots, N$, where $k_{\max}^{(i)}$ is the maximum of the transcription rate, $k_{\mathrm{d}}^{(i)}$ is a bounding constant (the threshold protein concentration at which the transcription rate is at half its maximum value), and the parameter β_i is the Hill coefficient and determines the steepness of the repression curve.

Deterministic stability. It is easy to see that Eq. 27 must have a unique positive equilibrium point, denoted by $(\mathbf{x}*, \mathbf{y}*)$, where $\mathbf{x}^* = (x_1^*, x_2^*, \ldots, x_N^*)$ and $\mathbf{y}^* = (y_1^*, y_2^*, \ldots, y_N^*)$, which are given by

$$\begin{aligned} x_1^* &= \frac{f_1}{\gamma_1}, \\ y_1^* &= \frac{b_1 f_1}{\tilde{\gamma}_1}, \\ x_i^* &= \frac{f_i\left(y_{i-1}^*\right)}{\gamma_i}, \\ y_i^* &= \frac{b_i f_i\left(y_{i-1}^*\right)}{\tilde{\gamma}_i}. \end{aligned} \quad (29)$$

for $i = 2, 3, \ldots, N$, where $b_i = K_i/\gamma_i$ represents the average copy number of protein Y_i produced by per transcript of mRNA X_i $(i = 1, 2, \ldots, N)$. The Jacobian matrix of Eq. 27 about $(\mathbf{x}^*, \mathbf{y}^*)$ is

$$\mathbf{J} = \begin{pmatrix} -\gamma_1 & & & & & & & \\ K_1 & -\tilde{\gamma}_1 & & & & & & \\ & \mathrm{d}f_2\left(y_1^*\right)/\mathrm{d}y_1 & -\gamma_2 & & & & & \\ & & K_2 & -\tilde{\gamma}_2 & & & & \\ & & & \ddots & \ddots & & & \\ & & & & \ddots & \ddots & & \\ & & & & & \mathrm{d}f_N\left(y_{N-1}^*\right)/\mathrm{d}y_{N-1} & -\gamma_N & \\ & & & & & & K_N & -\tilde{\gamma}_N \end{pmatrix}. \quad (30)$$

Notice that $2N$ possible eigenvalues of the matrix $\mathbf{J}$ are $-\gamma_1, \ldots, -\gamma_N, -\tilde{\gamma}_1, \ldots, -\tilde{\gamma}_N$, respectively. Thus, $(\mathbf{x}*, \mathbf{y}*)$ must be globally asymptotically stable.

Noise analysis. It is well known that mRNA molecules usually decay much faster than proteins, i.e., the ratio $\tilde{\gamma}_i/\gamma_i$ should be typically a small quantity for all $i = 1, 2, \ldots, N$ (6). This implies that the concentrations of proteins should be the slow variables compared with the concentrations of mRNAs. So, in mathematics and physics, the dynamical properties of the system are mainly determined by the dynamical behavior of the concentrations of proteins (12), i.e., Eq. 7.27 can be approximated as

$$\begin{aligned}\frac{\mathrm{d}y_1}{\mathrm{d}t} &= b_1 f_1 - \tilde{\gamma}_1 n_1, \\ \frac{\mathrm{d}y_i}{\mathrm{d}t} &= b_i f_i(y_{i-1}) - \tilde{\gamma}_i y_i \text{ for } i = 2, 3, \ldots, N\end{aligned} \tag{31}$$

with a uniquely globally asymptotically stable equilibrium $\mathbf{y}*$ that is given in Eq. 29. The Jacobian matrix of Eq. 31 about $\mathbf{y}^*$ is given by

$$\mathbf{A} = \begin{pmatrix} -\tilde{\gamma}_1 & & & \\ b_2 \frac{\mathrm{d}f_2(y_1^*)}{\mathrm{d}y_1} & -\tilde{\gamma}_2 & & \\ & \ddots & \ddots & \\ & & b_N \frac{\mathrm{d}f_N(y_{N-1}^*)}{\mathrm{d}y_{N-1}} & -\tilde{\gamma}_N \end{pmatrix} \tag{32}$$

with eigenvalues $-\tilde{\gamma}_1, \ldots, -\tilde{\gamma}_N$. Elf and Ehrenberg (12) have provided a nice mathematical proof of the elimination of fast variables in stochastic noise analysis. They showed that the elimination of fast variables can make the linear noise approximation more accurate near critical points.

Similar to the analysis in Subheading 2, let Ω be the system size. Then, the copy number of protein Y_i can be expressed as $n_i = \Omega y_i$ for $i = 1, 2, \ldots, N$, i.e., $\mathbf{n} = (n_1, n_2, \ldots, n_N)$. For the stochastic fluctuations in $\mathbf{n}$, the probabilities of having $\mathbf{n}$ can be described by a birth-and-death Markov process with events

$$\begin{aligned} n_1 &\xrightarrow{\Omega b_1 f_1} n + 1, \\ n_1 &\xrightarrow{\tilde{\gamma}_1 n_1} n_1 - 1, \\ n_i &\xrightarrow{\Omega b_i f_i(n_{i-1}/\Omega)} n_i + 1, \\ n_i &\xrightarrow{\tilde{\gamma}_i n_i} n_i - 1, \end{aligned}$$

for $i = 2, 3, \ldots, N$. Let $\phi(\mathbf{n}, t)$ denote the joint probability distribution. The master equation of $\phi(\mathbf{n}, t)$ is

$$\begin{aligned}\partial_t \phi(\mathbf{n}, t) = &\left[(E_1^+ - 1)\tilde{\gamma}_1 n_1 \phi + (E_1^- - 1)\Omega b_1 f_1 \phi\right] \\ &+ \sum_{i=2}^{N} \left[(E_i^+ - 1)\tilde{\gamma}_i n_i \phi + (E_i^- - 1)\Omega b_i f_i(n_{i-1}/\Omega)\phi\right],\end{aligned} \tag{33}$$

where the symbol $E_i^{\pm}$ represents the step operator. Similar also to the analysis in Subheading 2, let

$$\mathbf{n} = \Omega \mathbf{y} + \Omega^{1/2} \boldsymbol{\xi}, \tag{34}$$

where $\mathbf{y}(t)$ is the solution of Eq. 31, and $\xi = (\xi_1, \xi_2, \ldots, \xi_N)$ are the new variables associated with the fluctuations of $\mathbf{n}$. The joint probability distribution is now rewritten as the function of ξ, i.e., $\phi(\mathbf{n}, t) = \psi(\xi, t)$. Notice that $n_i \to n_i \pm 1 \Leftrightarrow \xi_i \to \xi_i \pm \Omega^{1/2}$. Then, the Taylor expansion of the step operator $E_i^{\pm}$ can be given by

$$E_i^{\pm} = 1 \pm \Omega^{-1/2} \frac{\partial}{\partial \xi_i} + \frac{1}{2} \Omega^{-1} \frac{\partial^2}{\partial \xi_i^2} \pm \cdots \tag{35}$$

for $i = 1, 2, \ldots, N$. The time derivative in Eq. 7.34 is taken with constants n_i, i.e.,

$$\frac{\mathrm{d}\xi_i}{\mathrm{d}t} = -\Omega^{-1/2} \frac{\mathrm{d}y_i}{\mathrm{d}t} \tag{36}$$

for $i = 1, 2, \ldots, N$. Hence

$$\begin{aligned} \partial_t \phi =& \frac{\partial \psi}{\partial t} - \sum_{i=1}^{N} \Omega^{-1/2} \frac{\partial \psi}{\partial \xi_i} \frac{\mathrm{d}y_i}{\mathrm{d}t} \\ =& \Omega \left(\Omega^{-1/2} \frac{\partial}{\partial \xi_1} + \frac{1}{2} \Omega^{-1} \frac{\partial^2}{\partial \xi_1^2} \right) \tilde{\gamma}_1 \left(y_1 + \Omega^{-1/2} \xi_1 \right) \psi \\ & + \Omega \left(-\Omega^{-1/2} \frac{\partial}{\partial \xi_1} + \frac{1}{2} \Omega^{-1} \frac{\partial^2}{\partial \xi_1^2} \right) b_1 f_1 \psi \\ & + \sum_{i=2}^{N} \Omega \left[\left(\Omega^{-1/2} \frac{\partial}{\partial \xi_i} + \frac{1}{2} \Omega^{-1} \frac{\partial^2}{\partial \xi_i^2} \right) \tilde{\gamma}_i \left(y_i + \Omega^{-1/2} \xi_i \right) \psi \right. \\ & \left. + \left(-\Omega^{-1/2} \frac{\partial}{\partial \xi_i} + \frac{1}{2} \Omega^{-1} \frac{\partial^2}{\partial \xi_i^2} \right) b_i f_i \left(y_{i-1} + \Omega^{-1/2} \xi_{i-1} \right) \psi \right], \end{aligned} \tag{37}$$

where the terms $f_i(y_{i-1} + \Omega^{-1/2} \xi_{i-1})$ are taken their Taylor expansion about $\xi_{i-1} = 0$, i.e.,

$$f_i(y_{i-1} + \Omega^{-1/2} \xi_{i-1}) = f_i(y_{i-1}) + \Omega^{-1/2} \frac{\mathrm{d}f_i(y_{i-1})}{\mathrm{d}y_{i-1}} \xi_{i-1} + \cdots. \tag{38}$$

for $i = 2, \ldots, N$. Collecting the terms in Eq. 37 of Ω^0, we have

$$\begin{aligned} \frac{\partial \psi(\xi, t)}{\partial t} =& \tilde{\gamma}_1 \frac{\partial}{\partial \xi_1} \xi_1 \psi + \frac{1}{2} (b_1 f_1 + \tilde{\gamma}_1 y_1) \frac{\partial^2 \psi}{\partial \xi_1^2} \\ & + \sum_{i=2}^{N} \left[-\frac{\partial}{\partial \xi_i} \left(-\tilde{\gamma}_i \xi_i + b_i \frac{\mathrm{d}f_i(y_{i-1})}{\mathrm{d}y_{i-1}} \xi_{i-1} \right) \psi + \frac{1}{2} (b_i f_i(y_{i-1}) + \tilde{\gamma}_i y_i) \frac{\partial^2 \psi}{\partial \xi_i^2} \right] \end{aligned} \tag{39}$$

with boundary conditions $\lim_{\xi\to\pm\infty}\psi = 0$ and $\lim_{\xi\to\pm\infty}\partial\psi/\partial\xi_i = 0$.

Using the boundary conditions of Eq. 39, we have

$$\frac{d\langle\xi_1\rangle}{dt} = \int_{-\infty}^{\infty}\cdots\int_{-\infty}^{\infty}\xi_1\frac{\partial\psi}{\partial t}d\xi_1\cdots d\xi_N = -\tilde{\gamma}_1\langle\xi_1\rangle,$$

$$\frac{d\langle\xi_i\rangle}{dt} = \int_{-\infty}^{\infty}\cdots\int_{-\infty}^{\infty}\xi_i\frac{\partial\psi}{\partial t}d\xi_1\cdots d\xi_N = -\tilde{\gamma}_i\langle\xi_i\rangle + b_i\frac{df_i\left(y_{i-1}^*\right)}{dy_{i-1}^*}\langle\xi_{i-1}\rangle$$

for $i = 2, 3, \ldots, N$,

$$\frac{d\langle\xi_1^2\rangle}{dt} = \int_{-\infty}^{\infty}\cdots\int_{-\infty}^{\infty}\xi_1^2\frac{\partial\psi}{\partial t}d\xi_1\cdots d\xi_N = -2\tilde{\gamma}_1\langle\xi_1^2\rangle + 2\tilde{\gamma}_1 y_1^*,$$

$$\begin{aligned}\frac{d\langle\xi_i^2\rangle}{dt} &= \int_{-\infty}^{\infty}\cdots\int_{-\infty}^{\infty}\xi_i^2\frac{\partial\psi}{\partial t}d\xi_1\cdots d\xi_N \\ &= -2\tilde{\gamma}_i\langle\xi_i^2\rangle + 2b_i\frac{df_i(y_{i-1}^*)}{dy_{i-1}^*}\langle\xi_{i-1}\xi_i\rangle + 2\tilde{\gamma}_i y_i^*\end{aligned}$$

for $i = 2, 3, \ldots, N$,

$$\begin{aligned}\frac{d\langle\xi_1\xi_i\rangle}{dt} &= \int_{-\infty}^{\infty}\cdots\int_{-\infty}^{\infty}\xi_1\xi_i\frac{\partial\psi}{\partial t}d\xi_1\cdots d\xi_N \\ &= -(\tilde{\gamma}_1 + \tilde{\gamma}_i)\langle\xi_1\xi_i\rangle + b_i\frac{df_i(y_{i-1}^*)}{dy_{i-1}}\langle\xi_1\xi_{i-1}\rangle\end{aligned}$$

for $i = 2, 3, \ldots, N$, and

$$\begin{aligned}\frac{d\langle\xi_i\xi_j\rangle}{dt} &= \int_{-\infty}^{\infty}\cdots\int_{-\infty}^{\infty}\xi_i\xi_j\frac{\partial\psi}{\partial t}d\xi_1\cdots d\xi_N \\ &= -\left(\tilde{\gamma}_i + \tilde{\gamma}_j\right)\langle\xi_i\xi_j\rangle + b_i\frac{df_i(y_{i-1}^*)}{dy_{i-1}}\langle\xi_{i-1}\xi_j\rangle \\ &\quad + b_j\frac{df_j\left(y_{j-1}^*\right)}{dy_{j-1}}\langle\xi_i\xi_{j-1}\rangle\end{aligned}$$

for $1<i<j\leqslant N$. Hence, the first moment dynamics of ξ can be given by

$$\frac{d\langle\xi\rangle^T}{dt} = \mathbf{A}\langle\xi\rangle^T, \tag{40}$$

where $\langle\xi\rangle = (\langle\xi_1\rangle, \langle\xi_2\rangle, \ldots, \langle\xi_N\rangle)$ and $\mathbf{A}$ is the Jacobian matrix of Eq. 31 about $\mathbf{y}^*$ (see Eq. 7.32), and the second moment dynamics of ξ

$$\frac{d\Xi}{dt} = \mathbf{A}\Xi + \Xi\mathbf{A}^T + \mathbf{B}, \tag{41}$$

where $\Xi = \left(\langle\xi_i\xi_j\rangle\right)_{N\times N}$ and

$$\mathbf{B} = \begin{pmatrix} 2\tilde{\gamma}_1 y_1^* & & & \\ & 2\tilde{\gamma}_2 y_2^* & & \\ & & \ddots & \\ & & & 2\tilde{\gamma}_N y_N^* \end{pmatrix}. \tag{42}$$

Notice that all eigenvalues of the Jacobian matrix $\mathbf{A}$ are negative. Thus, for Eq. 40, $\langle \xi \rangle = \mathbf{0}$ must be globally asymptotically stable. For the dynamics Eq. 41, the solution of $\mathbf{A}\Xi + \Xi\mathbf{A}^T + \mathbf{B} = 0$ can be expressed as in the matrix integral form

$$\Xi = \int_0^\infty e^{\mathbf{A}t}\mathbf{B}(e^{\mathbf{A}t})^T \mathrm{d}t \tag{43}$$

(the derivation is given in the Appendix). For simplicity, we assume that all proteins have the same degradation rate (24), i.e., $\tilde{\gamma}_i = \tilde{\gamma}$ for $i = 1, 2, \ldots, N$. Since the Jordan decomposition of the matrix $\mathbf{A}$ is $\mathbf{A} = \mathbf{PJP}^{-1}$, where

$$\mathbf{J} = \begin{pmatrix} -\tilde{\gamma} & 1 & & & \\ & -\tilde{\gamma} & 1 & & \\ & & \ddots & \ddots & \\ & & & -\tilde{\gamma} & 1 \\ & & & & -\tilde{\gamma} \end{pmatrix}$$

and

$$\mathbf{P} = \begin{pmatrix} & & & 1 \\ & & \frac{b_N}{\tilde{\gamma}}\frac{\mathrm{d}f_N(y_{N-1}^*)}{\mathrm{d}y_{N-1}} & \\ & \cdots & & \\ \prod_{j=1}^{N-1}\frac{b_{j+1}}{\tilde{\gamma}}\frac{\mathrm{d}f_{j+1}(y_j^*)}{\mathrm{d}y_j} & & & \end{pmatrix},$$

we have

$$\begin{aligned} e^{\mathbf{A}t} &= \mathbf{P}e^{\mathbf{J}t}\mathbf{P}^{-1} \\ &= \mathbf{P}e^{-\tilde{\gamma}t}\begin{pmatrix} 1 & t & \frac{t^2}{2!} & \cdots & \frac{t^{N-1}}{(N-1)!} \\ & 1 & t & \cdots & \frac{t^{N-2}}{(N-2)!} \\ & & \ddots & \ddots & \vdots \\ & & & 1 & t \\ & & & & 1 \end{pmatrix}\mathbf{P}^{-1}. \end{aligned}$$

After calculating the integral in Eq. 43, we obtain

$$\begin{aligned} \Xi_{ii} &= \langle \xi_i^2 \rangle \\ &= y_i^* + \sum_{j=1}^{i-1}\frac{[2(i-j)-1]!!}{[2(i-j)]!!}\frac{1}{\tilde{\gamma}^{2(i-j)}}\prod_{l=1}^{i-j}\left(b_{l+1}\frac{\mathrm{d}f_{l+1}(y_l^*)}{\mathrm{d}y_l}\right)^2 y_j^* \end{aligned} \tag{44}$$

for $i = 1, 2, \ldots, N$, where $m!! = 1 \times 3 \times 5 \times \cdots \times m$, if m is old and $m!! = 2 \times 4 \times 6 \times \cdots \times m$, if m is even, and

$$\Xi_{1i} = \langle \xi_1 \xi_i \rangle = \frac{1}{(2\tilde{\gamma})^{i-1}} \prod_{l=1}^{i-1} b_{l+1} \frac{\mathrm{d}f_{l+1}(y_l^*)}{\mathrm{d}y_l} y_1^* \tag{45}$$

for $i = 2, 3, \ldots, N$.

Similar to the analysis in Subheading 2, the expectation and variance of n_i can be expressed as $\langle n_i \rangle = \Omega y_i^*$ and $\sigma_i^2 = \Omega \langle \xi_i^2 \rangle$ for $i = 1, 2, \ldots, N$, and the covariance of n_i and n_j is $\mathrm{cov}(n_i, n_j) = \Omega \langle \xi_i \xi_j \rangle$ for $i \neq j$.. This implies that the noise of protein Υ_i is

$$\frac{\sigma_i^2}{\langle n_i \rangle^2} = \left[1 + \sum_{j=1}^{i-1} \frac{[2(i-j)-1]!!}{[2(i-j)]!!} \frac{1}{\tilde{\gamma}^{2(i-j)}} \prod_{l=1}^{i-j} \left(b_{l+1} \frac{\mathrm{d}f_{l+1}(y_l^*)}{\mathrm{d}y_l} \right)^2 \frac{y_j^*}{y_i^*} \right] \times \frac{1}{\langle n_i \rangle} \tag{46}$$

for $i = 1, 2, \ldots, N$, where the term $1/\langle n_i \rangle$ represents the intrinsic noise of protein Υ_i due to random births and deaths of individual molecules, and the term

$$\sum_{j=1}^{i-1} \frac{[2(i-j)-1]!!}{[2(i-j)]!!} \frac{1}{\tilde{\gamma}^{2(i-j)}} \prod_{l=1}^{i-j} \left(b_{l+1} \frac{\mathrm{d}f_{l+1}(y_l^*)}{\mathrm{d}y_l} \right)^2 \frac{y_j^*}{y_i^*} \frac{1}{\langle n_i \rangle}$$

the noise propagation called the extrinsic noise (9) due to fluctuations in reaction rates. It is easy to see that the noise of protein Υ_1 will be only due to its intrinsic noise (where the effect of stochastic fluctuations in mRNA X_1 on protein Υ_1 is ignored), i.e., $\sigma_1^2 / \langle n_1 \rangle^2 = 1/\langle n_1 \rangle$ and that the noise of protein Υ_i ($i = 2, 3, \ldots, N$) will be larger than intrinsic noise due to any single cascade stage. This means that the propagated noise can contribute to the stochastic fluctuations of downstream protein species. From Eq. 45, the normalized covariance of protein Υ_1 and protein Υ_i is

$$\frac{\mathrm{cov}(n_1, n_i)}{\langle n_1 \rangle \langle n_i \rangle} = \frac{1}{(2\tilde{\gamma})^{i-1}} \prod_{l=1}^{i-1} b_{l+1} \frac{\mathrm{d}f_{l+1}(y_l^*)}{\mathrm{d}y_l} \frac{1}{\langle n_i \rangle} \tag{47}$$

for $i = 2, 3, \ldots, N$. Normally, the normalized variance and normalized covariance are also called the self-correlation and correlation, respectively.

4. Results

Steady-state sensitivity. The ultrasensitive behavior refers to a situation in which the output is very sensitive to variation in the input, i.e., an ultrasensitive all-or-none response to graded inputs where very small changes in input stimuli switch the output between low and high levels. For a genetic cascade, Thattai and van Oudenaarden (24) noticed that the net transfer function of a multistage cascade can display an enhanced sensitivity over that of its single-stage transfer functions, producing sharper switching behavior as more cascade stages are added (see also ref. 28), and they argued that the idea thresholding device produces two distinct outputs: high when the input is above threshold and low when the input is below it.

For our model, when the system state is near the stable equilibrium $\mathbf{y}^*$, the function $f_i(y_{i-1})$ can be linearized about $\mathbf{y}^*$, i.e., $f_i(y_{i-1}) = \left(\mathrm{d}f_i\left(y_{i-1}^*\right)/\mathrm{d}y_{i-1}\right)y_{i-1} + c_i$, where $c_i = f_i\left(y_{i-1}^*\right) - \left(\mathrm{d}f_i\left(y_{i-1}^*\right)/\mathrm{d}y_{i-1}\right)y_{i-1}^*$ for $i = 2, 3, \ldots, N$. If we take $f_i(y_{i-1})$ as a Hill-type function (see Eq. 28), then we have

$$\frac{\mathrm{d}f_i(y_{i-1})}{\mathrm{d}y_{i-1}} = \frac{f_i(y_{i-1})^2\beta_i\left(y_{i-1}\big/k_{\mathrm{d}}^{(i)}\right)^{\beta_i-1}}{k_{\max}^{(i)}k_{\mathrm{d}}^{(i)}}$$

for $i = 2, 3, \ldots, N$. Clearly, form $f_i(y_{i-1}) = \left(\mathrm{d}f_i\left(y_{i-1}^*\right)/\mathrm{d}y_{i-1}\right) y_{i-1} + c_i$, $\mathbf{y}^*$ can be expressed as

$$\begin{aligned} y_1^* &= \frac{b_1 f_1}{\tilde{\gamma}}, \\ y_i^* &= y_1^* \prod_{j=2}^{i} \frac{b_j\left(\mathrm{d}f_j\left(y_{j-1}^*\right)/\mathrm{d}y_{j-1}\right)}{\tilde{\gamma}} \\ &\quad + \sum_{j=2}^{i}\left[\prod_{l=j+1}^{i}\frac{b_l\left(\mathrm{d}f_l\left(y_{l-1}^*\right)/\mathrm{d}y_{l-1}\right)}{\tilde{\gamma}}\right]\frac{b_j c_j}{\tilde{\gamma}} \end{aligned} \tag{48}$$

for $i = 2, 3, \ldots, N$. For the steady-state sensitivity, i.e., the sensitivity when the system state is near the stable equilibrium $\mathbf{y}^*$, the effect of y_1^* on y_i^* can be measured by

$$\frac{\mathrm{d}y_i^*}{\mathrm{d}y_1^*} = \prod_{j=2}^{i}\frac{b_j\left(\mathrm{d}f_j\left(y_{j-1}^*\right)/\mathrm{d}y_{j-1}\right)}{\tilde{\gamma}} \tag{49}$$

for $i = 2, 3, \ldots, N$.

As a special case, if $b_i = b$, $\frac{\mathrm{d}f_i(y_{i-1}^*)}{\mathrm{d}y_{i-1}} = f'$, and $c_i = c$ for all $i = 2, 3, \ldots, N$, then Eq. 48 can be rewritten as

$$y_i^* = \left(\frac{bf'}{\tilde{\gamma}}\right)^{i-1} y_1^* + \sum_{j=2}^{i} \left(\frac{bf'}{\tilde{\gamma}}\right)^{i-j} \frac{bc}{\tilde{\gamma}} \\ = \left(\frac{bf'}{\tilde{\gamma}}\right)^{i-1} \left(y_1^* - \frac{bc}{\tilde{\gamma} - bf'}\right) + \frac{bc}{\tilde{\gamma} - bf'} \tag{50}$$

for $i = 2, 3, \ldots, N$ with $\lim_{i\to\infty} y_i^* = bc/(\tilde{\gamma} - bf')$, if $|bf'| < \tilde{\gamma}$. Clearly, these properties can be easily characterized by a simple difference equation

$$z_i = \frac{bf'}{\tilde{\gamma}} z_{i-1} + \frac{bc}{\tilde{\gamma}} \tag{51}$$

for $i = 2, 3, \ldots, N$, where $z_i = y_i^*$ with $z_1 = y_1^* = bf_1/\tilde{\gamma}$.. It is easy to see that the unique fixed point of Eq. 51 is $bc/(\tilde{\gamma} - bf')$, and it is asymptotically stable, if and only if $|bf'| < \tilde{\gamma}$ (see Fig. 5). The dynamics properties of Eq. 51 imply that (1) y_i^* will equal exactly $bc/(\tilde{\gamma} - bf')$ for all possible $i = 2, 3, \ldots, N$, if $y_1^* = bc/(\tilde{\gamma} - bf')$; (2) for $|bf'| < \tilde{\gamma}$, y_i^* will converge to $bc/(\tilde{\gamma} - bf')$ with the increase of cascade stage i, if $y_1^* \neq bc/(\tilde{\gamma} - bf')$, and the convergence will be monotone, if $0 < bf' < \tilde{\gamma}$ (see Fig. 5a), or the convergence will be damped oscillatory around $bc/(\tilde{\gamma} - bf')$, if $-\tilde{\gamma} < bf' < 0$ (see Fig. 5b); (3) for $bf' > \tilde{\gamma}$, y_i^* will increases, or decreases, monotonously with the increase of cascade stage i, if $y_1^* > bc/(\tilde{\gamma} - bf')$, or $y_1^* < bc/(\tilde{\gamma} - bf')$ (see Fig. 5c); and (4) for $bf' < -\tilde{\gamma}$, $y_1^*, y_2^*, \ldots, y_N^*$ will be a divergent oscillatory sequence around $bc/(\tilde{\gamma} - bf')$, and its amplitude will increase with the increase of cascade stage i (see Fig. 5d). All of these results show clearly that for the situation with $|bf'| > \tilde{\gamma}$, the thresholding behavior of input y_1^* is quite evident since for large cascade stage i, the output y_i^* will be very sensitive to the sign of the difference $y_1^* - bc/(\tilde{\gamma} - bf')$, i.e., for the different signs, + and −, the fate of y_i^* will be very different. This corresponds exactly to a standard switch structure for the ultrasensitive all-or-none behavior. Naturally, the sensitivity of the sequence $y_1^*, y_2^*, \ldots, y_N^*$ can be measured well by the derivative $\mathrm{d}y_i^*/\mathrm{d}y_1^* = (bf'/\tilde{\gamma})^{i-1}$ for all $i = 2, 3, \ldots, N$. Obviously, if $|bf'| < \tilde{\gamma}$, then, for large cascade stage i, the effect of the input y_1^* on the output y_i^* will be very small; conversely, if $|bf'| > \tilde{\gamma}$, then, for large cascade stage i, y_i^* will be dependent sensitively on the thresholding behavior of y_1^*. The derivative $\mathrm{d}y_i^*/\mathrm{d}y_1^* = (bf'/\tilde{\gamma})^{i-1}$ reflects how the concentration of protein Υ_i responses to the small change in the concentration of protein Υ_1 when the system state is near the stable equilibrium y^*. Of course, this is only a very special case, but it reveals some important properties for why a protein cascade can be ultrasensitive.

Additivity of noise propagation. From Eq. 41, the noise of protein Υ_i, $\sigma_i^2/\langle n_i\rangle^2$, can be also expressed as

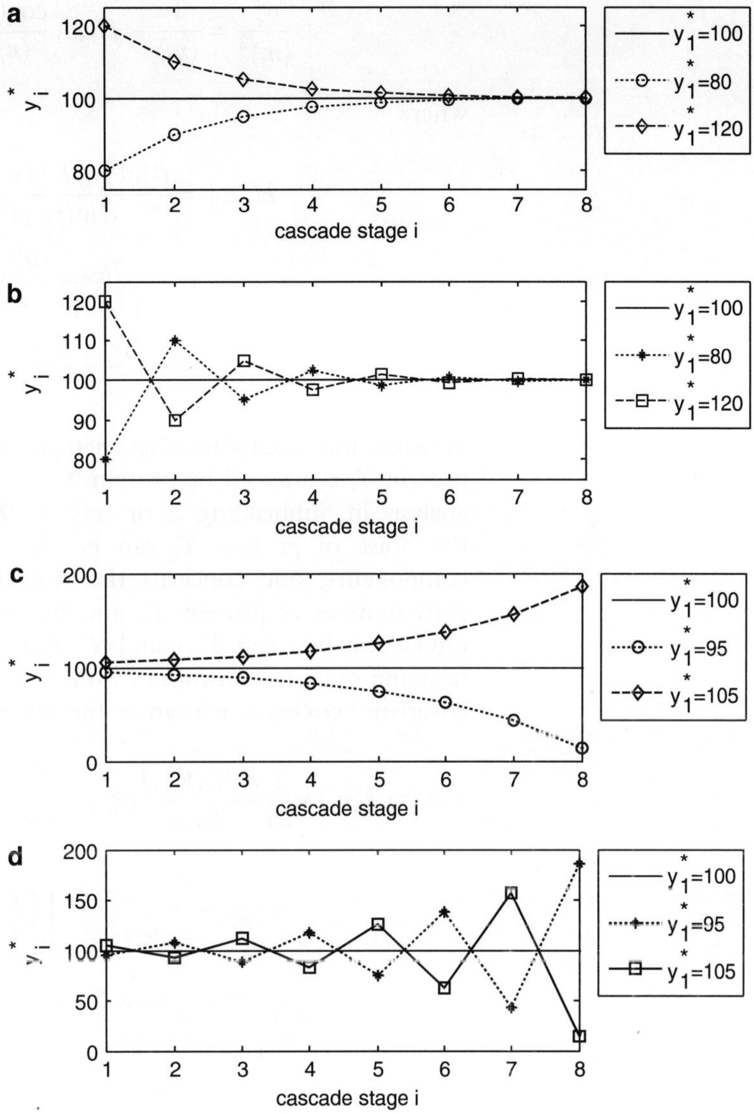

Fig. 5. The sequence $y_1^*, y_2^*, \ldots, y_N^*$ is plotted versus the cascade stage i. It shows clearly how the expected output levels dependent sensitively on the input level and the parameters, where (**a**) $bf' = 0.05,\ \ bc = 5,\ \ \tilde{\gamma} = 0.1,$ and $y_1^* = 80,\ 100,\ 120$; (**b**) $bf' = -0.05,\ \ bc = 15,\ \ \tilde{\gamma} = 0.1,$ and $y_1^* = 80,\ 100,\ 120$; (**c**) $bf' = 0.15,\ \ bc = -5,\ \ \tilde{\gamma} = 0.1,$ and $y_1^* = 95,\ 100,\ 105$; and (**d**) $bf' = -0.15,\ \ bc = 25,\ \ \tilde{\gamma} = 0.1,$ and $y_1^* = 95,\ 100,\ 105$. For all cases, we take $bc/(\tilde{\gamma} - bf') = 100$.

$$\frac{\sigma_i^2}{\langle n_i \rangle^2} = \frac{1}{\langle n_i \rangle} - H_{i,i-1} \frac{\mathrm{cov}(n_{i-1}, n_i)}{\langle n_{i-1} \rangle \langle n_i \rangle}, \tag{52}$$

where

$$\begin{aligned} H_{i,i-1} &= \left. \frac{\partial \ln(\tilde{\gamma} y_i / b_i f_i(y_{i-1}))}{\partial \ln(y_{i-1})} \right|_{\mathrm{y}=\mathrm{y}^*} \\ &= -\frac{y_{i-1}^*}{f_i(y_{i-1}^*)} \frac{\mathrm{d} f_i(y_{i-1}^*)}{\mathrm{d} y_{i-1}} \\ &= -\frac{b_i y_{i-1}^*}{\tilde{\gamma} y_i^*} \frac{\mathrm{d} f_i(y_{i-1}^*)}{\mathrm{d} y_{i-1}} \end{aligned} \tag{53}$$

measures how the balance between production and elimination of protein Υ_i is affected by protein Υ_{i-1} ($i = 2, 3, \ldots, N$) (see the analysis in Subheading 2 or refs. 1, 9, 16). This means that the noise of protein Υ_i can be decomposed into two basic components, one concerns the contribution of the expected copy number of protein Υ_i, and the other the contribution of interaction between Υ_{i-1} and Υ_i. On the other hand, to show how the noise is transmitted from upstream proteins to Υ_i, the covariance $\mathrm{cov}(n_{i-1}, n_i)$ can be further decomposed into

$$\begin{aligned} \mathrm{cov}(n_{i-1}, n_i) &= \frac{1}{2\tilde{\gamma}} \frac{b_i \mathrm{d} f_i(y_{i-1}^*)}{\mathrm{d} y_{i-1}} \sigma_{i-1}^2 + \sum_{j=2}^{i-1} \frac{[2(i-j)-3]!!}{[2(i-j)]!!} \\ &\quad \times \frac{1}{\tilde{\gamma}^{2(i-j)}} \frac{b_j \mathrm{d} f_j\left(y_{j-1}^*\right)}{\mathrm{d} y_{j-1}} \prod_{l=j}^{i-2} \left(b_{l+1} \frac{\mathrm{d} f_{l+1}(y_l^*)}{\mathrm{d} y_l} \right)^2 \sigma_j^2 \\ &= \sum_{j=1}^{i-1} \frac{[2(i-j)-3]!!}{[2(i-j)]!!} \frac{1}{\tilde{\gamma}^{2(i-j)}} \frac{b_j \mathrm{d} f_j\left(y_{j-1}^*\right)}{\mathrm{d} y_{j-1}} \\ &\quad \prod_{l=j}^{i-2} \left(b_{l+1} \frac{\mathrm{d} f_{l+1}(y_l^*)}{\mathrm{d} y_l} \right)^2 \sigma_j^2 \end{aligned} \tag{54}$$

for all $i = 2, 3, \ldots, N$. Thus, the variance σ_i^2 can be expressed as

$$\begin{aligned} \sigma_i^2 &= \langle n_i \rangle + \sum_{j=1}^{i-1} \frac{[2(i-j)-3]!!}{[2(i-j)]!!} \frac{1}{\tilde{\gamma}^{2(i-j)}} \\ &\quad \prod_{l=j}^{i-1} \left(b_{l+1} \frac{\mathrm{d} f_{l+1}(y_l^*)}{\mathrm{d} y_l} \right)^2 \sigma_j^2 \end{aligned} \tag{55}$$

i.e., the noise $\sigma_i^2 / \langle n_i \rangle^2$ can be rewritten as

$$\frac{\sigma_i^2}{\langle n_i \rangle^2} = \frac{1}{\langle n_i \rangle} + \sum_{j=1}^{i-1} \frac{[2(i-j)-3]!!}{[2(i-j)]!!} \frac{1}{\tilde{\gamma}^{2(i-j)}} \prod_{l=j}^{i-1} \left(b_{l+1} \frac{\mathrm{d}f_{l+1}(y_l^*)}{\mathrm{d}y_l} \right)^2 \left(\frac{y_j^*}{y_i^*} \right)^2 \frac{\sigma_j^2}{\langle n_j \rangle^2}$$
$$= \frac{1}{\langle n_i \rangle} + \frac{1}{2} H_{i,i-1}^2 \frac{\sigma_{i-1}^2}{\langle n_{i-1} \rangle^2} + \sum_{j=1}^{i-2} \frac{[2(i-j)-3]!!}{[2(i-j)]!!} \prod_{l=j}^{i-1} H_{l+1,l}^2 \frac{\sigma_j^2}{\langle n_j \rangle^2} \quad (56)$$

for $i = 2, 3, \ldots, N$. This reveals that the contributions of fluctuations in upstream proteins to the noise of protein Υ_i $(i = 2, 3, \ldots, N)$ should be additive, i.e., the noise of protein Υ_i can be expressed as a linear function of $\sigma_1^2/\langle n_1 \rangle^2$, $\sigma_2^2/\langle n_2 \rangle^2$, $\cdots$, $\sigma_{i-1}^2/\langle n_{i-1} \rangle^2$, where the terms $[2(i-j)-3]!!/[2(i-j)]!! \prod_{l=j}^{i-1} H_{l+1,l}^2$ for $j = 1, 2, \ldots, i-2$ and $H_{i,i-1}^2/2$ represent the effects of upstream proteins $\Upsilon_1, \Upsilon_2, \ldots, \Upsilon_{i-1}$ on protein Υ_i for noise propagation, respectively.

To show the output noise varies as the function of the input concentration and cascade length, we still consider only the special case with $b_i = b$, $\mathrm{d}f_i(y_{i-1}^*)/\mathrm{d}y_{i-1} = f'$, and $c_i = c$ for all $i = 2, 3, \ldots, N$. From Eqs. 46 and 47, we have

$$\frac{\sigma_i^2}{\langle n_i \rangle^2} = \left[1 + \sum_{j=1}^{i-1} \frac{(2j-1)!!}{(2j)!!} \left(\frac{bf'}{\tilde{\gamma}} \right)^{2j} \frac{y_{i-j}^*}{y_i^*} \right] \frac{1}{\langle n_i \rangle} \quad (57)$$

and

$$\frac{\mathrm{cov}(n_1, n_i)}{\langle n_1 \rangle \langle n_i \rangle} = \frac{1}{2^{i-1}} \left(\frac{bf'}{\tilde{\gamma}} \right)^{i-1} \frac{1}{\langle n_i \rangle} \quad (58)$$

for $i = 2, 3, \ldots, N$. Firstly, if $y_1^* = bc/(\tilde{\gamma} - bf')$, i.e., $y_i^* = y_1^*$ for all $i = 2, 3, \ldots, N$, then we must have

$$\frac{\sigma_i^2}{\langle n_i \rangle^2} = \left[1 + \sum_{j=1}^{i-1} \frac{(2j-1)!!}{(2j)!!} \left(\frac{bf'}{\tilde{\gamma}} \right)^{2j} \right] \frac{1}{\langle n_1 \rangle}.$$

This implies that

$$\frac{\sigma_i^2}{\langle n_i \rangle^2} < \frac{1}{2} \frac{1 - (bf'/\tilde{\gamma})^{2i}}{1 - (bf'/\tilde{\gamma})^2} \cdot \frac{1}{\langle n_1 \rangle}$$

with $|bf'| \neq \tilde{\gamma}$ since $(2j-1)!!/(2j)!! \leqslant 1/2$ for $j = 2, 3, \ldots, i-1$. Hence

$$\lim_{i \to \infty} \frac{\sigma_i^2}{\langle n_i \rangle^2} < \frac{1}{2} \frac{\tilde{\gamma}^2}{\tilde{\gamma}^2 - (bf')^2} \frac{1}{\langle n_1 \rangle}$$

if and only if $|bf'|<\tilde{\gamma}$, the noise $\sigma_i^2/\langle n_i\rangle^2$ will be bounded (23). On the other hand, notice also that

$$\frac{\sigma_i^2}{\langle n_i\rangle^2}-\frac{\sigma_{i-1}^2}{\langle n_{i-1}\rangle^2}=\frac{(2i-3)!!}{(2i-2)!!}\left(\frac{bf'}{\tilde{\gamma}}\right)^{2(i-1)}\frac{1}{\langle n_1\rangle}>0,$$

and that this difference will decreases with the increase of cascade stage i. Thus, the increase of $\sigma_i^2/\langle n_i\rangle^2$ will be monotone with the increase of cascade stage i, and it will converge to a limit for sufficiently large i, if $|bf'|<\tilde{\gamma}$ (see Fig. 6a).

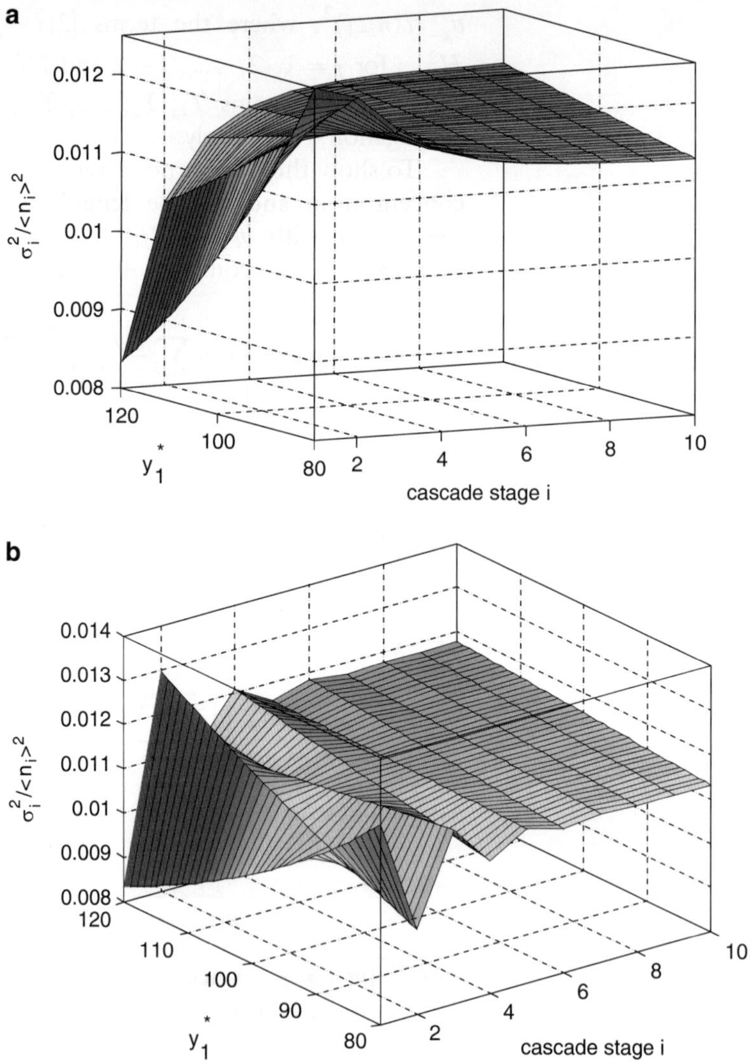

Fig. 6. The noises vary as the function of input concentration and cascade length. The *X*-axis denotes the cascade stage i, *Y*-axis the input concentration y_1^*, and *Z*-axis the noise. The parameters are taken as $bf'=0.05,\ bc=5$, and $\tilde{\gamma}=0.1$ in (**a**), and $bf'=-0.05,\ bc=15$, and $\tilde{\gamma}=0.1$ in (**b**).

Secondly, for the situation with $y_1^* \neq bc/(\tilde{\gamma} - bf')$, Eq. 57 can be rewritten as

$$\frac{\sigma_i^2}{\langle n_i \rangle^2} = \frac{1}{\langle n_i \rangle}\left[1 + \frac{1}{y_i^*}\left(y_1^* - \frac{bc}{\tilde{\gamma} - bf'}\right)\left(\frac{bf'}{\tilde{\gamma}}\right)^{-1}\sum_{j=1}^{i-1}\frac{(2j-1)!!}{(2j)!!}\left(\frac{bf'}{\tilde{\gamma}}\right)^{3j} + \frac{1}{y_i^*}\frac{bc}{\tilde{\gamma} - bf'}\sum_{j=1}^{i-1}\frac{(2j-1)!!}{(2j)!!}\left(\frac{bf'}{\tilde{\gamma}}\right)^{2j}\right]$$

for $i = 2, 3, \ldots, N$. Notice that the limit $\lim_{i\to\infty}\sigma_i^2/\langle n_i \rangle^2$ must exist, if $|bf'| < \tilde{\gamma}$ since

$$\sum_{j=1}^{i-1}\frac{(2j-1)!!}{(2j)!!}\left(\frac{bf'}{\tilde{\gamma}}\right)^{mj} < \frac{1}{2}\left(\frac{bf'}{\tilde{\gamma}}\right)^{m}\frac{1 - (bf'/\tilde{\gamma})^{m(j-1)}}{1 - (bf'/\tilde{\gamma})^{m}}$$

for $m = 2, 3$. Notice also that

$$\begin{aligned}\frac{\sigma_i^2}{\langle n_i \rangle^2} - \frac{\sigma_{i-1}^2}{\langle n_{i-1} \rangle^2} = {} & \left(\frac{1}{\langle n_i \rangle} - \frac{1}{\langle n_{i-1} \rangle}\right) + \left(\frac{1}{\langle n_i \rangle y_i^*} - \frac{1}{\langle n_{i-1} \rangle y_{i-1}^*}\right)\sum_{j=1}^{i-2}\frac{(2j-1)!!}{(2j)!!} \\ & \times \left(\frac{bf'}{\tilde{\gamma}}\right)^{2j}\left[\left(y_1^* - \frac{bc}{\tilde{\gamma} - bf'}\right)\left(\frac{bf'}{\tilde{\gamma}}\right)^{j-1} + \frac{bc}{\tilde{\gamma} - bf'}\right] \\ & + \frac{[2(i-1)-1]!!}{[2(i-1)]!!}\left(\frac{bf'}{\tilde{\gamma}}\right)^{2(i-1)} \\ & \times \left[\left(y_1^* - \frac{bc}{\tilde{\gamma} - bf'}\right)\left(\frac{bf'}{\tilde{\gamma}}\right)^{i-2} + \frac{bc}{\tilde{\gamma} - bf'}\right]\frac{1}{\langle n_i \rangle y_i^*}\end{aligned}$$

for $i = 2, 3, \ldots, N$. Thus, from the properties of the sequence $y_1^*, y_2^*, \ldots, y_N^*$, we have that (1) for $0 < bf' < \tilde{\gamma}$, if $y_1^* > bc/(\tilde{\gamma} - bf')$, then the noise $\sigma_i^2/\langle n_i \rangle^2$ will increase monotonously with the increase of cascade stage i (see Fig. 6a), and, conversely, if $y_1^* < bc/(\tilde{\gamma} - bf')$, then there must exist a positive integer i^* such that the noise $\sigma_i^2/\langle n_i \rangle^2$ will decrease monotonously with the increase of cascade stage i for all $i > i^*$, i.e., the noise can be attenuated if $y_1^* < bc/(\tilde{\gamma} - bf')$ (see also Fig. 6a); and (2) for $-\tilde{\gamma} < bf' < 0$, the noise $\sigma_i^2/\langle n_i \rangle^2$ will be damped oscillatory with the increase of cascade stage i (see Fig. 6b). These two results show clearly how the noise propagation depends sensitively on the thresholding behavior of y_1^* and cascade length. It is necessary to point out that since the variance of protein Y_1 equals exactly its expectation, where we ignore the effect of stochastic fluctuations in mRNA X_1 on protein Y_1), i.e., $\sigma_1^2 = \langle n_1 \rangle$, our results imply also that the thresholding behavior of input noise might be important in real system.

From the additivity of noise in propagation, the noise of protein Y_i can be also expressed as

$$\frac{\sigma_i^2}{\langle n_i\rangle^2} = \frac{1}{\langle n_i\rangle} + \frac{1}{2}\left(\frac{bf'}{\tilde{\gamma}}\right)^2 \left(\frac{y_{i-1}^*}{y_i^*}\right)^2 \frac{\sigma_{i-1}^2}{\langle n_{i-1}\rangle^2} + \sum_{j=1}^{i-2}\left[\frac{[2(i-j)-3]!!}{[2(i-j)]!!}\left(\frac{bf'}{\tilde{\gamma}}\right)^{2(i-j)} \prod_{l=j}^{i-1}\left(\frac{y_l^*}{y_{l+1}^*}\right)^2 \frac{\sigma_j^2}{\langle n_j\rangle^2}\right]$$

since $H_{i,i-1} = -(bf'/\tilde{\gamma})(y_{i-1}^*/y_i^*)$ for $i = 2, 3, \ldots, N$. This shows that if $|bf'/\tilde{\gamma}|<1$, then the effect of upstream protein Υ_j on the noise of protein Υ_i will decrease geometrically with the increase of $i-j$, i.e., the fact that the noise is additive can be reconciled with the fact that noise is bounded in magnitude.

A theoretical model for the stochastic fluctuations in a protein cascade is investigated in this section. For the steady-state sensitivity, we show the conditions that result in the ultrasensitive "all-or-none" behavior, and for the noise propagation, we show clearly that (1) for any one given protein species in this cascade, the contributions of fluctuations in upstream proteins to its noise should be additive and (2) the output noise level can vary as a function of the input concentrations and cascade length. These results not only provide a basic theoretical insight for understanding stochasticity and ultrasensitivity in a protein cascade, but also provide a possible theoretical explanation for the previous experimental studies (28, 29).

Appendix

Derivation of Eq. 43. To derive Eq. 43, we first introduce two theorems about the matrix equation in the following.

Theorem 1. The matrix equation $\mathbf{AX}+\mathbf{XB}=\mathbf{C}$ with $\mathbf{A},\mathbf{B},\mathbf{C}\in\mathbf{C}^{n\times n}$ has a solution $\mathbf{X}$, if and only if $\begin{bmatrix}\mathbf{A} & \mathbf{0}\\ \mathbf{0} & \mathbf{B}\end{bmatrix}$ and $\begin{bmatrix}\mathbf{A} & \mathbf{C}\\ \mathbf{0} & \mathbf{B}\end{bmatrix}$ are similar (30).

Theorem 2. If all the eigenvalues of $\mathbf{A}$ and $\mathbf{B}$ have nonnegative real parts, then the matrix equation $\mathbf{AX}+\mathbf{XB}=\mathbf{C}$ has a unique solution $X = -\int_0^\infty e^{\mathbf{A}t}\mathbf{C}e^{\mathbf{B}t}dt$ (30).

Notice that in Eqs. 40 and 41, the Jacobian matrix $\mathbf{A}$ satisfies *Theorem 1*. Thus, the matrix equation $\mathbf{A}\Xi+\Xi\mathbf{A}^T+\mathbf{B}=\mathbf{0}$ must have a solution. Notice also that all the possible eigenvalues of $\mathbf{A}$ are negative. Thus, from *Theorem 2*, we have Eq. 43.

References

1. Paulsson, J. (2005) Models of stochastic gene expression *Physics of Life Reviews* **2**, 157–175.
2. Ozbudark, E.M., Thattai, M., Kurster, I., Grossman, A.D., van Oudenaarden, A. (2002) Regulation of noise in the expression of a single gene *Nat Genet* **31**, 67–73.
3. Rigney, D.R. and Schieve, W.C. (1977) Stochastic model of linear, continuous protein-synthesis in bacterial populations *J Theor Biol* **69**, 761–766.
4. Berg , O.G. (1978) A model for statistical fluctuations of protein numbers in a microbial-population *J Theor Biol* **73**, 307–320.
5. McAdams , H.H. and Arkin, A. (1997) Stochastic mechanisms in gene expression *Proc Natl Acad Sci USA* **94**, 814–819.
6. Thattai , M. and A. van Oudenaarden. (2001) Intrinsic noise in gene regulatory networks *Proc Natl Acad Sci USA* **98**, 8614–8616.
7. Paulsson , J. and Ehrenberg, M. (2000) Random signal fluctuations can reduce random fluctuations in regulated components of chemical regulatory networks *Phys Rev Lett* **84**, 5447–5450.
8. Paulsson, J. and Ehrenberg, M. (2001) Intrinsic noise in gene regulatory networks *Q Rev Biophys* **34**, 1–59.
9. Paulsson, J. (2004) Summing up the noise in gene networks *Nature* **29**, 415–418.
10. Elowitz, M.B., Levine, A.J., Siggia, E.D., Swian, P.S. (2002) Stochastic gene expression in a single cell *Science* **297**, 183–186.
11. Blake, W.J., Kaern, M., Canto, C.R., Collins, J.J. (2003) Noise in eukaryotic gene expression *Nature* **422**, 633–637.
12. Elf, J.and Ehrenberg, M. (2003) Fast evaluation of fluctuations in biochemical networks with the linear noise approximation *Gemome Res* **13**, 2475–2484.
13. Kaern, M., Elston, T.C., Blake, W.J., Collins, J.J. (2005) Stochasticity in gene expression: from theories to phenotypes *Nat Rev Genet* **6**, 451–464.
14. Raj, A., van Oudenaarden, A. (2008) Nature, nurture, or chance: stochastic gene expression and its consequences *Cell* **135**, 216–226.
15. Becskei, A., Serrano, L. (2000) Engineering stability in gene networks by autoregulation *Nature* **405**, 590–593.
16. Tao, Y., Zheng, X-D., Sun, Y-H. (2007) Effect of feedback regulation on stochastic gene expression *J Theor Biol* **247**, 827–836.
17. Gardner, T.S., Cantor, C.R., Collins, J.J. (2001) Construction of a genetic toggle switch in Escherichia coli *Nature* **403**, 339–342.
18. Cherry, J.L. and Adler, F.R. (2000) How to make a biological switch *J Theor Biol* **203**, 117–133.
19. Shea, M., Ackers, G.K. (1985) The OR control system of bacteriophage lambda. A physical-chemical model for gene regulation *J Mol Biol* **181**, 211–230.
20. van Kampen, N.G. (1992) *Stochastic Process Theory in Physics and Chemistry* Amsterdam: North-Holland.
21. Tao, Y. (2004) Intrinsic and external noise in an auto-regulatory genetic network *J Theor Biol* **229**, 147–156.
22. Gillespie, D.T. (1977) Exact stochastic simulation of coupled chemical reactions *J Phys Chem* **81**, 2340–2361.
23. Gillespie, D.T. (2007) Stochastic simulation of chemical kinetics *Annu Rev Phys Chem* **58**, 35–55.
24. Thattai, M. and van Oudenaarden. A. (2002) Attenuation of noise in ultrasensitive signaling cascades *Biophys J* **82**, 2943–2950.
25. Lee , T.L., Rinaldi, N.J., Robert, F., Odom, D.T., Bar-Joseph, Z., Gerber, G.K., Hannett, N. M. , Harbison, C.T., Thompson, C.M., Simon, I. et al. (2002) Transcriptional regulatory networks in saccharomyces cerevisiae *Science* **298**, 799–804.
26. Shen-Orr , S.S., Milo, R., Mangan, S., Alon, U. (2002) Network motifs in the transcriptional regulation network of Escherichia coli *Nat Genet* **31**, 64–68.
27. Rosenfeld , N. and Alon, U. (2003) Response delays and the structure of transcription networks *J Mol Biol* **329**, 645–654.
28. Hooshangi, S., Thiberge, S., Weiss, R. (2005) Ultrasensitivity and noise propagation in a synthetic transcriptional cascade *Proc Natl Acad Sci USA* **102**, 3581–3586.
29. Pedraza, J.M. van Oudenaarden, A. (2005) Noise propagation in gene networks Science 307, 1965–1969.
30. Chen, G-N. (1990) Matrix Theory and Application Beijing: High Education Press, 1990 (in Chinese)

Chapter 8

Studying Adaptation and Homeostatic Behaviors of Kinetic Networks by Using MATLAB

Tormod Drengstig, Thomas Kjosmoen, and Peter Ruoff

Abstract

Organisms have the ability to counteract environmental perturbations and keep certain components within a cell homeostatically regulated. Closely related to homeostasis is the behavior of perfect adaptation where an organism responds to a step-wise perturbation by regulating some of its components, after a transient period, to their original pre-perturbation values. A particular interesting type of model relates to the so-called robust behavior where the homeostatic or perfect adaptation property is independent of the magnitude of the applied step-wise perturbation. It has been shown that this type of behavior is related to the control-theoretic concept of integral feedback (or integral control). Using downloadable MATLAB examples, we demonstrate how robust perfect adaptation sites can be identified in reaction kinetic networks by linearizing the system, applying the Laplace transform and inspecting the transfer function. We also show how the homeostatic set point in perfect adaptation is related to the presence of zero-order fluxes.

Key words: MATLAB, Kinetic networks, Metabolic control theory, Control engineering, Transfer functions, Control coefficients, Adaptation

1. Introduction

The capability of yeast and other organisms (and part of organsims) to adapt to environmental changes in nutrition (1–9), light (10–12), temperature (13, 14), or other stressors appear essential for an organism's fitness and survival. There are various adaptation modes (15, 16), which range from no adaptation at all, to partial adaptation, perfect adaptation, and overadaptation (Fig. 1).

There is a considerable interest in perfect adaptation which describes the response during a step-wise perturbation by maintaining some of the variables (concentration/fluxes) to their original pre-perturbation values. Perfect adaptation has been found, for example in bacterial (2–8) and eukaryotic (9) chemotaxis, osmoregulation in yeast (17), photoreceptor responses

Attila Becskei (ed.), *Yeast Genetic Networks: Methods and Protocols*, Methods in Molecular Biology, vol. 734,
DOI 10.1007/978-1-61779-086-7_8, © Springer Science+Business Media, LLC 2011

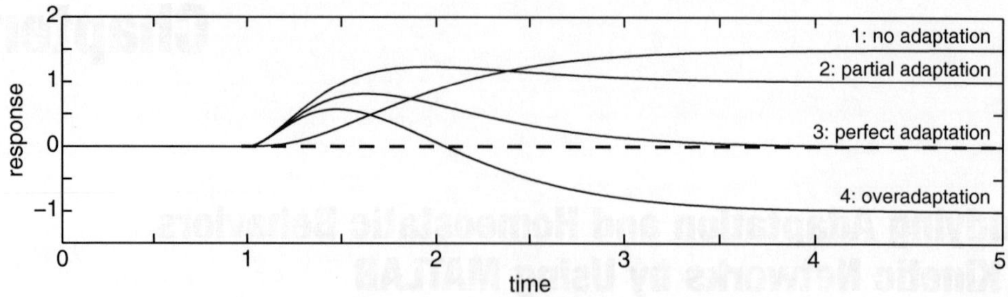

Fig. 1. Different adaptation behaviors of a system with respect to an applied step perturbation. Redrawn with permission from ref. (24).

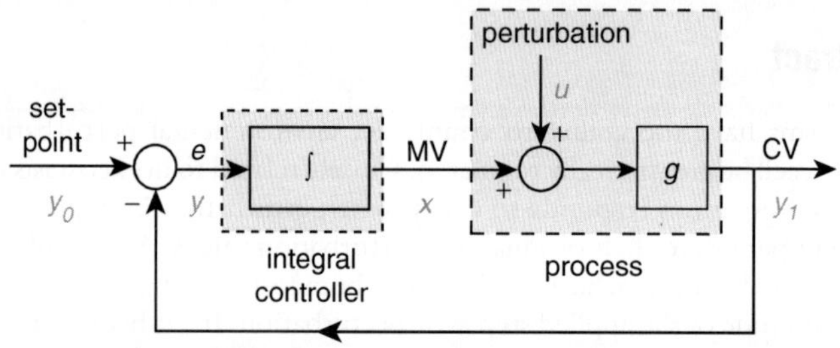

Fig. 2. Scheme of integral feedback/control of a perturbed system where the system output perfectly adapts to the setpoint. MV and CV are the manipulated and controlled variables, respectively. *e* denotes the error between the set point and controlled variable. *Gray symbols* represent the notation by Yi et al. (22). Redrawn with permission from ref. (25).

(10, 11), MAP-kinase regulations (18–20), as well as temperature homeostasis in circadian and ultradian rhythms (generally referred to as temperature compensation).

There are two ways in which to consider how perfect adaptation may be understood. In the first case, perfect adaptation is the result of a fine-tuning or balancing between rate parameters. This mode of adaptation is considered to be non-robust (21) because any change in a rate parameter will disrupt the balance and the adaptation behavior. In the second approach, perfect adaptation is the result of a network property, which does not need a fine-tuning in most of the parameters. Yi et al. (22) showed that this second form of adaptation can be described in terms of an integral feedback (also called integral control); a concept used in control theory (23).

Figure 2 illustrates the principles of integral feedback regulation. The error *e* between a reference signal (set point) and the output (CV, controlled variable) is integrated, processed together with a perturbation and then fed back again to calculate the error again. In this way a closed loop is generated, where the output value of the system converges to the set point and error *e* approaches zero. While the integral feedback principle is quite general, it is not obvious how to identify potential (robust) perfect adaptation sites in kinetic networks and how set points can be interpreted in terms of a molecular mechanism.

In the following section we will illustrate, by using MATLAB, how robust perfect adaptation sites can be identified (24) and how zero-order fluxes can become important in defining set points of homeostatic controllers (25).

2. Predicting Robust Perfect Adaptation Sites in Reaction Kinetic Networks

The procedure for identifying robust perfect adaptation sites consists of the following steps, which are described in detail in the next section, together with the MATLAB commands.

1. Define a state space model of the reaction kinetic network in question. This is a set of coupled first-order differential equations. The input variables on which step-wise perturbations are performed are generally the rate constants k_n, where n is an index for the reaction associated with k_n. The output variables are either the concentrations of the molecular components or reaction velocities (fluxes) or a combination of them (see below).
2. Perform a linearization of the network model (if possible). In some cases, it is difficult or impossible to find analytically a linearized model.
3. Laplace-transform the linearized network. In kinetics we generally think of a (perturbed) rate constant k_n or a concentration I_m of species m to be a function of time t, such as $k_n(t)$ or $I_m(t)$. The Laplace transform $F(s)$ of function $f(t)$ is defined as

$$F(s) = \mathcal{L}\{f(t)\} = \int_0^\infty e^{-st} f(t)\,dx, \tag{1}$$

   and becomes a function of the complex-valued s- or frequency space. The advantage of working in s-space is that the differential equations are transformed into algebraic equations, which are often easier to analyze and handle.
4. Calculate the transfer functions $H_{y_p,k_n}(s) = \Delta y_p(s)/\Delta k_n(s)$ between a small change in the Laplace transformed input elements $\Delta k_n(s)$ (i.e., rate constants) and the corresponding small change in the Laplace transformed output elements $\Delta y_p(s)$ (e.g., concentration or fluxes). Note that we use the same symbol for both time- and Laplace-domain signals. The transfer function elements $H_{y_p,k_n}(s)$ are part of the transfer function matrix $\mathbf{H}(s)$ from the vector of input elements to the

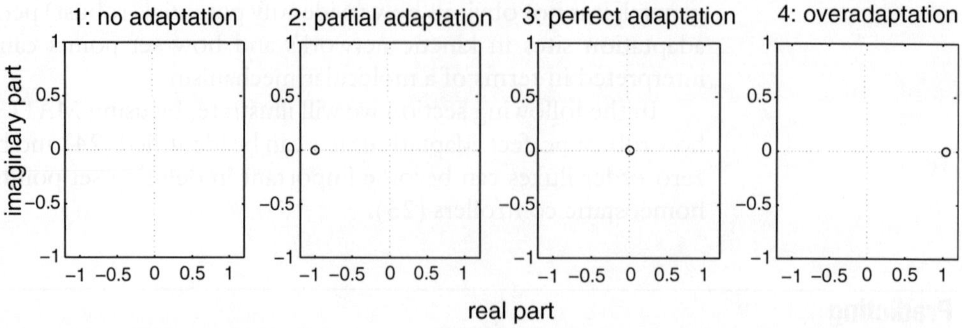

Fig. 3. The different adaptation behaviors in Fig. 1 are described by the solution of $n(s) = 0$. (1) When $n(s)$ is a constant, no adaptation exist; (2) partial adaptation is observed when the solution of $n(s) = 0$ lies in the left half of the complex s-plane; (3) the system shows perfect adaptation when the solution of $n(s) = 0$ lies in the origo; (4) overadaptation is observed when the solution of $n(s) = 0$ lies in the right half of the complex s-plane. The four transfer functions are described by $H_i(s) = n_i(s)/d(s), \; i = 1,\ldots,4$ with the denominator $d(s) = (0.2s + 1)(0.35s + 1)(0.45s + 1)$ and numerators: $n_1(s) = 1.5, \; n_2(s) = (s + 1), \; n_3(s) = s, n_4(s) = s - 1$. Redrawn with permission from ref. (24).

vector of output elements. In general, the elements of the transfer function matrix are written as:

$$H_{y_p,k_n}(s) = \frac{\Delta y_p(s)}{\Delta k_n(s)} = \frac{K \cdot \prod_{r=1}^{n} \left(-\frac{1}{z_r}s + 1\right)}{\prod_{q=1}^{m} \left(-\frac{1}{p_q}s + 1\right)} = \frac{n(s)}{d(s)}, \quad (2)$$

where z_r are defined as the transfer function's zeros, p_q are the poles, and K is the gain.

As indicated by Eq. (2), $H_{y_p,k_n}(s)$ is described as the ratio between two polynomials, the numerator-polynomial $n(s)$ and the denominator-polynomial $d(s)$. The solution to $n(s) = 0$ (i.e., the position of the zeros z_r in the complex plane) indicates the type of adaptation behavior. Figure 3 shows the different $n(s)$ polynomials that relate to the four adaptation types shown in Fig. 1. When, for a given $H_{y_p,k_n}(s)$, $n(s)$ has a zero in origo of the s-plane regardless of the values of any of the rate constants, the output shows robust perfect adaptation with respect to a stepwise increase of the rate constant considered as input. For a more detailed discussion about the influence of the transfer function's poles on adaptation kinetics we refer to ref. (24).

2.1. Detailed Outline of the Principles

Consider a reaction kinetic network with M chemical components (I_m) and N reaction steps, where each step n is associated with a rate constant k_n. The network can be stimulated by changing one of the rate constants (k_n) by means of a step function. Such a stimulation may occur due to a signal coming from a receptor acting specifically on k_n. In this respect, the rate constants are considered to be time-dependent. The kinetics of the network are described by the rate equations for each chemical component I_m:

$$\frac{dI_m(t)}{dt} = f_m(k_1(t), \ldots, k_N(t), I_1(t), \ldots, I_M(t)). \quad (3)$$

From this model we define P outputs, described as y_p (the ***model output***), which are the different properties of the network we want to investigate. For instance, these outputs can be concentrations $I_m(t)$, fluxes $J_n(t)$, or other network properties, which depend on concentrations and/or rate constants. Hence, the P nonlinear output models are given by:

$$y_p(t) = g_p(k_1(t), \ldots, k_N(t), I_1(t), \ldots, I_M(t)). \quad (4)$$

In order to find the ***transfer function matrix*** $\mathbf{H}(s)$ from the input (changes in rate constants) to the output candidates y_p, Eqs. (3) and (4) are first linearized around the (unperturbed) steady-state (ss) values $\mathbf{I}_{ss} = [I_1, \ldots, I_M]$, $\mathbf{y}_{ss} = [y_1, \ldots, y_P]$, and the pre-perturbation values of the rate constants $\mathbf{k}_{ss} = [k_1, \ldots, k_N]$ (note the independence of time t of the vector elements to indicate steady-state values), giving the following linear state-space model:

$$\Delta\dot{\mathbf{I}}(t) = A \cdot \Delta\mathbf{I}(t) + B \cdot \Delta\mathbf{k}(t), \quad (5)$$

$$\Delta\mathbf{y}(t) = C \cdot \Delta\mathbf{I}(t) + D \cdot \Delta\mathbf{k}(t), \quad (6)$$

where, $\Delta\mathbf{k}(t) = [\Delta k_1(t), \ldots, \Delta k_N(t)]^T$, $\Delta\mathbf{I}(t) = [\Delta I_1(t), \ldots, \Delta I_M(t)]^T$ and $\Delta\mathbf{y}(t) = [\Delta y_1(t), \ldots, \Delta y_P(t)]^T$ are vectors of small deviations around $\mathbf{k}_{ss}$, $\mathbf{I}_{ss}$, and $\mathbf{y}_{ss}$, respectively. The $M \times M$ ***state*** matrix A, the $M \times M$ ***input matrix*** B, the $P \times M$ ***output matrix*** C, and the $P \times N$ ***feed-through matrix*** D are defined as

$$A_{ij} = \frac{\partial f_i}{\partial I_j}|_{ss}, \quad (7)$$

$$B_{ij} = \frac{\partial f_i}{\partial k_j}|_{ss}, \quad (8)$$

$$C_{ij} = \frac{\partial g_i}{\partial I_j}|_{ss}, \quad (9)$$

$$D_{ij} = \frac{\partial g_i}{\partial k_j}|_{ss}. \quad (10)$$

Laplace-transforming the linearized model in Eqs. (5) and (6), gives the $(P \times N)$ ***transfer function*** matrix $\mathbf{H}(s)$ as (23):

$$\mathbf{H}(s) = C(sI - A)^{-1}B + D, \quad (11)$$

where I is the $M \times M$ identity matrix. $\mathbf{H}(s)$ describes the relationship between ***a small change*** in all possible inputs, i.e., the array of Laplace-transformed rate constants $\Delta\mathbf{k}(s) = [\Delta k_1(s),$

…, $\Delta k_N(s)]^{\mathrm{T}}$ and the resulting *changes* in all possible outputs, i.e., $\Delta \mathbf{y}(s) = [\Delta y_1(s), \ldots, \Delta y_P(s)]^{\mathrm{T}}$.

2.2. Calculating Control Coefficients

In metabolic control analysis (26–29) sensitivities are generally calculated as dimension-independent control or sensitivity coefficients:

$$C_{k_n}^{y_p} = \frac{\partial \log y_p}{\partial \log k_n}. \tag{12}$$

These sensitivity coefficients can also be calculated in *s* -(frequency) domain. The relationship between the frequency-dependent transfer function matrix and the frequency-dependent control coefficient matrix is found to be (30)

$$\mathbf{C}_{\mathbf{k}}^{\mathbf{y}}(s) = \mathbf{H}(s) \bullet \frac{\mathbf{k}_{ss}}{\mathbf{y}_{ss}}, \tag{13}$$

where the steady-state control coefficient matrix becomes

$$\mathbf{C}_{\mathbf{k}}^{\mathbf{y}} = \mathbf{H}(0) \bullet \frac{\mathbf{k}_{ss}}{\mathbf{y}_{ss}} \tag{14}$$

by using element-wise multiplication, or the so-called Hadamard matrix multiplication "•" (31). An alternative approach to relate control/sensitivity coefficients to their *s*-dependent counterpart was described by Ingalls (32).

2.3. Illustrating the Principles

We will use motif M1 shown in Eq. (15). below to illustrate the MATLAB commands used to calculate the transfer functions and control coefficients.

$$\xrightarrow{k_1} \mathrm{I}_1 \underset{k_{-2}}{\overset{k_2}{\rightleftarrows}} \mathrm{I}_2 \xrightarrow{k_3} \tag{15}$$

The rate equations as in Eq. (3) become

$$\frac{dI_1(t)}{dt} = \dot{I}_1(t) = k_1(t) - k_2(t)I_1(t) + k_{-2}(t)I_2(t), \tag{16}$$

$$\frac{dI_2(t)}{dt} = \dot{I}_2(t) = k_2(t)I_1(t) - k_3(t)I_2(t) - k_{-2}(t)I_2(t). \tag{17}$$

Since concentration is the model output, we get $y_1(t) = I_1(t)$ and $y_2(t) = I_2(t)$ as described in Eq. (4). Defining the order of the reaction constants as $\Delta\mathbf{k}(t) = [\Delta k_1(t), \Delta k_2(t), \Delta k_{-2}(t), \Delta k_3(t)]^T$ gives the following MATLAB code for the implementation of Eqs. (3), (4), and (7)–(10),

```
% file I122.m
clear all
close all
```

```
syms k1 k2 km2 k3 I1 I2

% differential equations
d_I1 = k1 - k2*I1 + km2*I2;
d_I2 =      k2*I1 - km2*I2 - k3*I2;

% ouput equations
y1 = I1;y2 = I2;

% system matrix A
A11=diff(d_I1,I1);A12=diff(d_I1,I2);
A21=diff(d_I2,I1);A22=diff(d_I2,I2);
A=[A11 A12;A21 A22];

% input matrix B
B11=diff(d_I1,k1);B12=diff(d_I1,k2);
B13=diff(d_I1,km2);B14=diff(d_I1,k3);
B21=diff(d_I2,k1);B22=diff(d_I2,k2);
B23=diff(d_I2,km2);B24=diff(d_I2,k3);
B=[B11 B12 B13 B14;B21 B22 B23 B24];

C11 = diff(y1,I1);C12 = diff(y1,I2);
C21 = diff(y2,I1);C22 = diff(y2,I2);
C=[C11 C12;C21 C22];

D11=diff(y1,k1);D12=diff(y1,k2);
D13-diff(y1,km2);D14=diff(y1,k3);
D21=diff(y2,k1);D22=diff(y2,k2);
D23=diff(y2,km2);D24-diff(y2,k3);
D=[D11 D12 D13 D14;D21 D22 D23 D24];
```

which produces the following matrices for the linearized state-space model:

$$A = \begin{bmatrix} -k_2 & k_{-2} \\ k_2 & -(k_3 + k_{-2}) \end{bmatrix}, \quad B = \begin{bmatrix} 1 & -I_1 & I_2 & 0 \\ 0 & I_1 & -I_2 & -I_2 \end{bmatrix},$$

$$C = \begin{bmatrix} 1 & 0 \\ 0 & 1 \end{bmatrix}, \qquad D = \begin{bmatrix} 0 & 0 & 0 & 0 \\ 0 & 0 & 0 & 0 \end{bmatrix}.$$

The MATLAB code for the transfer function matrix $\mathbf{H}(s)$ using Eq. (11) is shown below:

```
% identity matrix sI
syms s
sI = eye(2)*s;
% calculating the transfer function
Hs = C*inv(sI-A)*B+D;
```

The results can be presented in MATLAB workspace by typing `pretty(simple(Hs))` or more readable as

$$\mathbf{H}(s) = \frac{1}{s^2 + s(k_2 + k_3 + k_{-2}) + k_3 k_2} \cdot \begin{bmatrix} s + k_3 + k_{-2} & -I_1(s + k_3) & I_2(s + k_3) & -I_2 k_{-2} \\ k_2 & I_1 s & I_2 s & -I_2(s + k_2) \end{bmatrix}. \quad (19)$$

We see that the transfer function depends on the steady-state value of I_1 and I_2. Since these values depends on the rate constants, we first calculate and thereafter insert these expressions into the transfer function making the transfer function only dependent upon rate constants. The MATLAB commands used here are `solve` and `subs`

```
% state space calculations
ss = solve(d_I1,I1,d_I2,I2);
I1 = ss.I1;
I2 = ss.I2;
Hs1=subs(Hs,{'I1','I2'},{I1,I2})
```

where `ss` is a structure with two elements, i.e., the steady-state expressions for I_1 and I_2:

$$I_1 = \frac{k_{-2}k_1 + k_1 k_3}{k_3 k_2}, \quad I_2 = \frac{k_1}{k_3}. \quad (20)$$

By typing `pretty(simple(Hs1))` in MATLAB workspace, we get the rate constant-dependent transfer function `Hs1` as:

```
>> pretty(simple(Hs1))
[s + k3 + km2   k1 (k3 + km2) (s + k3)  k1 (s + k3)   km2 k1]
[-----------, - ----------------------, -----------, - ------]
[    %1              %1 k2 k3             %1 k3        %1 k3]
[                                                            ]
[ k2          k1 (k3 + km2) s          k1 s        (s + k2) k1]
[--,          ---------------,       - -----,    - -----------]
[ %1             %1 k2 k3             %1 k3         %1 k3    ]
                 2
         %1 := s  + s k3 + s km2 + k2 s + k2 k3
```

Based on this, it is now possible to investigate the step response, frequency response (i.e., Bode plot), and the location of poles and zeros. First the symbolic rate constants have to replaced by the actual values (e.g., $k_1 = 1$, $k_2 = 2$, $k_{-2} = 3$, $k_3 = 4$.)

```
Hs2 = subs(Hs1,{'k1','k2','km2','k3'},{1,2,3,4})
s = zpk('s')
Hs3=eval(simplify(Hs2)) % canceling overlapping poles/
zeros
figure;step(Hs3)
```

```
figure;bode(Hs3)

figure
subplot(2,4,1);pzmap(Hs3(1,1))
subplot(2,4,2);pzmap(Hs3(1,2))
subplot(2,4,3);pzmap(Hs3(1,3))
subplot(2,4,4);pzmap(Hs3(1,4))

subplot(2,4,5);pzmap(Hs3(2,1))
subplot(2,4,6);pzmap(Hs3(2,2))
subplot(2,4,7);pzmap(Hs3(2,3))
subplot(2,4,8);pzmap(Hs3(2,4))
```

producing the following results (Figs. 4–6).

2.4. Determination of Control Coefficients

Using the steady-state expressions for the outputs, i.e., $y_1 = I_1$ and $y_2 = I_2$, we use Eq. (13) to calculate the frequency dependent and Eq. (14) to calculate the steady-state concentration control matrices using the following MATLAB code,

```
k_y= [k1/I1 k2/I1 km2/I1 k3/I1
k1/I2 k2/I2 km2/I2 k3/I2];
C= Hs.*k_y;  % elementwise multiplication
```

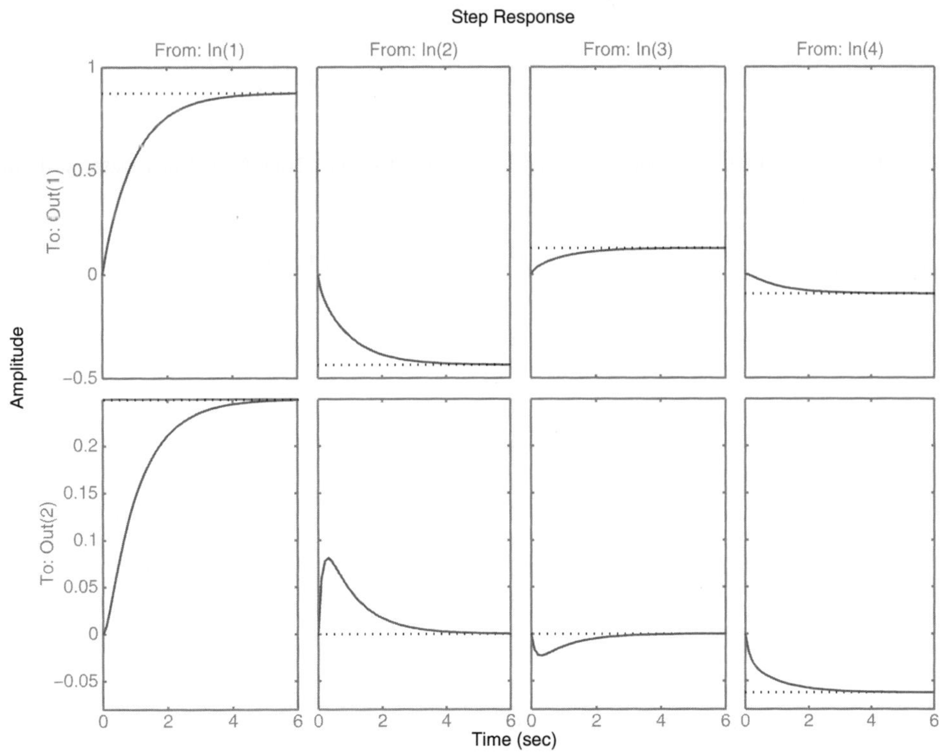

Fig. 4. MATLAB generated plot showing the result of a step response on rate constants k_1 (In(1)), k_2 (In(2)), k_{-2} (In(3)), and k_3 (In(4)) with the resulting time-behavior of concentrations (amplitude) of I_1 (Out(1)) and I_2 (Out(2)). Please note that by default MATLAB implicitly assumes that the time scale by which rate constants are defined is given in seconds (sec).

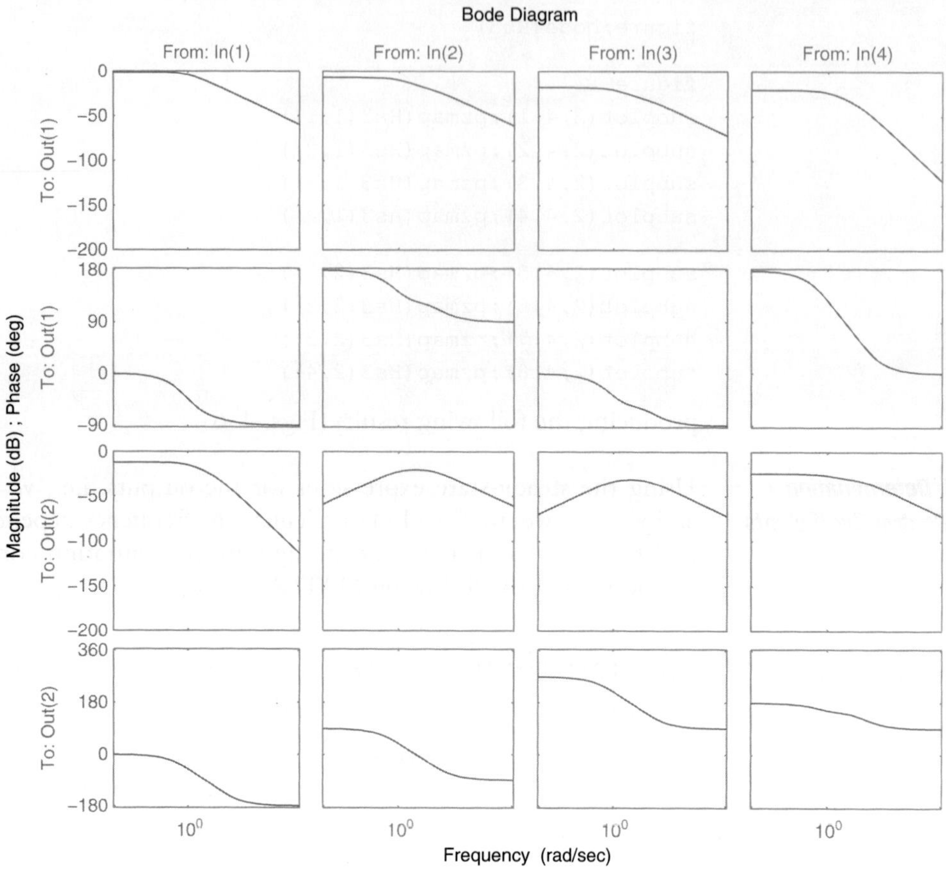

Fig. 5. MATLAB generated Bode plot showing the magnitude of the outputs (in dB) and the output's phases (in degrees) as a function of the logarithm of the frequency (radians/second).

```
C1 = subs(C,{'I1','I2'},{I1,I2})
Css = simplify(subs(C1,'s',0)
```

The results are presented below

$$\mathbf{C}_{\mathbf{k}}^{\mathbf{y}}(s) = \mathbf{H}(s) \bullet \frac{\mathbf{k}_{ss}}{\mathbf{y}_{ss}} = \frac{1}{s^2 + s(k_2 + k_3 + k_{-2}) + k_3 k_2} \begin{bmatrix} \frac{(s+k_3+k_{-2})k_2 k_3}{k_3+k_{-2}} & -(s+k_3)k_2 & \frac{(s+k_3)k_{-2}k_2}{k_3+k_{-2}} & \frac{-k_{-2}k_2k_3}{k_3+k_{-2}} \\ k_2 k_3 & (k_3+k_{-2})s & -k_{-2}s & -(s+k_2)k_3 \end{bmatrix}, \tag{21}$$

and the steady-state concentration control coefficient matrix is

$$\mathbf{C}_{\mathbf{k}}^{\mathbf{y}} = \begin{bmatrix} 1 & -1 & \frac{k_{-2}}{k_{-2}+k_3} & -\frac{k_{-2}}{k_{-2}+k_3} \\ 1 & 0 & 0 & -1 \end{bmatrix}. \tag{22}$$

Applying the numerical values to the rate constants

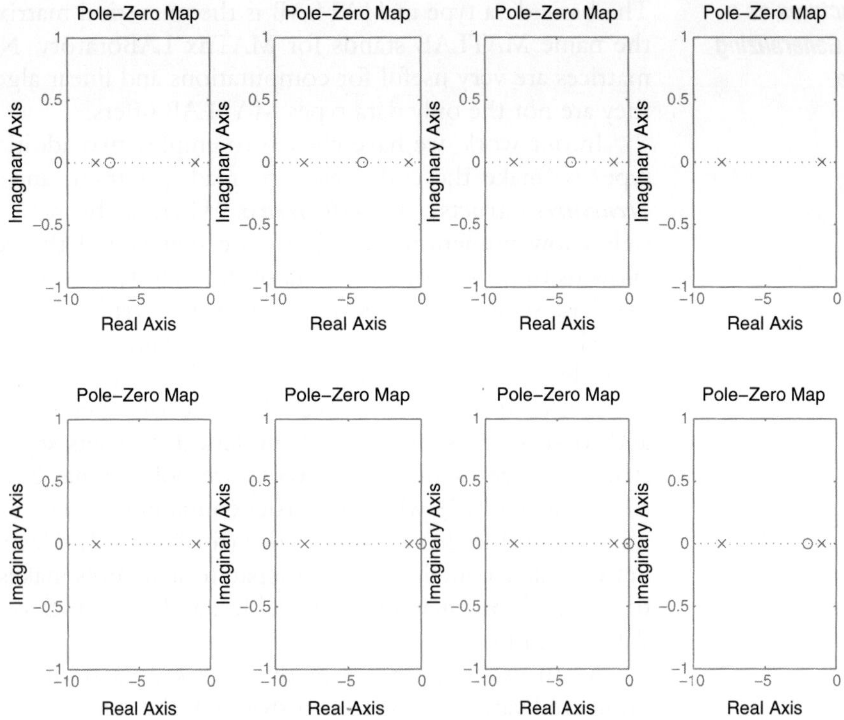

Fig. 6. MATLAB generated plot showing the zeros ($n(s) = 0$) as *circles* and poles ($d(s) = 0$) as *crosses* for each element in the transfer function matrix $\mathbf{H}(s)$.

```
C2 = subs(C1,{'k1','k2','km2','k3'},{1,2,3,4})
C3 = eval(simplify(C2)) % freq. dependent CC
C4 = dcgain(C3); % steady state CC
```

produce the following results in MATLAB:

```
>> C4
C4 =
1.0000    -1.0000    0.4286    -0.4286
1.0000          0         0    -1.0000
```

In the same manner as for the transfer-function matrix, step responses, Bode plots, and pole/zero plots can now be found for $\mathbf{C_k}^{\mathbf{y}}(s)$ in Eq. (21).

The summation theorem applied to either of the concentration-control coefficient matrices (i.e., the frequency-dependent matrix in Eq. (21) or the steady-state matrix in Eq. (22)) gives (summed over all N reactions):

$$\sum_{all\ N} \mathbf{C}_{\mathbf{k}}^{\mathbf{y}}(s) = \sum_{all\ N} \mathbf{C}_{\mathbf{k}}^{\mathbf{y}} = \begin{bmatrix} 0 \\ 0 \end{bmatrix},$$

which is easily verified by summing the rows in the MATLAB results for `C4` shown above.

2.5. Structures as a Tool for Generalizing the Code

The basic data type in MATLAB is the numerical matrix; in fact, the name MATLAB stands for MATrix LABoratory. Numerical matrices are very useful for computations and linear algebra, but they are not the only data types MATLAB offers.

In our work, we have chosen to employ two additional data types to make the code easier to read, program, and manage: *Structures* (structs) and *cell arrays*. Whereas the basic matrices only allow numerical data, both the structs and the cell arrays allow us to group together data of different types such as scalars, matrices, symbols, strings, and indeed even structs and cell arrays.

The structs are variables that have named *fields*, making it possible to create hierarchies of named variables, matrices, etc. We have used these structs to easily model our chemical networks, and to keep the symbolic and numerical data sets separate. The struct that represents the networks symbolically has been named `s` (for "symbolic"), while the struct containing the corresponding numerical values has been named `v` (for "values"). Having all the network data contained in two separate structures makes it trivial to save and load new sets of variables, and keep track of multiple data sets simultaneously.

As an example, the rate constants k_n, and indeed all other numbered variables, have been stored in the structs as cell arrays. The symbolic rate constants are accessed by using `s.k{n}`, and the corresponding values by using `v.k{n}`. Using cell arrays for such variables means we can quickly, not to mention independently of the current particular chemical network, check how many rate constants the network has. This allows us to write code that is more dynamic and does not need to be tailor made for each network to be evaluated, i.e., we can easily add more inputs or outputs and use the same framework.

In our work, we use the network models to study one or more of the following input/output relationships in our search for robust/nonrobust perfect adaptation:

- Rate constants/concentration
- Rate constants/fluxes
- Temperature/concentration
- Temperature/fluxes

We specify the equations of the network in the `s` struct and the values for which we want to evaluate the network in the `v` struct.

For the example in motif Eq. (15), i.e., rate constant/concentration relationship, the network specification together with the differential equations will look like

```
v.intermediates = 2;
v.rate_constants = 4;
```

```
v.inputs = v.rate_constants;
v.outputs = v.intermediates;

v.k = {1, 2, 3, 4};

% differential equations
s.d_I{1} = s.k{1} - s.k{2}*s.I{1};
s.d_I{2} =          s.k{2}*s.I{1} - s.k{3}*s.I{2};
```

The generic code (used for all motifs) for calculation of e.g., the system matrix A and the output matrix $\mathbf{C}$ in the s struct becomes

```
% system matrix A
for kk = 1:v.intermediates
    for jj = 1:v.intermediates
        s.d_I{kk}.dI{jj} = diff(s.d_I{kk}, s.I{jj});
        s.A(kk,jj) = s.d_I{kk}.dI{jj};
    end
end

% output matrix C
for kk = 1:v.outputs
    for jj = 1:v.intermediates
        s.dy{kk}.dI{jj} = diff(s.y{kk},s.I{jj});
        s.C(kk,jj) = s.dy{kk}.dI{jj};
    end
end
```

By using the structs and cell arrays, the specification for each motif needs approximately 20 individual lines of code, whereas the generic calculation for transfer function, control coefficients, search for nonrobust perfect adaptation and others are programmed over approximately 1,000 lines of code.

3. Defining the Set Point in a Homeostatic Controller

Homeostasis is another aspect of how to view (robust) perfect adaptation of a controlled compound A. To see how the set point in the integral feedback scheme (Fig. 2) can be defined in kinetic terms, we consider in Fig. 7 an homeostatic inflow controller (25), where species A is under negative stabilizing (33) feedback control by species E_{adapt}. We assume that A is synthesized by zero-order process with rate constant k_{synth} and is subject to unpredictable inflow perturbations by (varying) rate constant k_{pert}. Enzyme E_{tr} transforms A into another species, while enzyme E_{adapt} induced by A (through k_{adapt}) removes/degrades A. Enzyme E_{set} removes or inactivates E_{adapt}. Concentrations of E_{tr} and E_{set}

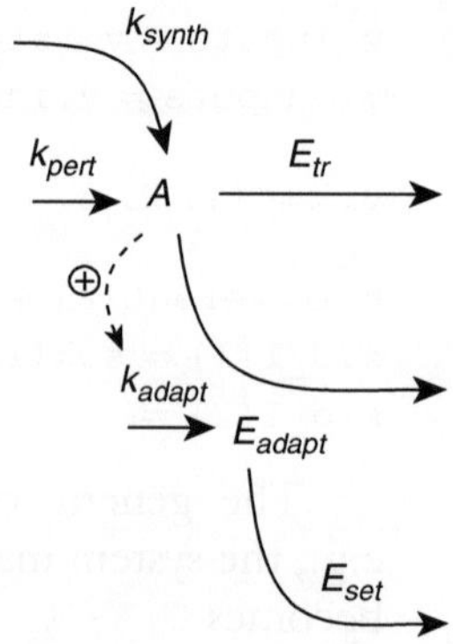

Fig. 7. Homeostatic inflow controller keeping robust homeostasis in *A*. Redrawn with permission from ref. (25).

are considered to be constant. All enzymatic reactions are described by standard Michaelis–Menten kinetics

$$v = \frac{V^E_{max} \cdot S}{K^E_M + S}, \tag{24}$$

where v is the reaction velocity, S denotes the concentration of substrate, K^E_M is the Michaelis constant, and V^E_{max} is the maximum velocity described by $V^E_{max} = k^E_{cat} \cdot E$ with turnover number k^E_{cat} and enzyme concentration E.

The rate equations are:

$$\frac{dA}{dt} = k_{pert} + k_{synth} - \frac{V^{E_{adapt}}_{max} A}{K^{E_{adapt}}_M + A} - \frac{V^{E_{tr}}_{max} A}{K^{E_{tr}}_M + A}, \tag{25}$$

$$\frac{dE_{adapt}}{dt} = k_{adapt} A - \frac{V^{E_{set}}_{max} E_{adapt}}{K^{E_{set}}_M + E_{adapt}}. \tag{26}$$

Equation (26) defines the error between the set point in A-homeostasis, A_{set}, and the actual value in A by comparing Eq. (26) with the equation

$$\frac{dE_{adapt}}{dt} = k_{adapt}(A - A_{set}), \tag{27}$$

which gives the following A_{set}:

$$A_{set} = \frac{V^{E_{set}}_{max}}{k_{adapt}} \cdot \frac{E_{adapt}}{K^{E_{set}}_M + E_{adapt}}. \tag{28}$$

Equation (28) indicates that $V^{E_{set}}_{max}/k_{adapt}$ is an upper bound for A_{set} and robust homeostasis in A with the set point

$$A_{set} = \frac{V^{E_{set}}_{max}}{k_{adapt}} \tag{29}$$

is achieved when $K_M^{E_{set}} \ll E_{adapt}$, i.e., when there is a strong binding between E_{adapt} and its processing enzyme E_{set} leading to zero-order kinetics in the removal/inactivation of E_{adapt} (25).

To solve the rate equations (25) and (26) two m-files are created and put in the path of MATLAB. The first file `LShifc.m` contains initial concentrations to the dynamical variables `y(i)`, values to the rate parameters `k(i)`, the method of integration, and the simulation time. The file can also include plotting instructions as shown here:

```
%le: LShifc.m


clear all

% dynamic variables
% y(1) <-> A
% y(2) <-> E_adapt

%rate constants/rate parameters
% k(1) <-> k1
% k(2) <-> kcat(E_adapt)
% k(3) <-> KM (E_adapt)
% k(4) <-> k_adapt
% k(5) <-> kcat (E_set)
% k(6) <-> KM (E_set)
% k(7) <-> k_synth
% k(8) <-> kcat(E_tr)
% k(9) <-> KM (E_tr)
% k(10) <-> E_set
% k(11) <-> E_tr

% define rate constant values [k(1) k(2)..... k(10)]
ks=[0.1  1.0  2.0  3.0  6.0e+6  1.0e-6  1.0  0.01  5.0
5.0e-7  0.1];

% simulation time
t=[0,50];

% initial concentrations
y0=[1.0   0.03];

% options for numerical integration
options = odeset('RelTol',0.000001,'MaxStep',0.01);
% solve model
[T Y]=ode15s(@hifc,t,y0,options,ks);

% making Figure 1
figure(1),
subplot(2,1,1),plot(T,Y(:,2),'-',T,Y(:,1),'-');
xlabel('time, ␣au');
```

```
ylabel('concentration, au');
hold on
grid on
legend('E_{adapt}','A');
hold off

subplot(2,1,2),plot(Y(:,1),Y(:,2),'-');
xlabel('A-concentration, au');
ylabel('E_{adapt}-concentration, au');
title(['inflow homeostatic controller'])
hold on
legend('E_{adapt}-A phase plane');
hold on
grid on
hold off
```

The second file `hifc.m` defines symbolically the rate equations:

```
%le: hifc.m
function dy=hifc(t,y,k)
dy=zeros(2,1);
dy(1)=k(1)-k(2)*y(1)*y(2)/(k(3)+y(1))+k(7)-k(8)*k(11)
*y(1)/(k(9)+y(1));
dy(2)=k(4)*y(1)-k(5)*y(2)*k(10)/(k(6)+y(2));
```

The model is run by placing the files `LShifc.m` and `hifc.m` somewhere in MATLAB's path, typing `LShifc` in the MATLAB console, and hitting the RETURN key. Figure 8 shows the adaptation behavior of the inflow controller with the initial concentrations and rate constants given above.

3.1. Harmonic Oscillations in Homeostatic Controllers

Interestingly, the negative feedback in the A–E_{adapt} homeostatic system can lead to harmonic oscillations when the binding between A and E_{adapt} becomes strong (leading to low $K_M^{E_{adapt}}$ values) and, additionally, the removal of A by transforming enzyme E_{tr} is negligible (either by a large $K_M^{E_{tr}}$ value and/or by a low $V_{max}^{E_{tr}}$ value). In this case, the rate equations (25) and (26) can be combined and lead to the harmonic oscillator equation

$$\frac{\ddot{A}}{k_{cat}^{E_{adapt}} \cdot k_{adapt}} + A = A_{set} = \frac{V_{max}^{E_{set}}}{k_{adapt}}, \tag{30}$$

indicating that A shows harmonic oscillations around A_{set} with a period length P given by

$$P = \frac{2\pi}{\sqrt{k_{cat}^{E_{adapt}} \cdot k_{adapt}}}. \tag{31}$$

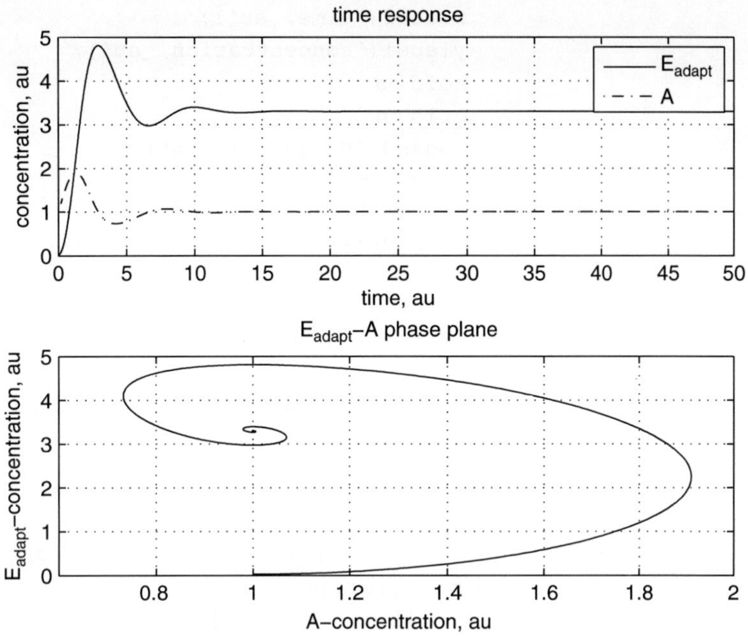

Fig. 8. MATLAB generated plot showing (robust) adaptation in *A*.

To observe these oscillations, we make a slight change in the $K_M^{E_{adapt}}$ `(k(3))` value from `2.0` to `1.0e-6` by appending the following code in `LShifc.m`:

```
%file: LShifc.m
...
% repeat calculations, but now with low k(3) (KM (E_adapt)) value...
% define rate constant values [k(1) k(2)..... k(10)]
ks=[0.1 1.0 1.0e-6 3.0 6.0e+6 1.0e-6 1.0 0.01 5.0 5.0e-7 0.1];

% simulation time
t=[0,50];

% initial concentrations
y0=[1.0   0.03];

% options for numerical integration
options = odeset('RelTol',0.000001,'MaxStep',0.01);
% solve model
[T Y]=ode15s(@hifc,t,y0,options,ks);

% making Figure 2
figure(2),
subplot(2,1,1),plot(T,Y(:,2),'-',T,Y(:,1),'-');
```

```
xlabel('time, au');
ylabel('concentration, au');
hold on
grid on
legend('E_{adapt}','A');
hold off

subplot(2,1,2),plot(Y(:,1),Y(:,2),'-');
xlabel('A-concentration, au');
ylabel('E_{adapt}-concentration, au');
title(['inflow homeostatic controller'])
hold on
legend('E_{adapt}-A phase plane');
hold on
grid on
hold off
```

This change in $K_M^{E_{adapt}}$ generates harmonic oscillations in A and E_{adapt}, see Fig. 9.

These type of oscillations have been considered to occur in the negative-feedback regulation of the p53-Mdm2 system (34, 35), where p53 is considered to be bound by Mdm2 to an upper (sub-apoptotic) level (36). Recent experimental findings using a

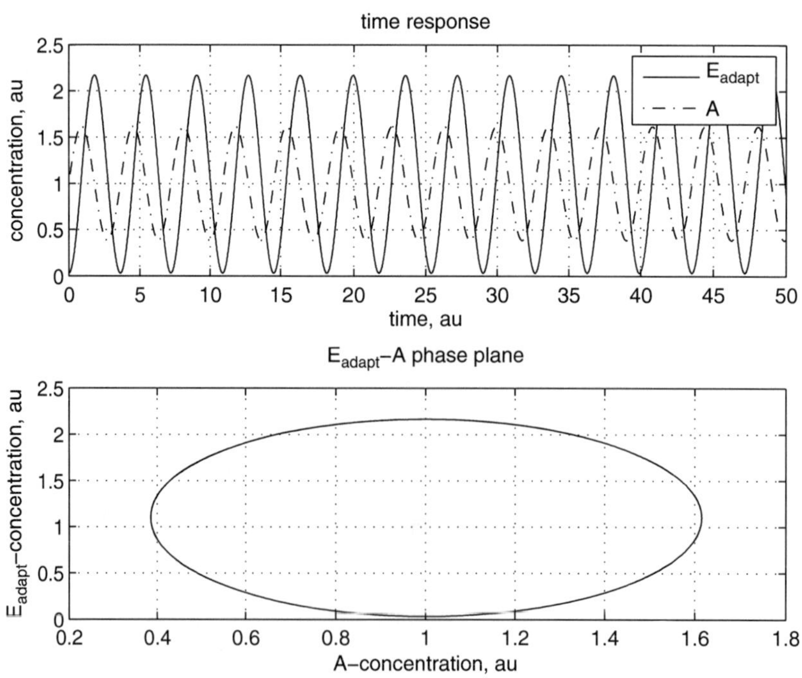

Fig. 9. Harmonic oscillations generated in the homeostatic inflow controller. Due to the harmonic character of the oscillations no limit-cycle is observed but multiple trajectories in phase space occur (only one is shown) which depend on the initial concentrations.

synthetic–natural hydrid oscillator of the p53 network showed indeed the presence of a major harmonic component (37). Another interesting aspect of oscillations arising in homeostatic controllers may be related to the pulsatile manner of how hormones are released leading to homeostatic control of important metabolites (38).

4. Supplementary Information

MATLAB files `I122.m`, `LShifc.m`, and `hifc.m` described in the text can be downloaded from http://bioinfo.ux.uis.no/adapt.zip.

References

1. Grylls, F. S., and J. S. Harrison, 1956. Adaptation of yeast to maltose fermentation. *Nature* **178**, 1471–2.
2. Berg, H. C., and P. M. Tedesco, 1975. Transient response to chemotactic stimuli in Escherichia coli. *Proc Natl Acad Sci U S A* **72**, 3235–9.
3. Alon, U., M. G. Surette, N. Barkai, and S. Leibler, 1999. Robustness in bacterial chemotaxis. *Nature* **397**, 168–71.
4. Bray, D., 2002. Bacterial chemotaxis and the question of gain. *Proc Natl Acad Sci U S A* **99**, 7–9.
5. Mello, B. A., and Y. Tu, 2003. Perfect and near-perfect adaptation in a model of bacterial chemotaxis. *Biophys J* **84**, 2943–56.
6. Berg, H. C., 2004. E. coli in Motion. Springer-Verlag, New York.
7. Mello, B. A., and Y. Tu, 2007. Effects of adaptation in maintaining high sensitivity over a wide range of backgrounds for Escherichia coli chemotaxis. *Biophys J* **92**, 2329–37.
8. Hansen, C. H., R. G. Endres, and N. S. Wingreen, 2008. Chemotaxis in Escherichia coli : A Molecular Model for Robust Precise Adaptation. *PLoS Computational Biology* **4**, 0014–0027.
9. Levchenko, A., and P. A. Iglesias, 2002. Models of eukaryotic gradient sensing: application to chemotaxis of amoebae and neutrophils. *Biophys J* **82**, 50–63.
10. Ratliff, F., H. K. Hartline, and W. H. Miller, 1963. Spatial and temporal aspects of retinal inhibitory interaction. *J Opt Soc Am* **53**, 110–20.
11. He, Q., and Y. Liu, 2005. Molecular mechanism of light responses in Neurospora: from light-induced transcription to photoadaptation. *Genes Dev* **19**, 2888–99.
12. Walters, R. G., 2005. Towards an understanding of photosynthetic acclimation. *J Exp Bot* **56**, 435–47.
13. Arthur, H., and K. Watson, 1976. Thermal adaptation in yeast: growth temperatures, membrane lipid, and cytochrome composition of psychrophilic, mesophilic, and thermophilic yeasts. *J Bacteriol* **128**, 56–68.
14. Margesin, R., 2009. Effect of temperature on growth parameters of psychrophilic bacteria and yeasts. *Extremophiles* **13**, 257–62.
15. Asthagiri, A. R., and D. A. Lauffenburger, 2000. Bioengineering models of cell signaling. *Annu Rev Biomed Eng* **2**, 31–53.
16. Koshland, J., D. E., A. Goldbeter, and J. B. Stock, 1982. Amplification and adaptation in regulatory and sensory systems. *Science* **217**, 220–5.
17. Muzzey, D., C. A. Gomez-Uribe, J. T. Mettetal, and A. van Oudenaarden, 2009. A systems-level analysis of perfect adaptation in yeast osmoregulation. *Cell* **138**, 160–71.
18. Asthagiri, A. R., C. M. Nelson, A. F. Horwitz, and D. A. Lauffenburger, 1999. Quantitative relationship among integrin-ligand binding, adhesion, and signaling via focal adhesion kinase and extracellularsignal-regulated kinase 2. *J Biol Chem* **274**, 27119–27.
19. Hao, N., M. Behar, T. C. Elston, and H. G. Dohlman, 2007. Systems biology analysis of G protein and MAP kinase signaling in yeast. *Oncogene* **26**, 3254–66.
20. Mettetal, J. T., D. Muzzey, C. Gomez-Uribe, and A. van Oudenaarden, 2008. The Frequency Dependence of Osmo-Adaptation in Saccha- romyces cerevisiae. *Science* **319**, 482–4.
21. Hong, C. I., E. D. Conrad, and J. J. Tyson, 2007. A proposal for robust temperature

compensation of circadian rhythms. *Proc Natl Acad Sci U S A* **104**, 1195–200.

22. Yi, T. M., Y. Huang, M. I. Simon, and J. Doyle, 2000. Robust perfect adaptation in bacterial chemotaxis through integral feedback control. *Proc Natl Acad Sci U S A* **97**, 4649–53.
23. Wilkie, J., M. Johnson, and K. Reza, 2002. Control Engineering. An Introductory Course. Palgrave, New York.
24. Drengstig, T., H. R. Ueda, and P. Ruoff, 2008. Predicting Perfect Adaptation Motifs in Reaction Kinetic Networks. *J Phys Chem B* **112**, 16752–16758.
25. Ni, X. Y., T. Drengstig, and P. Ruoff, 2009. The control of the controller: molecular mechanisms for robust perfect adaptation and temperature compensation. *Biophys J* **97**, 1244–53.
26. Kacser, H., and J. A. Burns, 1979. Molecular democracy: who shares the controls? *Biochem Soc Trans* 7, 1149–60.
27. Burns, J. A., A. Cornish-Bowden, A. K. Groen, R. Heinrich, H. Kacser, J. W. Porteous, S. M. Rapoport, T. A. Rapoport, J. W. Stucki, J. M. Tager, R. J. A. Wanders, and H. V. Westerhoff, 1985. Control analysis of metabolic systems. *Trends Biochem Sci* **19**, 16.
28. Heinrich, R., and S. Schuster, 1996. The Regulation of Cellular Systems. Chapman and Hall, New York.
29. Fell, D., 1997. Understanding the Control of Metabolism. Portland Press, London and Miami.
30. Drengstig, T., T. Kjosmoen, and P. Ruoff. On the Relationships between Sensitivity Coefficients and Transfer Functions in Reaction Kinetic Networks. To be published.
31. Lutkepohl, H., 1996. Handbook of matrices. John Wiley & Sons.
32. Ingalls, B. P., 2004. A Frequency Domain Approach to Sensitivity Analysis of Biochemical Networks. *J. Phys. Chem. B* **108**, 1143–1152.
33. Beckskei, A., and L. Serrano, 2000. Engineering stability in gene networks by autoregulation. *Nature* **405**, 261–74.
34. Lahav, G., N. Rosenfeld, A. Sigal, N. Geva-Zatorsky, A. J. Levine, M. B. Elowitz, and U. Alon, 2004. Dynamics of the p53-Mdm2 feedback loop in individual cells. *Nat Genet* **36**, 147–50.
35. Geva-Zatorsky, N., N. Rosenfeld, S. Itzkovitz, R. Milo, A. Sigal, E. Dekel, T. Yarnitzky, Y. Liron, P. Polak, G. Lahav, and U. Alon, 2006. Oscillations and variability in the p53 system. *Mol Syst Biol* **2**, 2006 0033.
36. Jolma, I. W., X. Y. Ni, L. Rensing, and P. Ruoff, 2010. Harmonic Oscillations in Homeostatic Controllers: Dynamics of the p53 Regulatory System. *Biophys J* **98**, 743–752.
37. Toettcher, J. E., C. Mock, E. Batchelor, A. Loewer, and G. Lahav, 2010. A synthetic-natural hybrid oscillator in human cells. *Proc Natl Acad Sci U S A* **107**, 17047–52.
38. Chadwick, D. J., and J. A. Goode, 2000. Mechanisms and Biological Significance of Pulsatile Hormone Secretion. Wiley, New York.

Chapter 9

Biochemical Systems Analysis of Signaling Pathways to Understand Fungal Pathogenicity

Jacqueline Garcia, Kellie J. Sims, John H. Schwacke, and Maurizio Del Poeta

Abstract

Over the past decade, researchers have recognized the need to study biological systems as integrated systems. While the reductionist approaches of the past century have made remarkable advances of our understanding of life, the next phase of understanding comes from systems-level investigations. Additionally, biology has become a data-intensive field of research. The introduction of high throughput sequencing, microarrays, high throughput proteomics, metabolomics, and now lipidomics are producing significantly more data than can be interpreted using existing methods. The field of systems biology brings together methods from computer science, modeling, statistics, engineering, and biology to explore the volumes of data now being produced and to develop mathematical representations of metabolic, signaling, and gene regulatory systems. Advances in these methods are allowing biologists to develop new insights into the complexities of life, to predict cellular responses and treatment outcomes, and to effectively plan experiments that extend our understanding. In this chapter, we are providing the basic steps of developing and analyzing a small S-system model of a biochemical pathway related to sphingolipid metabolism in the regulation of virulence of the human fungal microbial pathogen *Cryptococcus neoformans (Cn)*.

Key words: *Cryptococcus neoformans*, Fungal infection, Melanin, Sphingolipid, Protein kinase C, Diacylglycerol, Cell wall, Computational analysis, Model

1. Introduction

In recent years, most biologists have recognized that reductionism alone cannot explain every cellular biological process and now admit that integralism or pluralism must accompany reductionism to fully explain biological phenomena. One key issue in the reductionism process is the use of appropriate methodologies in the study of the phenomenon of interest. For instance, methodologies that study molecules in the time and space of the living cell should be preferable to and complement those that study the molecules in vitro. Outside of this critical cellular context, the richness

Attila Becskei (ed.), *Yeast Genetic Networks: Methods and Protocols*, Methods in Molecular Biology, vol. 734,
DOI 10.1007/978-1-61779-086-7_9, © Springer Science+Business Media, LLC 2011

of integrated cell behavior may be lost. Systems biology can dramatically improve the selection of such experimental strategies.

Since mathematical models can be used to theoretically predict the patterns of enzymatic activity that could lead to an observed steady state phenotype, it is possible, in principle, to determine which enzyme would have the greatest effect in achieving that phenotype. Therefore, systems biology can aid in selecting which enzyme would have to be altered and in what manner to obtain the phenotype of interest. In other words, mathematical models can be used as valuable tools for exhaustive prescreening studies for all kinds of scenarios and for creating novel hypotheses that are then to be tested in the laboratory. Computational modeling can help in selecting the experiments most likely to disprove the hypothesis and further improve our conceptual model of the system under study.

Cellular systems frequently employ cascade mechanisms to facilitate the transduction of external signals and activation of transcription factors that regulate the expression of specific genes in response to specific signals. Cascade pathways and signaling in *Cryptococcus neoformans* (*Cn*) has been extensively studied (1–17). In our studies, we found that an enzyme in the fungal sphingolipid pathway, inositol phosphoryl ceramide synthase 1 (Ipc1), controls the signaling cascade leading to the production of melanin (18). In particular, Ipc1 produces inositol-containing sphingolipids (e.g., inositol phosphoryl ceramide or IPC) and diacylglycerol (DAG) (19), which, in *Cn*, activates protein kinase C1 (Pkc1) (20). Although previous studies showed that DAG does not activate Pkc1 of other fungal species, such as *Candida albicans* (*Ca*) (21) and *Saccharomyces cerevisiae* (*Sc*) (22), we found that this was not the case for the *Cn* Pkc1. In *Cn*, Pkc1 activation occurs through the C1 domain of Pkc1 since deletion of this domain reduces its activation by DAG (20). The Ipc1–DAG–Pkc1 pathway appears to drive laccase to its proper location (cell wall) so that it can transform L-Dopamine into melanin in the outer leaflet of the cell wall (Fig. 1).

In this chapter, we demonstrate the development of a simple mathematical model of the regulation of melanin by the sphingolipid pathway in the pathogenic microorganisms *C. neoformans*. This microbe is an environmental fungus that, upon inhalation, can cause a life-threatening meningoencephalitis, especially in immunocompromised patients (23). *Cn* produces a black melanin pigment deposited on the outer cell wall that protects the fungus from the environment and from the host immune response (24–27). Melanin deficient mutants are not pathogenic (28–30), thus it became important to define how its production is regulated.

Mathematical modeling of biochemical and regulatory systems is a well-developed field of research, and there are numerous strategies that could be employed to implement the sphingolipid activated production of melanin in *Cn*. We choose Biochemical

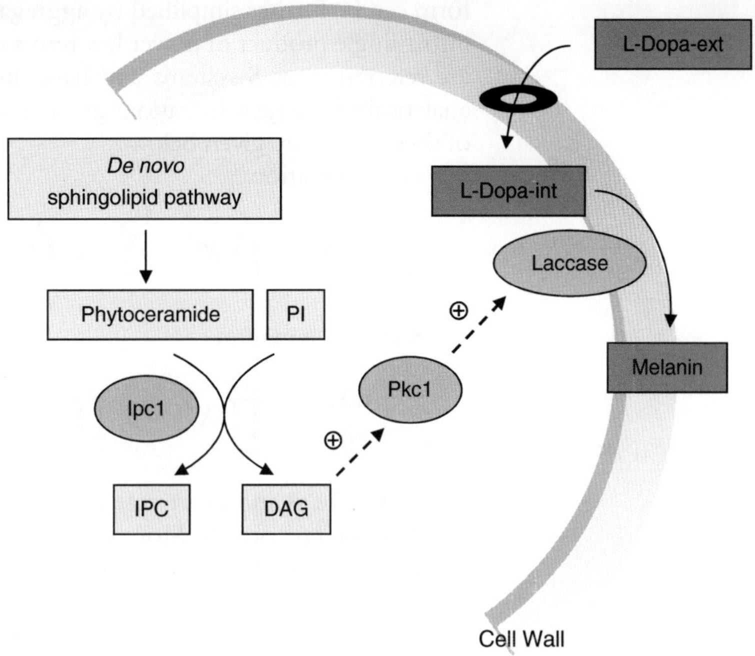

Fig. 1. *Signaling pathway regulating melanogenesis in Cryptococcus neoformans (Cn) by sphingolipids.* Diacylglycerol (DAG) produced by Ipc1 activates Pkc1 through the C1 domain of Pkc1. Activation of Pkc1 maintains the structure of the cell wall, which enables laccase to produce melanin granules deposited in the cell wall. Melanin production is required for pathogenesis of *Cn*. *PI* phosphatidylinositol, *Ipc1* inositol phosphoryl ceramide synthase 1, *Pkc1* protein kinase C 1, *L-DOPA-ext* L-Dopamine extracellular, *L-DOPA-int* L-Dopamine intracellular.

Systems Theory (BST) as developed by Savageau in the 1960s (31, 32) and applied in numerous modeling efforts since. BST employs a power-law formalism derived from a Taylor series approximation to each process rate law in logarithmic coordinates. The resulting canonical representation greatly simplifies the construction of models, provides a rigorous basis for analyzing the stability and performance of the system, and has been successfully applied to metabolic pathways, gene regulatory networks, and signal transduction systems. BST has been successfully used to model a variety of pathways in a variety of organisms (33–35), including metabolic pathways in *S. cerevisiae* (36, 37) and *Cn* (38, 39). In the BST framework, the system's dynamic behavior is captured in a set of differential equations, where the rates of individual processes are given as the product of power laws. Each time varying quantity of interest, termed a dependent variable, is described by one of these equations. When each of the underlying processes (reaction, transport, expression, etc.) is described individually, the model is termed a Generalized Mass Action (GMA) system owing to its mathematical similarity to Mass Action systems. In many cases, this form can be further simplified by aggregating influxes and effluxes into a single product of power law terms each. Models in this form are referred to as S-systems and have distinct computational and analytical advantages. Equations giving the canonical forms for each of these forms are given below

GMA equation

$$\frac{dX_i}{dt} = \sum_{k=1}^{m} \alpha_{ik} \prod_{j=1}^{n} X_j^{g_{ijk}} - \sum_{k=1}^{m} \beta_{ik} \prod_{j=1}^{n} X_j^{h_{ijk}} \quad i = 1, 2, \ldots, n$$

S-system equation

$$\frac{dX_i}{dt} = \alpha_i \prod_{j=1}^{n+m} X_j^{g_{ij}} - \beta_i \prod_{j=1}^{n+m} X_j^{h_{ij}} \quad i = 1, 2, \ldots, n$$

Here, the αs and βs are referred to as rate constants and the gs and hs as kinetic orders. More formally, these gs and hs are called ***apparent kinetic orders*** to differentiate them from the kinetic orders familiar to those describing process rates using mass action kinetics. For a more detailed explanation of the theory behind BST models, see appendix below and Voit (35).

Since this demonstration model is based on our previous experiments, model analyses show that under alteration of its parameters the simulations strictly reflect the results obtained in the previous experiments. Indeed, any mathematical model crucially depends on solid experimental data. However, once a model is established, it becomes a rich tool for analyses that are often unattainable with wet experimentation. For instance, the model can simulate the effect of Pkc1 downregulation on cell wall integrity and laccase location at the cell wall; these predictions could help to design additional experiments that specifically address this hypothesis. This chapter walks the reader through the basic steps of developing and analyzing a small S-system model of a biochemical pathway related to sphingolipid metabolism in *Cn*.

2. Methods

2.1. The Modeling Process

Regardless of which mathematical method is chosen to model a biological system, the same general process is followed to develop and test a model. Each step is discussed in greater detail in the appendix below, but briefly here are the main steps. The crucial first step consists of defining the pathway to be modeled and deciding which components are altered by the system (dependent variables) and which remain unchanged (independent variables). The second step is to write the system equations; each dependent variable represented by one differential equation that calculates how the amount of the associated molecular species changes. The equations are first written symbolically in terms of the dependent and independent variables and kinetic parameters and then numerically once all parameters have been estimated. Next, the quality and robustness of the model is assessed by calculating the local stability at steady state and the logarithmic gains and sensitivities of the variables and parameters. Last, simulations of known behavior are used to examine the dynamics of the model in response to perturbations. Reasonable responses indicate that the model is ready for predictive simulations. The modeling process is not strictly linear but iterative with successive rounds of experimentation and refinement as determined by the results of the analysis.

2.2. Graphical Model Design

As described in the introduction and illustrated in Fig. 1, our pathway of interest is a signaling cascade that promotes melanin production in *Cn*. After listing the components of the pathway (see Table 1), the most crucial step is to create a drawing or map of the biological system to be modeled, as it is from this map that the equations are written. This map connects the real biological system with the mathematical analysis. Thus, the map should be as accurate as possible based on the published literature and the researchers knowledge and should include the level of detail desired for the given problem. This map is similar to the conceptual drawings of biological pathways often found in textbooks or journal articles, but there are some simple guidelines to insure consistency between model maps and to help avoid ambiguity or confusion.

The system map is represented as a network graph with two basic elements: nodes and directed edges. Nodes typically represent a pool of material, such as metabolites, cofactors, signaling molecules, proteins, enzymes, or genes. Nodes may represent ***dependent*** (time varying) or ***independent*** (fixed) variables. For example, a canonical reaction, where the substrate is transformed into a product (e.g., L-DOPA-int and melanin) by an enzyme (laccase) has two dependent and one independent variable; the substrate and product concentrations are changed by the reaction, but the enzyme typically is not. Each dependent variable has an equation that describes the influx and efflux of that variable, while independent variables have a constant value. Some typical independent variables are enzymes and cofactors.

Solid edges are used to indicate a flow or conversion of material and must connect to nodes. Single-headed arrows denote irreversible reactions and double-headed arrows indicate reversible reactions. If a different enzyme catalyzes the reverse reaction, then two single-headed arrows pointing in opposite directions are used. Several variations of flux arrows are possible depending on the number of substrates, products, enzymes, and cofactors involved in the reaction; see Sims et al. for more examples (40).

Table 1
Model variables with initial values

Type	Symbol	Variable name	Role	Initial value	Reference
Dependent	X_1	IPC[a]	Metabolite	1 mol%[b]	(56)
	X_2	DAG[c]	Metabolite	38 pmol/nmol Pi	(18)
	X_3	L-DOPA-int[d]	Metabolite	1[e]	(42)
	X_4	Melanin	Metabolite	1[f]	(42)
Independent	X_5	Phytoceramide	Metabolite	3.8 pmol/nmol Pi	(18)
	X_6	PI[g]	Metabolite	4.54 mol%	(56)
	X_7	Ipc1[h]	Enzyme	35 pmol/min/mg	(18)
	X_8	Pkc1[i]	Signaling molecule	53.5 pmol/min/mg[j]	(18)
	X_9	L-DOPA-ext[k]	Metabolite	10^6 nM	(42)
	X_{10}	Transport	Process	3.5 nmol/min/mg of cells[l]	(43)
	X_{11}	Laccase	Enzyme	1,500 pmol/min/10^7 cells	(18)

[a] Inositol phosphoryl ceramide
[b] The total membrane concentration of IPC is 1 mol% under normal conditions during exponential growth, value reported by Wu et al. (56). Mol% is equivalent to the concentration of sphingoid base or phosphatidate/concentration of total phospholipid
[c] Diacylglycerol
[d] L-Dopamine intracellular
[e] *C. neoformans* grown in l-3,4-dihydroxyphenylalanine (L-DOPA) external (42). The L-DOPA internal concentration is assumed equal to the relative melanin contents
[f] This concentration refers to the relative melanin contents whose value is equal to 1 (42)
[g] Phosphatidylinositol
[h] Inositol phosphoryl ceramide synthase 1
[i] Protein kinase C 1
[j] Serine/threonine kinase whose specific activity in the absence of lipids is reported as 31.5 pmol/min/mg. In the presence of the DAG subspecies, its activity was increased by 1.7 fold (18)
[k] L-Dopamine extracellular
[l] V_{max}

Edges are also used to indicate the flow of information, i.e., signals, from one variable that regulate some process in the model. In this case, the arrows are dashed and may have a positive or negative sign to indicate activation or inhibition, respectively. Lack of a sign on a dashed arrow is considered activation. Information flow arrows connect a node to an edge associated with a flow of material (solid arrow). Often these dashed arrows are used to indicate the relationship between an enzyme and the reactions that it modulates or the inhibition of one metabolite on some step in the pathway.

2.3. Equation Formulation, Symbolic, and Numeric

Once satisfied that the system map is accurate and includes all details relevant to the model, the differential equations can be set up from the map. It is helpful to first write the symbolic equations of the system. These indicate all the pertinent components that affect each dependent variable, but no specific numerical values, such as metabolite concentration are used. After the symbolic equations are complete, parameter values are estimated and plugged in to create the numeric equations that are used for further analysis.

It is mathematically convenient to substitute symbols for the proper names of the associated molecular species. They are represented by X_i for dependent and X_j for independent variables with the respective subscript. By convention, the numeration is consecutive beginning with $i = 1$ to n for the dependent variables $j = n + 1$ to $n + m$ for the independent variables. Table 1 lists the components of the sample system along with the symbolic name and initial value. The fluxes between metabolites are also given symbolic names of the form $v_{i,j}$, where i and j indicate the two nodes the flux flows between. Next, for each dependent variable X_i, we identify the other variables and signals that influence its influx and efflux. A symbolic differential equation is then written for each dependent variable using either the S-system or GMA method of flux representation.

For example, Fig. 2 shows the network map with symbolic notation for the signaling cascade Ipc1–DAG–Pkc1 described in the introduction. In the first half of the cascade, the enzyme Ipc1, (X_7) transfers the phosphoryl-inositol moiety from phosphatidylinositol (PI), (X_6) to phytoceramide, (X_5) forming IPC, (X_1) and DAG, (X_2)

In the second half of the system, melanin (X_4) is synthesized from an internal concentration of L-DOPA-int, (X_3) via laccase, (X_{11}). In *Cn*, melanin (X_4) is produced in the presence of

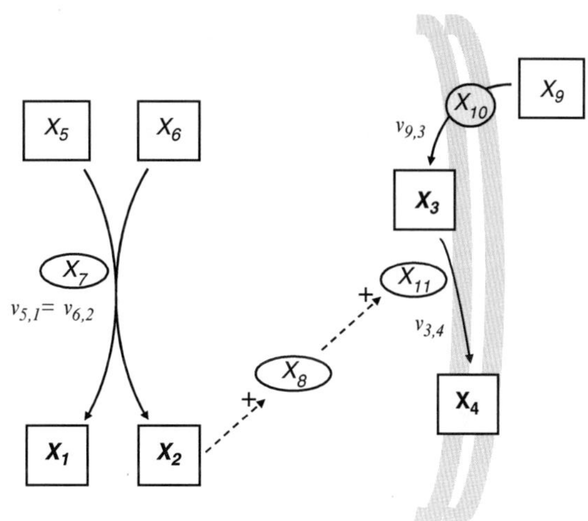

Fig. 2. *Network map of the production of melanin via the signaling cascade: Ipc1–-DAG–Pkc1*. Based on the pathway described in Fig. 1., the system has four dependent variables X_1, X_2, X_3, X_4 and seven independent variables X_5, X_6, X_7, X_8, X_9, X_{10}, and X_{11}. Metabolites are shown as *boxes* with dependent variables in *bold*, enzymes as *ovals*, and the transport process as a *circle*. *Solid arrows* indicate flux and *dashed arrows* indicate that the variable has an effect on the system. Positive signals are indicating activation.

phenolic substrates, such as L-DOPA-ext, (X_9) (41, 42) that are actively transported into the cell (43). This transport process is identified in this relatively simple model as the variable X_{10}.

These two metabolic pathways are connected by two signals. First, DAG, (X_2) released from the production of IPC activates the enzyme Pkc1, (X_8), which then activates laccase, (X_{11}), stimulating the increased production of melanin.

This example contains four dependent variables, the metabolites: X_1, X_2, X_3, and X_4. Their synthesis is affected by their respective precursors and their degradation depends only on their own concentration. Note that the production of melanin requires numerous precursors and reactions, but has been simplified in this example as a direct substrate of L-DOPA-int. For each X_i, all components whether dependent or independent that have a part in its synthesis are aggregated in V_i^+, the influx function. Similarly, all variables dependent or independent that have a part in the degradation of X_i are aggregated in V_i^-, the efflux function. For the system shown in Fig. 2, the influx and efflux functions for each dependent variable can be represented as follows.

2.3.1. Mass Balance Equations

$$\frac{dX_1}{dt} = V_1^+(X_5, X_6, X_7) - V_1^-(X_1)$$

$$\frac{dX_2}{dt} = V_2^+(X_5, X_6, X_7) - V_2^-(X_2)$$

$$\frac{dX_3}{dt} = V_3^+(X_9, X_{10}) - V_3^-(X_2, X_3, X_8, X_{11})$$

$$\frac{dX_4}{dt} = V_4^+(X_2, X_3, X_8, X_{11}) - V_4^-(X_4)$$

The next step is to flesh out these influx and efflux functions. While the actual function that governs an enzymatic reaction may sometimes be described using Michaelis–Menten kinetic rate laws, often the exact mechanism is not known. This is when using the power law formalism has a great advantage. We can rewrite these flux functions as symbolic S-system equations as shown below, where each of the fluxes is given as a product of power-law terms.

2.3.2. Symbolic Equations

$$\frac{dX_1}{dt} = \alpha_1 X_5^{g_{1,5}} X_6^{g_{1,6}} X_7^{g_{1,7}} - \beta_1 X_1^{h_{1,1}}$$

$$\frac{dX_2}{dt} = \alpha_2 X_5^{g_{2,5}} X_6^{g_{2,6}} X_7^{g_{2,7}} - \beta_2 X_2^{h_{2,2}}$$

$$\frac{dX_3}{dt} = \alpha_3 X_9^{g_{3,9}} X_{10}^{g_{3,10}} - \beta_3 X_2^{h_{3,2}} X_3^{h_{3,3}} X_8^{h_{3,8}} X_{11}^{h_{3,11}}$$

$$\frac{dX_4}{dt} = \alpha_4 X_2^{g_{4,2}} X_3^{g_{4,3}} X_8^{g_{4,8}} X_{11}^{g_{4,11}} - \beta_4 X_4^{h_{4,4}}$$

2.4. Parameter Estimation

Once the symbolic equations are defined, it is time to provide numeric values for each parameter. Most often, this procedure occurs "bottom-up" by estimating the parameters for individual reactions based on kinetic characterizations of the associated enzymes. Recently, efforts have also focused on estimating these parameters from time series measurements of the dependent variables, the so-called top-down approach (44–47). Here, we employ the "bottom-up" approach and focus on using kinetic data for each reaction or process in the model. This step utilizes published information about the enzymes and reactions, such as *Km*, *Ki*, and V_{max}, although there are times when the needed data are unavailable. In such cases, custom experiments may need to be conducted, parameter values may be estimated from characterizations in other organisms, or default "guesstimates" may be used for a limited number of parameters.

On initial inspection, our system appears to have 8 rate constants and 19 kinetic orders that need to be calculated; however, there are constraints on the system that equate some parameters, thus reducing the total number to be determined. For example, the synthesis of X_1 and X_2 occur from the same process so V_1^+ and V_2^+ are equivalent. Also, the precursor–product relationship conserves the flux of a reaction, thus the efflux V_3^- must equal the influx of V_4^+. This means that the rate constants and kinetic orders of each of these relationships must be equal resulting in the following constraints on the parameters.

$$\alpha_1 = \alpha_2; g_{1,5} = g_{2,5}; g_{1,6} = g_{2,6}; g_{1,7} = g_{2,7}$$
$$\beta_3 = \alpha_4; h_{3,2} = g_{4,2}; h_{3,8} = g_{4,8}; h_{3,3} = h_{4,3}; h_{3,11} = g_{4,11}$$

2.4.1. Kinetic Orders

In BST, the kinetic order of a variable indicates the influence of that variable on the flux in which the variable appears and is derived from the slope of the rate function when expressed in logarithmic coordinates. Kinetic orders influence the stability of the system and the logarithmic gains, and sensitivities of the dependent variables (35). Several methods are available for the estimation of kinetic orders, including estimation from time series data (as in the top-down approach), estimation from kinetic data of individual enzymes, or through approximations to established rate laws for well-characterized processes. Kinetic orders are frequently determined directly from experimental data by estimating the slope in a log–log plot of rate versus concentration data (34). When a rate law is available to describe a process of interest, it is also possible to calculate kinetic orders using partial derivatives as shown below.

$$g_{i,j} = \frac{\partial V_i}{\partial X_j} \frac{X_j}{V_i}$$

This expression is derived from the partial derivative of the log rate V_i with respect to the log of the variable of interest X_j. For example, the simple Michaelis–Menten rate law produces the following kinetic order with respect to the substrate X_i.

Given

$$g_{i,j} = \left[\frac{\partial}{\partial X_i}\left(\frac{V_{\max} \times X_i}{Km + X_i}\right)\right] \times \frac{X_i}{\left(\frac{V_{\max} \times X_i}{Km + X_i}\right)}$$

$$= \frac{Km}{Km + X_i}$$

For the first example, consider the enzyme Ipc1, X_7 of the sphingolipid pathway and its substrates X_5 and X_6 which are assumed to be independent variables and thus are constants. However, the products of this reaction can change; so the kinetic orders of V_1^+ and V_2^+ are associated with the IPC and DAG. Estimations of the kinetic orders $g_{1,5}$ and $g_{1,6}$ are illustrated in the appendix below. To determine the kinetic order of the enzyme, we assume a direct proportionality between activity and concentration of enzyme. Differentiation gives a value of 1. Recalling the constraints on parameters discussed previously, we get $g_{1,7} = g_{2,7} = 1$.

Next, we assume a simple kinetic Michaelis–Menten rate law for the laccase reaction from which we compute the kinetic orders $h_{3,3}$ and $g_{3,9}$. First, $h_{3,3}$ quantifies the effect of internal L-DOPA on its own degradation through laccase. This enzyme exhibits a constant kinetic for DOPA with a $Km = 0.59$ mM (48). After differentiation and substitution of measured values, we get the following:

$$h_{3,3} = \frac{Km}{Km + X_4} = \frac{0.59}{0.59 + 1} = 0.3710$$

Similarly, the kinetic order $g_{3,9}$ is obtained by differentiation of V_3^+ with respect to X_9. The kinetic order $g_{3,9}$ reflects the effect of L-DOPA-ext on L-DOPA-int via transport through the cell wall with a kinetic constant $Km = 0.45$ mM (48) as shown below.

$$g_{3,9} = \frac{Km}{Km + X_9} = \frac{0.45}{0.45 + 1} = 0.3103$$

We note that in the calculation of these kinetic orders, we supply a value for one or more of the system variables (X_4 and X_9 in the examples above). The power-law derived in this manner is an approximation to the underlying rate law (Michaelis–Menten in this case), more formally it is a first order Taylor approximation in logarithmic coordinates. As such, we must choose a point around which to make this approximation. This becomes what is called the ***operating point*** and is often chosen to match the nominal conditions around which the system is expected to operate. At this operating point, both the rate and the slope of the power law approximation match that of the rate law being approximated.

2.4.2. Rate Constants

Rate constants represent the speed of the processes. Their values can be calculated with data for the *V*max, metabolite and enzyme concentrations along with the calculated kinetic orders (49). This is accomplished by setting the power law flux term equal to the original rate law at the operating point and solving for the rate constant this guarantees that the velocity of the original rate law and the power law approximation match.

For example, the rate constant associated with the formation of the variable X_1 is determined from the set of values for the flux rate, concentrations, and kinetic orders:

$$\alpha_1 = \frac{V_1^+}{X_5^{g_{1,5}} X_6^{g_{1,6}} X_7^{g_{1,7}}} = \frac{12.29}{3.8^{0.2621} \times 4.54^{0.5241} \times 35} = 0.1119$$

From the parameter constraints, we know that the rate constant associated with the formation of the variable X_1 is the same value as the rate constant associated with the formation of the variable X_2. Thus, $\alpha_2 = 0.1119$. Similar calculations give the remaining rate constants.

Once all the values have been calculated for the kinetic orders and rate constants, those values replace the symbols to produce the numeric equations that are then ready for analysis and testing. For several detailed examples of parameter estimation, see (35) and (40).

2.4.3. S-System Representation

$$\frac{dX_1}{dt} = 0.1119\, X_5{}^{0.2621} X_6^{0.5241}\, X_7 - 12.29 X_1$$

$$\frac{dX_2}{dt} = 0.1119\, X_5{}^{0.2621}\, X_6^{0.5241}\, X_7 - 0.3234\, X_2$$

$$\frac{dX_3}{dt} = 0.9474e - 2\, X_9^{0.3103} X_{10} - 0.7915e - 6\, X_2\, X_3^{0.3710} X_8\, X_{11}$$

$$\frac{dX_4}{dt} = 0.7915e - 6 X_2\, X_3^{0.3710} X_8\, X_{11} - 2.41\,3\, X_4$$

2.5. Model Analysis

The system of equations that have been developed are now tested for steady state values, eigenvalues, and sensitivity or logarithmic gains. The computations associated with these analyses are somewhat more complicated; fortunately, these steps have been automated and are available in at least two freely available software packages; PLAS http://www.dqb.fc.ul.pt/docentes/aferreira/plas.html (50) and PLMaddon http://www.sbi.uni-rostock.de/plmaddon (51).

At this time, it is useful to employ a software package that has been developed specifically for this task, such as PLAS. The equations and initial values (see Table 1) for the variables are entered into the software. The software can then automatically compute the steady state of the system by simultaneously setting the system of differential equations to zero. For S-systems the system of equations can be solved directly after logarithmic transformation. When the model is characterized using the GMA model, the software must use numerical integration techniques as a closed form for the steady state is not available. See appendix below and (35) for more details. Our system reaches steady state as shown in Table 2.

Along with the steady state, it is necessary to check the eigenvalues of the system which indicates the local stability of the system. Stability is an indication of the system's behavior following small perturbations from the steady state. If the system eventually returns to the steady state following a perturbation it is considered stable, a desirable property of our model. Stability is assessed by examining the eigenvalues of a linear approximation to the nonlinear system, constructed at the steady state. When the real parts of all the eigenvalues are negative, the steady state of the system is considered stable. Our system has negative real parts (see Table 2) and is thus stable. Also, the imaginary parts are all zero, indicating that the system does not oscillate as it returns to the steady state following a perturbation. Next, we check the sensitivity of the parameters and independent variables. Again, the available software package implements the required calculations. Sensitivities and logarithmic gains indicate how much the steady state values of dependent variables or fluxes of the system change when a parameter or independent variable is changed by a small amount. These measures are interpreted as relative changes. For example, a sensitivity of 5 means that a 1% change in the parameter or independent variable cause a 5% change in the steady

Table 2
***Steady state and stability assessment using PLAS software.* Eigenvalues for the S-system model (Fig. 2) of cascade Ipc1–DAG–Pkc1–laccase**

	Steady state		Eigenvalues	
Variable	**Value**	**Flux**	**Re**	**Im**
X_1	1	12.29	−12.29	0
X_2	38	12.29	−2.41	0
X_3	1	2.41	−0.90	0
X_4	1	2.41	−0.32	0

Table 3
Influence of the rate constant on the metabolite concentrations of the model in Fig. 2

		Metabolite concentration				Flux of metabolite			
Equation	**Rate constant**	X_1	X_2	X_3	X_4	$V(X_1)$	$V(X_2)$	$V(X_3)$	$V(X_4)$
IPC[a]	α_1	1	–	–	–	1	–	–	–
	β_1	−1	–	–	–	–	–	–	–
DAG[b]	α_2	–	1	−2.7	–	–	1	–	–
	β_2	–	−1	2.7	–	–	–	–	–
L-DOPA int[c]	α_3	–	–	2.7	1	–	–	1	1
	β_3	–	–	−2.7	−1	–	–	–	−1
Melanin	α_4	–	–	–	1	–	–	–	1
	β_4	–	–	–	−1	–	–	–	–

[a]Inositol phosphoryl ceramide synthase 1
[b]Diacylglycerol
[c]L-Dopamine intracellular

state or flux. Positive sensitivities indicate a change in the same direction, whereas negative sensitivities indicate opposite directions of change. Robust models have mostly small sensitivities; unusually, large values (e.g., >10) suggests that something may be amiss with the model or importantly, that the node in question may play a key role in the pathway.

Several types of sensitivities are calculated, but each with the dependent variables and the fluxes of the system; sensitivity with respect to kinetic orders (Table 3) and to rate constants (Table 4), and logarithmic gains of the independent variables (Table 5). As can be seen in the tables, our system has low log gains and mostly low to medium sensitivities of the kinetic orders. The primary exception is the kinetic order of laccase with respect to L-DOPA-int which equals −19.71.

The results presented show that the relatively simple model presented in the Fig. 2 is self-consistent with a steady state that is stable. The sensitivities are relatively small suggesting that this model is robust. Note that this model is only a preliminary analysis and consists of only four dependent variables. Additional variables and pathways could be included in the system, which would lead to a more detailed analysis of the formation of melanin and its regulation by sphingolipid metabolism in *Cn*. Now, we can simulate a perturbation to study the dynamics of the system.

2.6. Model Dynamics

Since the analysis of our system was favorable, we can now perform simulations to see how the system behaves dynamically. This is done using the software package by adding a statement that changes a variable's value at a specific time and then returns that

Table 4
Logarithmic gains of the independent variables with respect to metabolites concentration and with respect to fluxes of the model in Fig. 3. The metabolite most influenced by changes in the independent variables is L-Dopamine (L-DOPA)-int. The influence of the fluxes on metabolite concentrations shows that almost all the magnitudes are less than 1

			Metabolite concentration				Flux of metabolite			
Equation	**Variable**	**Kinetic order**	X_1	X_2	X_3	X_4	$V(X_1)$	$V(X_2)$	$V(X_3)$	$V(X_4)$
IPC	PhytoCer[a]	$g(1,5)$	0.35	–	–	–	0.35	–	–	–
	PI[b]	$g(1,6)$	0.79	–	–	–	0.79	–	–	–
	Ipc1[c]	$g(1,7)$	3.56	–	–	–	3.56	–	–	–
	IPC[d]	$h(1,1)$	–	–	–	–	–	–	–	–
DAG	PhytoCer	$g(2,5)$	–	0.35	−0.94	–	–	0.35	–	–
	PI	$g(2,6)$	–	0.79	−2.14	–	–	0.79	–	–
	Ipc1	$g(2,7)$	–	3.56	−9.58	–	–	3.56	–	–
	DAG[e]	$h(2,2)$	–	−3.64	9.80	–	–	–	–	–
L-DOPA int	L-DOPA-ext[f]	$g(3,9)$	–	–	11.56	4.29	–	–	4.29	4.29
	Transport	$g(3,10)$	–	–	3.38	1.25	–	–	1.25	1.25
	DAG	$h(3,2)$	–	–	−9.80	−3.64	–	–	–	−3.64
	L-DOPA-int[g]	$h(3,3)$	–	–	–	–	–	–	–	–
	Pkc1[h]	$h(3,8)$	–	–	−10.73	−3.98	–	–	–	−3.98
	Laccase	$h(3,11)$	–	–	−19.71	−7.31	–	–	–	−7.31
Melanin	DAG	$g(4,2)$	–	–	–	3.64	–	–	–	3.64
	L-DOPA-int	$g(4,3)$	–	–	–	–	–	–	–	–
	Pkc1	$g(4,8)$	–	–	–	3.98	–	–	–	3.98
	Laccase	$g(4,11)$	–	–	–	7.31	–	–	–	7.31
	Melanin	$h(4,4)$	–	–	–	–	–	–	–	–

[a]Phytoceramide
[b]Phosphatidylinositol
[c]Inositol phosphoryl ceramide synthase 1
[d]Inositol phosphoryl ceramide
[e]Diacylglycerol
[f]L-Dopamine extracellular
[g]L-Dopamine intracellular
[h]Protein kinase C 1

variable to its initial value. This is typically designed to simulate the presentation and removal of a stimulus as might be accomplished in a laboratory experiment but may also be applied to any sort of perturbation to the system. For example, Fig. 3 illustrates the dynamics when Ipc1p activity is decreased by 85% at 1 min and then returned to its initial value at 5 min. Another example is shown in Fig. 4, where the concentration of DAG is decreased by 85% at 1 min. These two simulation show that the system behaves as expected if either of the enzymes are decreased and then returns to steady state. Further scenarios could be tested as well, such as increasing the input of PI and or phytoceramide.

Table 5
Sensitivity of the kinetic orders on the metabolite concentrations and fluxes of the model in Fig. 2. The largest negative influence is on L-Dopamine (L-DOPA) concentration, X_3. This metabolite responds to a change in the kinetic order associate to degradation $h_{3,11}$; increase its concentration when this parameter decreases. Additionally, $S(X_4, g_{4,11})$ indicates an increase in melanin concentration, X_4 when its own synthesis increases

Independent variable		Metabolite concentration				Flux of metabolite			
		X_1	X_2	X_3	X_4	$V(X_1)$	$V(X_2)$	$V(X_3)$	$V(X_4)$
Phytoceramide	X_5	0.26	0.26	−0.71	–	0.26	0.26	–	–
PI[a]	X_6	0.52	0.52	−1.41	–	0.52	0.52	–	–
Ipc1[b]	X_7	1.00	1.00	−2.70	–	1.00	1.00	–	–
Pkc1[c]	X_8	–	–	−2.70	–	–	–	–	–
L-DOPA-ext[d]	X_9	–	–	0.84	0.31	–	–	0.31	0.31
Transport	X_{10}	–	–	2.70	1.00	–	–	1.00	1.00
Laccase	X_{11}	–	–	−2.70	–	–	–	–	–

[a]Phosphatidylinositol
[b]Inositol phosphoryl ceramide synthase 1
[c]Protein kinase C 1
[d]L-Dopamine extracellular

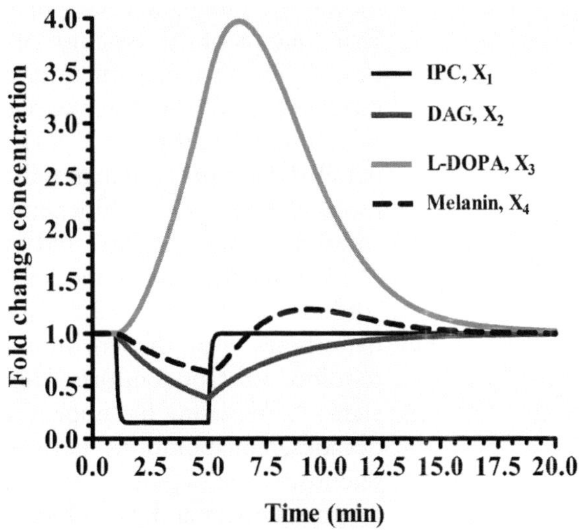

Fig. 3. *Decrease of the enzyme Ipc1 by 85%.* The results show that X_1 and X_2 decrease and then reach the steady state. X_3 increases rapidly and significantly (approximately fourfold) and then back to the steady state. X_4 exhibits a slight S shape before reaching the steady state.

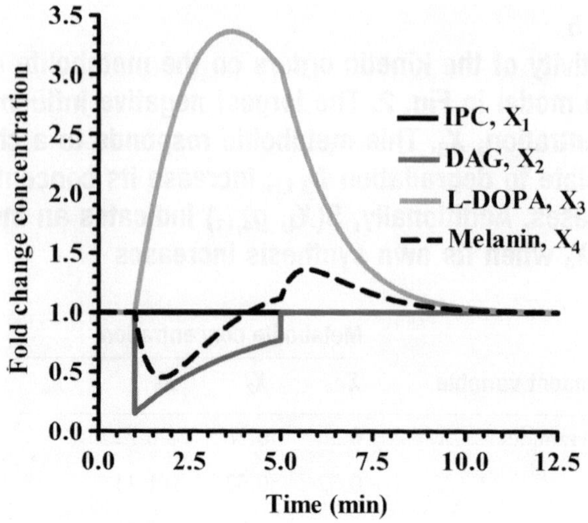

Fig. 4. *Decrease of the metabolite DAG by 85%*. The results show an increase of X_3 by 3.36 fold and a decrease of X_4. This perturbation does not affect X_1. After the perturbation, all variables reach quickly their initial values.

2.7. Validation of the Model

Once the mathematical model has been established it becomes essential to perform laboratory experimentations to validate its accurateness. For instance, the downregulation or/and deletion of *IPC1* or/and *PKC1* genes by homologous recombination should produce mutant strains that make less or no melanin. We would expect IPC and DAG lipid measurements to be decreased in the mutant in which Ipc1 is downregulated. Also, in this mutant, Pkc1 enzymatic activity should be decreased. Experiments of this type not only help to prove (or disprove) the model but also help in finding additional components of the model (e.g., cell wall genes regulated by Pkc1 in the regulation of cell wall integrity) (52, 53). In the Chapter 16, we provide a detailed description of materials and methods used for performing molecular biology and biochemistry studies in *Cn* that can be used to validate theories generated by systems biology.

Ultimately, reliable models are used as tools for prescreening studies for different kind of scenario and for creating novel hypotheses. But the creation of reliable mathematical models requires substantial efforts from both biologists and mathematicians. As modeling has improved significantly during the past few decades, collaborations between biological and computational scientists have begun to show that their effort reveal insights for a better understanding of biological processes that neither biologists nor mathematicians could have obtained without each other. Thus, the development of mathematical models should be seen as a tool that can analyze the system in different ways, complementing laboratory experimentations.

3. Analytical Methods

In the sections that follow, we provide a more complete description of the analytical methods involved in the modeling of melanin regulation by the sphingolipid pathway.

3.1. Methods of System Characterization

Within the framework of BST, models are usually constructed with either the S-system or GMA system representation. In special cases such as ours, where no branch points are present, the S- and GMA representations are equivalent. In the more general case, the S-system representation can be constructed so as to be equivalent to the GMA model at the operating point by aggregating all incoming or outgoing fluxes for each dependent variable into one incoming and one outgoing flux (please see examples in (35)).

Power-Law Formalism
Synergistic System

$$\frac{dX_i}{dt} = \alpha_i \prod_{j=1}^{n+m} X_j^{g_{ij}} - \beta_i \prod_{j=1}^{n+m} X_j^{h_{ij}} \quad i = 1, 2, \ldots, n$$

Generalized Mass Action

$$\frac{dX_i}{dt} = \sum_{k=1}^{m} \alpha_{ik} \prod_{j=1}^{n} X_j^{g_{ijk}} - \sum_{k=1}^{m} \beta_{ik} \prod_{j=1}^{n} X_j^{h_{ijk}} \quad i = 1, 2, \ldots, n$$

Two kinetic representations are related within the power-law formalism (33, 54). The aggregation for Synergistic System (S-System) has one differential equation for each dependent variable X_i with one term for the accumulation or synthesis and another term for the degradation. In GMA, each equation may contain one, two or more terms. In both aggregations, the derivatives of the variables with respect to time t are dX_i/dt. Each term contains all variables that affect the process that the term represents. The multiplicative rate constants α and β could be zero but not negative and the state variables X_j are positive. The exponential parameters are the kinetic order g and h that can be positive or negative real numbers; the subscript i enumerates the equations of the process, k refers to the process number of production or degradation (in GMA only), n refers to the number of dependent variables, and m to the number of independent variables.

3.1.1. S-System

The explanations below correspond with the simple model for the Ipc1–diacylglycerol–Pkc–laccase–melanin pathway. The general equation that describes the biological changes with respect to time can be written as

$$\frac{dX_i}{dt} = V_i^+ - V_i^-, \quad i = 1, \ldots, n,$$

where V_i^+ is a function that contains all the variables (dependent X_i and independent X_j) that influence the synthesis of the given X_i while V_i^- is a function of all variables related with degradation of X_i. This can be written as

$$\frac{dX_i}{dt} = V_i^+ (X_1, X_2, X_3, X_n, X_{n+1}, \ldots, X_{n+m}) - V_i^- \times (X_1, X_2, X_3, X_n, X_{n+1}, \ldots, X_{n+m})$$

According with the general properties of biochemical system, a good representation of V_i^+ and V_i^- is a product of power-law functions of the variables that directly influence the production (V_i^+) or degradation (V_i^-) of the quantity X_i, n is the number of dependent variables, i corresponds to the dependent variable subscript which typically ranges from 1 to n. Each function can be written as a power-law function, as shown below

$$V_i^+ (X_1, X_2, X_{3,} X_n, X_{n+1}, \ldots, X_{n+m}) = \alpha_i \left(X_1^{g1} X_2^{g2} X_3^{g3} X_n^{gn}, X_{n+1}^{gn+1}, \ldots, X_{n+m}^{gn+m} \right)$$

$$V_i^- (X_1, X_2, X_{3,} X_n, X_{n+1}, \ldots, X_{n+m}) = \beta_i (X_1^{h1} X_2^{h2} X_3^{h3} X_n^{hn}, X_{n+1}^{hn+1} \ldots, X_{n+m}^{hn+m})$$

The parameters α and β are positive real numbers called the rate constants and g and h are kinetic orders and can take on positive or negative values. α and g are parameters related with the synthesis of X_i, whereas β and h are related to the degradation of X_i. Putting these two equations together and writing in compact form gives the canonical S-system representation.

$$\frac{dX_i}{dt} = \alpha_i \prod_{j=1}^{n+m} X_j^{g_{ij}} - \beta_i \prod_{j=1}^{n+m} X_j^{h_{ij}} \quad i = 1, 2, \ldots, n$$

The S-system in our model contains four equations. Each term contains all the dependent and independent variables that have direct effect on the associated degradation or production process. Also each variable in each term has one exponent called the kinetic order. The kinetic order in the synthesis term is typically labeled g and the kinetic order in the degradation term labeled h. Each first subscript i in the kinetic order g or h refers to the dependent variable of the equation, and the second subscript j refers to the variable of the exponent. The rate constants α_i and β_i have one subscript that identifies the equation in question. The S-system equations for the model Fig. 3 are the following:

$$\frac{dX_1}{dt} = V_1^+ - V_1^- = \alpha_1 X_5^{g_{1,5}} X_6^{g_{1,6}} X_7^{g_{1,7}} - \beta_1 X_1^{h_{1,1}}$$

$$\frac{dX_2}{dt} = V_2^+ - V_2^- = \alpha_2 X_5^{g_{2,5}} X_6^{g_{2,6}} X_7^{g_{2,7}} - \beta_2 X_2^{h_{2,2}}$$

$$\frac{dX_3}{dt} = V_3^+ - V_3^- = \alpha_3 X_9^{g_{3,9}} X_{10}^{g_{3,10}} - \beta_3 X_2^{h_{3,2}} X_3^{h_{3,3}} X_8^{h_{3,8}} X_{11}^{h_{3,11}}$$

$$\frac{dX_4}{dt} = V_4^+ - V_4^- = \alpha_4 X_2^{g_{4,2}} X_3^{g_{4,3}} X_8^{g_{4,8}} X_{11}^{g_{4,11}} - \beta_4 X_4^{h_{4,4}}$$

$$X_5, X_6, X_7, X_8, X_9, X_{10}, X_{11} = \text{constant}$$

3.1.2. GMA

As with the S-system, a GMA system contains one differential equation for each dependent variable. However, each of these equations may include a sum of any number of terms. Typically, the number of terms in an equation is related to the number of reactions in which the associated dependent variable is involved. Each term has a rate constant α_{ik} associated with each synthesis (processes where X_i is a product) and other rate constant β_{ik} associated with each degradation (processes where X_i is a reactant). In some instances, we relax the constraint on the signs of α_{ik} and β_{ik} giving a single sum with rate constant γ_{ik}. Each term contains all dependent and independent variables that directly affect the process of synthesis or degradation of X_i that the term represents. The index i identifies the dependent variable, j the variable influencing the process, and k gives the index of the production or degradation process ($k = 1,\ldots,m$). Each variable X_j is raised to its kinetic order g_{ijk} and h_{ijk} (sometimes denominated as f_{ijk}).

$$\frac{dX_i}{dt} = \sum_{k=1}^{m} \alpha_{ik} \prod_{j=1}^{n} X_j^{g_{ijk}} - \sum_{k=1}^{m} \beta_{ik} \prod_{j=1}^{n} X_j^{h_{ijk}} \quad i = 1, 2, \ldots, n$$

An important advantage, GMA representation permits the identification of each component and each process to be expressed explicitly, retaining the original the stoichiometry of influxes and effluxes. However, a significant inconvenience of GMA representation is that it does not permit the easy calculation of the steady state solution as does the S-system representation.

S- and GMA systems are closely related and, as in our example, sometimes equivalent. Models using the GMA representation explicitly represent each process in the system and preserve the system stoichiometry. S-systems are often constructed from a GMA model and thus require an additional step, the aggregation of fluxes. Despite the fact that the GMA system explicitly represents each process, research indicates that the S-system representation permits error compensation and approximates the branches of traditional rate laws more exactly than GMA representation, although S-system can introduce discrepancies in flux stoichiometry. S-systems have a distinct computational advantage in the availability of a closed form solution for the steady state. This can be a significant advantage in optimization problems and in cases where a large parameter space is being explored. In this example, the model includes four differential equations for the dependent variables X_1, X_2, X_3, and X_4 each with one production and one degradation term. In this case, the GMA equations coincide with the S-system equations as our model system has no branch points.

The GMA equations are:

$$\frac{dX_1}{dt} = \alpha_{1,5} X_5^{g_{1,5,5}} X_6^{g_{1,6,5}} X_7^{g_{1,7,5}} - \beta_{1,1} X_1^{h_{1,1,1}} X_2^{h_{1,2,1}}$$

$$\frac{dX_2}{dt} = \alpha_{2,5} X_5^{g_{2,5,5}} X_6^{g_{2,6,5}} X_7^{g_{2,7,5}} - \beta_{2,2} X_1^{h_{2,1,2}} X_2^{h_{2,2,2}}$$

$$\frac{dX_3}{dt} = \alpha_{3,9} X_9^{g_{3,9,9}} X_{10}^{g_{3,10,9}} - \beta_{3,3} X_2^{h_{3,2,3}} X_3^{h_{3,3,3}} X_8^{h_{3,8,3}} X_{11}^{h_{3,11,3}}$$

$$\frac{dX_4}{dt} = \alpha_{4,3} X_2^{g_{4,2,3}} X_3^{g_{4,3,3}} X_8^{g_{4,8,3}} X_{11}^{g_{4,11,3}} - \beta_{4,4} X_4^{h_{4,4,4}}$$

3.2. Kinetic Order Estimation

Kinetic orders can be estimated using several, different methods but frequently these values are obtained directly from experimental data, or from any mathematical representation of such data. In some particular cases, the slope in a log–log plot of rate versus concentration data gives the corresponding kinetic order directly (34). At steady state, when the net flux through a dependent variable is zero, the influx and efflux terms must be equal. Thus, a given exponent g (or h) can be computed via partial differentiation of V with respect to X and multiplied by the ratio of X and V all evaluated at the operating point. The expression is formulated as:

$$g_{i,j} = \frac{\partial V_i}{\partial X_j} \frac{X_j}{V_i}.$$

The kinetic orders for the influence of a reactant, derived from a Michaelis–Menten rate law are between 0 and 1, where a value of 0.5 is obtained when the operating point is such that the substrate concentration is equal to the Km of the enzyme.

For example, the production of IPC (X_1) is represented by the flux $v_{5,1}$. The process that contributes to this flux is assumed to be irreversible, and the bisubstrate reaction includes phytoceramide, X_5 and phosphatidylinositol, X_6. The enzyme Ipc1 exhibits

Michaelis–Menten kinetics with a $Km = 1.35$ mol% for phytoceramide, $Km = 5$ mol% for DAG (48). The following equation includes substrate phytoceramide, X_5 and phosphatidylinositol, X_6:

$$v_{5,1} = \left(V_{\max} \times \frac{X_5}{(1.35 + X_5)} \times \frac{X_6}{(5 + X_6)} \right)$$

Derived from $V_{\max}$, the flux was calculated from the specific activity $V_{\text{Ipc1}} = 35$ pmol/min/mg (18). In this simple model, we are assuming that 1 L contains 1 mg of protein.

This gives the following equation for the flux $v_{5,1}$.

$$V_1^+ = v_{5,1} = \left(35 \times \frac{X_5}{(1.35 + X_5)} \times \frac{X_6}{(5 + X_6)} \right) \quad (1)$$

The kinetic order $g_{1,5}$ is then computed through partial differentiation of V_1^+ with respect to X_5 which yields:

$$g_{1,5} = \left[\frac{\partial}{\partial X_5} \left(35 \times \frac{X_5}{(Km_{(X_5)} + X_5)} \times \frac{X_6}{(Km_{(X_6)} + X_6)} \right) \right] \times \frac{X_5}{V_1^+}$$

The partial derivative using the initial concentrations of the substrates is

$$= \left(\frac{35}{(1.35 + 3.8)} \times \frac{X_6}{(5 + X_6)} - \frac{35 \times 3.8}{(1.35 + 3.8)^2} \times \frac{X_6}{(5 + X_6)} \right) \times \frac{X_5}{V_1^+}$$

and the kinetic order is then calculated as

$$g_{1,5} = 0.8475 \times \frac{3.8}{12.29} = 0.2621,$$

where 12.29 is produced from solving Eq. 1 using the initial concentrations of X_5 and X_6. The kinetic order $g_{1,6}$ is obtained in the same fashion, but its partial differentiation is with respect to X_6 resulting in $g_{1,6} = 0.5241$.

3.3. Software Implementation

As mentioned previously, software packages such as PLAS, greatly simplify the computations associated with model analysis and simulation. With PLAS, and other software packages, the user must typically provide (1) a model description, in this case, the system of differential equations for each of the dependent variables, (2) the operating point given as a list of values for the dependent and independent variables, normally a steady state of the system, (3) equations needed to translate the predicted system response to that of the experimental measurement system, and when simulating the system, (4) a set of initial conditions, starting time, end time, and reporting time interval. The sample PLAS input given below provides these for our model. Here, each of the differential equations is given as `X1'`, `X2'`, `X3'`, and `X4'` where the "'" indicates that this equation gives the first derivative of X_i and the expression is given as a sum of power-law terms ("^" indicates exponentiation). The operation point (and steady state) are given by the lines `X1=1` to `X11=1500` and the start time, end time, and reporting interval are given by `t0`, `t1`, and `hr`. Additional details on the PLAS model syntax and options are given in the software documentation. Below is the PLAS code for our model.

```
   X1 ' = .1119780350157*X5^.2621359223298*X7^1.
*X6^.5241090146750-12.29000020354*X1^1.
   X2 ' = .1119780350157*X5^.2621359223298*X7^1.
*X6^.5241090146750-.3234210579878*X2^1.
   X3  '  =  .9474646868175e-2*X9^.3103448275861
*X10^1.-.7915373351201e-6*X11^1.*X8^1.*X2^1.
*X3^.3710
   X4  '  =  .7915373351201e-6*X11^1.*X8^1.*X2^1.
*X3^.3710-2.413793103448*X4^1.
   && X5 X6 X7 X8 X9 X10 X11
   !! XX1 XX2 XX3 XX4 Ib
   !! XXINDEP Ib
   '// Dependent and independent variables' = 1 .. 11
   X1 = 1
   X2 = 38
   X3 = 1
   X4 = 1
   X5 = 3.8
   X6 = 4.54
   X7 = 35
   X8 = 53.5
   X9 = 1000000
   X10 = 3.5
   X11 = 1500
   XX1 = X1
   XX2 = 1/38*X2
   XX3 = X3
   XX4 = X4
   // Times
   t0 = 0
   hr = .1
   tf = 150
```

3.4. Steady State Solution

At steady state, the time rate of change for all dependent variables must be 0 and thus all of the equations in the S-system model (or GMA model) must equal 0. For S-systems, the resulting expression equates a difference of two power-law terms to 0. Moving the degradation term to the opposite side and taking logarithms results in a system of equations linear in the log concentrations. Given the *g*s, *h*s, αs, and βs, this system of equations can be solved for the system steady state (32).

The PLAS software executes these calculations and provides the computed steady state. The software additionally evaluates a linearized model at the computed steady state and gives us the eigenvalues for that model from which we can determine the stability of the system. In this model, the real parts are negative; the steady state is locally stable indicating that the system returns to the steady state following small perturbations.

3.5. Logarithmic Gains

We often wish to predict how the system responds to an increase or decrease in one of the independent variables to understand, for example, how over expression of an enzyme or increase in available nutrient might change the steady state of the system. Logarithmic gains indicate this relation of change between the dependent concentration X_i and the independent concentration X_j. The logarithmic gains characterize the propagation of biochemical signals throughout the system (55). These systemic properties are obtained by a single analytical solution of the steady state equations within the framework of the S-system representation (33).

Logarithmic gains can be used to understand changes in either dependent variable steady states or changes in steady state fluxes. The expression for a flux gain is given by the equation

$$L(V_i, X_j) = \left(\frac{\partial V_i}{\partial X_j}\right)\left(\frac{\partial X_j}{\partial V_i}\right) = \frac{\partial l(\log V_i)}{\partial l(\log X_j)}$$

$$i = 1, \ldots, n;\ j = n + 1, \ldots, n + m$$

A similar equation can be used to compute dependent variable logarithmic gains. If the resulting log gain is greater than 0, this implies amplification of the original signal; a magnitude less than 0 indicates attenuation. If the log gain is positive, this indicates that the changes of the independent and dependent variable are in the same direction, both increase and both decrease. If the log gain is negative, this indicates that the changes are in opposite directions.

3.6. Sensitivities

Sensitivities, like logarithmic gains, provide a measure of how the system steady state concentrations and steady state fluxes change with changes to rate constants and kinetic orders. Again, these values provide a relative indication of effect.

3.6.1. Sensitivities of the Rate Constant Parameters on the Metabolites

Metabolite sensitivities with respect to a rate constant indicates a relative change in the steady state dependent concentration X_i in response to changes in the rate constant, calculated by differentiation. For S-systems, some sensitivities are linked by the structure of the equation system. Increasing a production rate constant is mathematically equivalent to decreasing the corresponding degradation rate constant. Therefore, the sensitivity of X_i with respect to α is equivalent but with negative sign to the sensitivity of X_i with respect β. This can be expressed as the following equations giving the relative change in a dependent concentration X_i with respect to relative change in the rate constants α and β.

$$S\left(X_i, \beta_j\right) = \left(\frac{\partial X_i}{\partial \beta_j} \frac{\beta_j}{X_i}\right) = \frac{\partial(\log X_i)}{\partial\left(\log \beta_j\right)} \qquad i, j = 1, 2, \ldots, n$$

$$S(X_i, \alpha_j) = \left(\frac{\partial X_i}{\partial \alpha_j} \frac{\alpha_j}{X_i}\right) = \frac{\partial(\log X_i)}{\partial(\log \alpha_j)} \qquad i, j = 1, 2, \ldots, n$$

3.6.2. Sensitivities of the Rate Constant Parameters on the Fluxes

The equations for the sensitivities of the rate constants with respect to the fluxes are

$$S\left(V_i, \beta_j\right) = \left(\frac{\partial V_i}{\partial \beta_j} \frac{\beta_j}{V_i}\right) = \frac{\partial(\log V_i)}{\partial\left(\log \beta_j\right)} \qquad i, j = 1, \ldots, n$$

$$S(V_i, \alpha_j) = \left(\frac{\partial V_i}{\partial \alpha_j} \frac{\alpha_j}{V_i}\right) = \frac{\partial(\log V_i)}{\partial(\log \alpha_j)} \qquad i, j = 1, \ldots, n$$

The details of these derivations of these equations can be found in (35). The PLAS program includes procedures to calculate the logarithmic gains and sensitivities.

3.6.3. Sensitivities of the Kinetic Order Parameters on the Metabolite

This sensitivity shows a relative change in X_i given a relative change in a kinetic order g_{ij}. This influence is given by a magnitude that correspond to $S(X_i, g_{ij})$.

The sensitivities with respect to kinetic orders are:

$$S(X_i, g_{jk}) = \left(\frac{\partial X_i}{\partial g_{jk}} \frac{g_{jk}}{X_i}\right) = \frac{\partial(\log X_i)}{\partial(\log g_{jk})} \qquad i = 1, 2, \ldots, n;$$

$$j = n+1, \ldots, m$$

$$S(X_i, h_{jk}) = \left(\frac{\partial X_i}{\partial h_{jk}} \frac{h_{jk}}{X_i}\right) = \frac{\partial(\log X_i)}{\partial(\log h_{jk})} \qquad i = 1, 2, \ldots, n;$$

$$j = n+1, \ldots, m$$

3.6.4. Sensitivities of the Kinetic Order Parameters on the Fluxes

The change that can be generated in a flux when a kinetic order is changed is defined in the following fashion:

$$S(V_i, g_{jk}) = \left(\frac{\partial V_i}{\partial g_{jk}} \frac{g_{jk}}{V_i}\right) = \frac{\partial(\log V_i)}{\partial(\log g_{jk})} \qquad i = 1, 2, \ldots, n;$$

$$j = n+1, \ldots, m$$

$$S(V_i, h_{jk}) = \left(\frac{\partial V_i}{\partial h_{jk}} \frac{h_{jk}}{V_i}\right) = \frac{\partial(\log V_i)}{\partial(\log h_{jk})} \quad i = 1, 2, \ldots, n;$$
$$j = n + 1, \ldots, m$$

3.7. Advantages of S-System Representation

In this analysis, we have chosen to use a modeling representation developed from BST, in particular the S-systems representation. Choosing this framework brings a wealth of theory, numerous examples from the literature, established methods for the analysis of biochemical systems, and freely available software implementing the required calculations. The S-system representation provides some additional advantages. The steady state in S-systems can be expressed in linear equations that govern the local behavior of the intact biological system (32). The formalism is consistent with biologically relevant allometric relationships that quantitatively characterize the relative growth among the parts of the biological systems. S-system equations allow explicit symbolic determination of conditions for local stability and have been shown to represent the behavior of many biological systems with sufficient accuracy. For further study, we recommend the textbook from Voit (35).

Acknowledgments

This work was supported by Grants AI56168 and AI72142 (to M. D.P) and was conducted in a facility constructed with support from the National Institutes of Health, Grant Number C06 RR015455 from the Extramural Research Facilities Program of the National Center for Research Resources. Kellie J Sims is funded by Grant 5K12GM081265-03, an Institutional Research and Academic Career Development Award (IRACDA) program from NIGMS. John H. Schwacke is supported in part by a contract from the National Institutes of Health, National Heart Lung and Blood Institute (NHLBI NO1-HV-28181). Dr. Maurizio Del Poeta is a Burroughs Wellcome New Investigator in Pathogenesis of Infectious Diseases.

References

1. Alspaugh, J. A., Pukkila-Worley, R., Harashima, T., Cavallo, L. M., Funnell, D., Cox, G. M., Perfect, J. R., Kronstad, J. W., and Heitman, J. (2002) Adenylyl cyclase functions downstream of the Galpha protein Gpa1 and controls mating and pathogenicity of *Cryptococcus neoformans*. *Eukaryot. Cell* **1**, 75–84.
2. Chang, Z. L., Netski, D., Thorkildson, P., and Kozel, T. R. (2006) Binding and internalization of glucuronoxylomannan, the major capsular polysaccharide of Cryptococcus

neoformans, by murine peritoneal macrophages. *Infect Immun* **74**, 144–51.
3. Cramer, K. L., Gerrald, Q. D., Nichols, C. B., Price, M. S., and Alspaugh, J. A. (2006) Transcription factor Nrg1 mediates capsule formation, stress response, and pathogenesis in Cryptococcus neoformans. *Eukaryot Cell* **5**, 1147–56.
4. D'Souza, C. A., Alspaugh, J. A., Yue, C., Harashima, T., Cox, G. M., Perfect, J. R., and Heitman, J. (2001) Cyclic AMP–dependent protein kinase controls virulence of the fungal pathogen *Cryptococcus neoformans*. *Mol Cell Biol* **21**, 3179–91.
5. Kerry, S., TeKippe, M., Gaddis, N. C., and Aballay, A. (2006) GATA transcription factor required for immunity to bacterial and fungal pathogens. *PLoS One* **1**, e77.
6. Klengel, T., Liang, W. J., Chaloupka, J., Ruoff, C., Schroppel, K., Naglik, J. R., Eckert, S. E., Mogensen, E. G., Haynes, K., Tuite, M. F., Levin, L. R., Buck, J., and Muhlschlegel, F. A. (2005) Fungal adenylyl cyclase integrates CO2 sensing with cAMP signaling and virulence. *Curr. Biol.* **15**, 2021–6.
7. Ko, Y. J., Yu, Y. M., Kim, G. B., Lee, G. W., Maeng, P. J., Kim, S., Floyd, A., Heitman, J., and Bahn, Y. S. (2009) Remodeling of global transcription patterns of Cryptococcus neoformans genes mediated by the stress–activated HOG signaling pathways. *Eukaryot Cell* **8**, 1197–217.
8. Langfelder, K., Streibel, M., Jahn, B., Haase, G., and Brakhage, A. A. (2003) Biosynthesis of fungal melanins and their importance for human pathogenic fungi. *Fungal Genet Biol* **38**, 143–58.
9. Maeng, S., Ko, Y. J., Kim, G. B., Jung, K. W., Floyd, A., Heitman, J., and Bahn, Y. S. (2010) Comparative Transcriptome Analysis Reveals Novel Roles of the Ras and Cyclic AMP Signaling Pathways in Environmental Stress Response and Antifungal Drug Sensitivity in Cryptococcus neoformans. *Eukaryot Cell* **9**, 360–78.
10. McQuiston, T., Luberto, C., and Del Poeta, M. (2010) The role of host sphingosine kinase 1 (SK1) in the lung response against cryptococcosis. *Infect Immun.*
11. Nichols, C. B., Ferreyra, J., Ballou, E. R., and Alspaugh, J. A. (2009) Subcellular localization directs signaling specificity of the Cryptococcus neoformans Ras1 protein. *Eukaryot Cell* **8**, 181–9.
12. O'Meara, T. R., Norton, D., Price, M. S., Hay, C., Clements, M. F., Nichols, C. B., and Alspaugh, J. A. (2010) Interaction of Cryptococcus neoformans Rim101 and protein kinase A regulates capsule. *PLoS Pathog* **6**, e1000776.
13. Palmer, D. A., Thompson, J. K., Li, L., Prat, A., and Wang, P. (2006) Gib2, a novel Gbeta–like/RACK1 homolog, functions as a Gbeta subunit in cAMP signaling and is essential in Cryptococcus neoformans. *J Biol Chem* **281**, 32596–605.
14. Shea, J. M., and Del Poeta, M. (2006) Lipid signaling in pathogenic fungi. *Curr Opin Microbiol* **9**, 352–8.
15. Siafakas, A. R., Sorrell, T. C., Wright, L. C., Wilson, C., Larsen, M., Boadle, R., Williamson, P. R., and Djordjevic, J. T. (2007) Cell wall–linked cryptococcal phospholipase B1 is a source of secreted enzyme and a determinant of cell wall integrity. *J Biol Chem* **282**, 37508–14.
16. Wang, P., Perfect, J. R., and Heitman, J. (2000) The G–protein beta subunit *GPB1* is required for mating and haploid fruiting in *Cryptococcus neoformans*. *Mol Cell Biol* **20**, 352–62.
17. Waugh, M. S., Vallim, M. A., Heitman, J., and Andrew Alspaugh, J. (2003) Ras1 controls pheromone expression and response during mating in Cryptococcus neoformans. *Fungal Genet Biol* **38**, 110–21.
18. Heung, L. J., Luberto, C., Plowden, A., Hannun, Y. A., and Del Poeta, M. (2004) The sphingolipid pathway regulates protein kinase C 1 (Pkc1) through the formation of diacylglycerol (DAG) in *Cryptococcus neoformans*. *J. Biol. Chem.* **279**, 21144–53.
19. Luberto, C., Toffaletti, D. L., Wills, E. A., Tucker, S. C., Casadevall, A., Perfect, J. R., Hannun, Y. A., and Del Poeta, M. (2001) Roles for inositol–phosphoryl ceramide synthase 1 (*IPC1*) in pathogenesis of *C. neoformans*. *Genes Dev.* **15**, 201–12.
20. Heung, L. J., Kaiser, A. E., Luberto, C., and Del Poeta, M. (2005) The role and mechanism of diacylglycerol–protein kinase C1 signaling in melanogenesis by *Cryptococcus neoformans*. *J. Biol. Chem.* **280**, 28547–55.
21. Paravicini, G., Mendoza, A., Antonsson, B., Cooper, M., Losberger, C., and Payton, M. A. (1996) The *Candida albicans PKC1* gene encodes a protein kinase C homolog necessary for cellular integrity but not dimorphism. *Yeast* **12**, 741–56.
22. Watanabe, M., Chen, C. Y., and Levin, D. E. (1994) Saccharomyces cerevisiae PKC1 encodes a protein kinase C (PKC) homolog with a substrate specificity similar to that of mammalian PKC. *J Biol Chem* **269**, 16829–36.
23. Casadevall, A., and Perfect, J. R. (1998) *Cryptococcus neoformans*, ASM Press, Washington, DC, 381–405.
24. Perfect, J. R. (2005) Cryptococcus neoformans: a sugar–coated killer with designer

genes. *FEMS Immunol Med Microbiol* **45**, 395–404.
25. Wang, Y., Aisen, P., and Casadevall, A. (1995) *Cryptococcus neoformans* melanin and virulence: mechanism of action. *Infect. Immun.* **63**, 3131–6.
26. Mednick, A. J., Nosanchuk, J. D., and Casadevall, A. (2005) Melanization of Cryptococcus neoformans affects lung inflammatory responses during cryptococcal infection. *Infect Immun* **73**, 2012–9.
27. Nosanchuk, J. D., Rosas, A. L., and Casadevall, A. (1998) The antibody response to fungal melanin in mice. *J Immunol* **160**, 6026–31.
28. Kwon-Chung, K. J., Polacheck, I., and Popkin, T. J. (1982) Melanin-lacking mutants of *Cryptococcus neoformans* and their virulence for mice. *J. Bacteriol.* **150**, 1414–21.
29. Salas, S. D., Bennett, J. E., Kwon-Chung, K. J., Perfect, J. R., and Williamson, P. R. (1996) Effect of the laccase gene *CNLAC1*, on virulence of *Cryptococcus neoformans. J Exp Med* **184**, 377–86.
30. Noverr, M. C., Williamson, P. R., Fajardo, R. S., and Huffnagle, G. B. (2004) CNLAC1 is required for extrapulmonary dissemination of Cryptococcus neoformans but not pulmonary persistence. *Infect. Immun.* **72**, 1693–9.
31. Savageau, M. A. (1969) Biochemical systems analysis. II. The steady-state solutions for an n-pool system using a power-law approximation. *J Theor Biol* **25**, 370–9.
32. Savageau, M. A. (1969) Biochemical systems analysis. I. Some mathematical properties of the rate law for the component enzymatic reactions. *J Theor Biol* **25**, 365–9.
33. Sorribas, A., and Savageau, M. A. (1989) Strategies for representing metabolic pathways within biochemical systems theory: reversible pathways. *Math Biosci* **94**, 239–69.
34. Shiraishi, F., and Savageau, M. A. (1992) The tricarboxylic acid cycle in Dictyostelium discoideum. I. Formulation of alternative kinetic representations. *J Biol Chem* **267**, 22912–8.
35. Voit, E. O. (2000) Computational Analysis of Biochemical System. A practical guide for biochemists and Molecular Biologists., Cambridge University Press.
36. Alvarez-Vasquez, F., Sims, K. J., Cowart, L. A., Okamoto, Y., Voit, E. O., and Hannun, Y. A. (2005) Simulation and validation of modelled sphingolipid metabolism in *Saccharomyces cerevisiae. Nature* **433**, 425–30.
37. Alvarez-Vasquez, F., Sims, K. J., Hannun, Y. A., and Voit, E. O. (2004) Integration of kinetic information on yeast sphingolipid metabolism in dynamical pathway models. *J Theor Biol* **226**, 265–91.
38. Garcia, J., Shea, J., Alvarez-Vasquez, F., Qureshi, A., Luberto, C., Voit, E. O., and Del Poeta, M. (2008) Mathematical modeling of pathogenicity of *Cryptococcus neoformans. Molecular System Biology* **4**, 183–95.
39. Macura, N., Zhang, T., and Casadevall, A. (2007) Dependence of macrophage phagocytic efficacy on antibody concentration. *Infect Immun* **75**, 1904–15.
40. Sims, K. J., Alvarez-Vasquez, F., Voit, E. O., and Hannun, Y. A. (2007) A guide to biochemical systems modeling of sphingolipids for the biochemist. *Methods Enzymol* **432**, 319–50.
41. Wang, Y., Aisen, P., and Casadevall, A. (1996) Melanin, melanin "ghosts," and melanin composition in *Cryptococcus neoformans. Infect Immun* **64**, 2420–4.
42. Wang, Y., and Casadevall, A. (1996) Susceptibility of melanized and nonmelanized *Cryptococcus neoformans* to the melanin-binding compounds trifluoperazine and chloroquine. *Antimicrob Agents Chemother* **40**, 541–5.
43. Polacheck, I., Hearing, V. J., Kwon-Chung, K. J., Csukai, M., Chen, C. H., De Matteis, M. A., and Mochly Rosen, D. (1982) Biochemical studies of phenoloxidase and utilization of catecholamines in Cryptococcus neoformans. *J Bacteriol* **150**, 1212–20.
44. Chou, I. C., and Voit, E. O. (2009) Recent developments in parameter estimation and structure identification of biochemical and genomic systems. *Math Biosci* **219**, 57–83.
45. Goel, G., Chou, I. C., and Voit, E. O. (2006) Biological systems modeling and analysis: a biomolecular technique of the twenty-first century. *J Biomol Tech* **17**, 252–69.
46. Goel, G., Chou, I. C., and Voit, E. O. (2008) System estimation from metabolic time-series data. *Bioinformatics* **24**, 2505–11.
47. Voit, E. O., and Almeida, J. (2004) Decoupling dynamical systems for pathway identification from metabolic profiles. *Bioinformatics* **20**, 1670–81.
48. Fischl, A. S., Liu, Y., Browdy, A., and Cremesti, A. E. (2000) Inositolphosphoryl ceramide synthase from yeast. *Methods Enzymol.* **311**, 123–30.
49. Savageau, M. A. (1975) Optimal design of feedback control by inhibition: dynamic considerations. *J Mol Evol* **5**, 199–222.
50. Ferreira, A. (2000), pp. Power Law Analysis and Simulation (PLAS). http://www.dqb.fc.ul.pt/docentes/aferreira/plas.html.
51. Vera, J., Sun, C., Oertel, Y., and Wolkenhauer, O. (2007) PLMaddon: a power-law module for the Matlab SBToolbox. *Bioinformatics* **23**, 2638–40.

52. Gerik, K. J., Bhimireddy, S. R., Ryerse, J. S., Specht, C. A., and Lodge, J. K. (2008) PKC1 is essential for protection against both oxidative and nitrosative stresses, cell integrity, and normal manifestation of virulence factors in the pathogenic fungus Cryptococcus neoformans. *Eukaryot Cell* 7, 1685–98.
53. Gerik, K. J., Donlin, M. J., Soto, C. E., Banks, A. M., Banks, I. R., Maligie, M. A., Selitrennikoff, C. P., and Lodge, J. K. (2005) Cell wall integrity is dependent on the PKC1 signal transduction pathway in Cryptococcus neoformans. *Mol Microbiol* **58**, 393–408.
54. Voit, E. O., and Savageau, M. A. (1987) Accuracy of alternative representations for integrated biochemical systems. *Biochemistry* **26**, 6869–80.
55. Savageau, M. A. (1971) Parameter sensitivity as a criterion for evaluating and comparing the performance of biochemical systems. *Nature* **229**, 542–4.
56. Wu, W. I., McDonough, V. M., Nickels, J. T., Jr., Ko, J., Fischl, A. S., Vales, T. R., Merrill, A. H., Jr., and Carman, G. M. (1995) Regulation of lipid biosynthesis in Saccharomyces cerevisiae by fumonisin B1. *J Biol Chem* **270**, 13171–8.

Chapter 10

Clustering Change Patterns Using Fourier Transformation with Time-Course Gene Expression Data

Jaehee Kim

Abstract

To understand the behavior of genes, it is important to explore how the patterns of gene expression change over a period of time because biologically related gene groups can share the same change patterns. In this study, the problem of finding similar change patterns is induced to clustering with the derivative Fourier coefficients. This work is aimed at discovering gene groups with similar change patterns which share similar biological properties. We developed a statistical model using derivative Fourier coefficients to identify similar change patterns of gene expression. We used a model-based method to cluster the Fourier series estimation of derivatives. We applied our model to cluster change patterns of yeast cell cycle microarray expression data with alpha-factor synchronization. It showed that, as the method clusters with the probability-neighboring data, the model-based clustering with our proposed model yielded biologically interpretable results. We expect that our proposed Fourier analysis with suitably chosen smoothing parameters could serve as a useful tool in classifying genes and interpreting possible biological change patterns.

Key words: Fourier coefficient, *K*-means clustering, Model-based clustering, Silhouette width, Yeast cell cycle data

1. Introduction

Time course experiments can be classified into two main categories termed as periodic and developmental. Periodic time courses include natural biological processes whose temporal profiles follow regular patterns. Examples are cell cycles with regulated genes to have periodic expression patterns in (1) and (2). In developmental time course experiments, gene expression levels are measured at successive times during a developmental process, for example, during the natural growth and development of, or following a treatment applied.

Methods for identifying the genes of interest to the experimenter are required to find genes which change over time, or

Attila Becskei (ed.), *Yeast Genetic Networks: Methods and Protocols*, Methods in Molecular Biology, Vol. 734,
DOI 10.1007/978-1-61779-086-7_10,

genes which change differently over time between two or more biological conditions. This task can be viewed as a "filtering" of the genes to remove those which are not of interest, and clustering genes for validation and further characterization. Questions of interest to an investigator might concern the temporal profiles of genes for one biological condition, such as a desire to identify cell-cycle regulated genes. Alternatively, interest might focus on comparison between gene profiles across two or more conditions. The identification of temporally changing or differentially changing genes not only gives insight into the biological processes under study, but also provides a way of selecting a subset of genes for further analysis. The identification of differentially expressed genes narrows down the number of genes for further analysis.

Clustering genes with similar temporal profiles is commonly the next phase. This is done in the belief that the genes with similar temporal profiles may well be involved in similar biological processes. Grouping genes that share similar expression profiles into clusters is usually the first step in understanding the huge amount of DNA microarray data associated with complicated biological networks. However, most research on gene clustering has been performed with the observed expression data, while ignoring the change patterns.

Due to the differences in the initial levels of background noise in the experiment, difference values or derivatives need to be used as a measure of change. Also a basic premise is that the genes sharing similar change profiles may be functionally related or co-regulated. As such, microarray derivative data provide further insight into gene–gene interactions, gene functions, and pathways. Derivative functions also provide statistical convenience in that (1) functions with a constant amount of difference have the same derivatives and (2) difference values give information about their changes as well as about their original functions.

We propose to use Fourier coefficients in clustering expression patterns and change patterns. Fourier coefficients have several advantages over other methods. Some of these advantages are (1) the dimension of a data set can be reduced to several Fourier coefficients, (2) the estimated Fourier coefficients give information about the underlying function and enable automatic estimation of the change pattern function, (3) the Fourier coefficient estimation does not depend strongly on the covariance structure, and (4) the sample Fourier coefficients asymptotically follow the multivariate normal distributions.

There has been a considerable amount of research into discovering patterns using clustering and testing (3–7). Time-course gene expression data are often measured to study dynamic biological systems and gene regulatory networks. Smoothing away noise-induced wiggles with the Fourier series has been studied by some researchers (8–11). There are other approaches for identifying genes (12–16) including partial least squares (PLS) regression and B-splines matching. A comprehensive review (17) was presented about time series expression data analysis. Unlike *K*-means or hierarchical clustering, model-based clustering is a clustering approach considering probability distribution. The performance and successful applications of model-based clustering are provided (18–20).

We propose a new method for clustering change patterns with derivative Fourier coefficients. We will primarily focus on the Fourier method as gene profiles and demonstrate the usefulness of the Fourier analysis and model-based clustering. In order to provide application of our method, yeast gene expression data is analyzed resulting in interpretable genes.

2. Fourier Series

A brief development of Fourier series is provided without proofs. Detailed expositions of Fourier methods are provided by Tolstov (21) and Stein and Shakarchi (22). Lestrel (23) discussed applications of Fourier descriptors in biological sciences.

Based on the trigonometric system

$$\{1, \cos x, \sin x, \cos 2x, \sin 2x, \ldots\}, \qquad (1)$$

a trigonometric series of the form

$$a_0 + (a_1 \cos x + b_1 \sin x) + (a_2 \cos 2x + b_2 \sin 2x) + \cdots$$

is said to be a Fourier series if the constants $a_0, a_1, b_1, a_2, b_2, \ldots$ satisfy the following relations:

$$\begin{aligned} a_0 &= \frac{1}{2\pi}\int_{-\pi}^{\pi} f(x)\mathrm{d}x \\ a_j &= \frac{1}{2\pi}\int_{-\pi}^{\pi} f(x)\cos(jx)\mathrm{d}x, \quad j = 1, 2, \ldots, \infty \end{aligned} \qquad (2)$$

and

$$b_j = \frac{1}{2\pi}\int_{-\pi}^{\pi} f(x)\sin(jx)\mathrm{d}x, \quad j = 1, 2, \ldots, \infty.$$

The Fourier series is said to correspond to the function $f(x)$. This correspondence, in discrete form, is shown as

$$f(x) \sim a_0 + \sum_{j=1}^{\infty} a_j \cos jx + b_j \sin jx. \qquad (3)$$

One useful property is that they are pairwise orthogonal since

$$\int_{-\pi}^{\pi} \sin(jx)\cos(lx)\mathrm{d}x = 0, \tag{4}$$

$$\int_{-\pi}^{\pi} \cos(jx)\cos(lx)\mathrm{d}x = \begin{cases} 0, & j \neq l \\ \pi, & j = l \neq 0 \\ 2\pi, & j = l = 0 \end{cases}$$

and

$$\int_{-\pi}^{\pi} \sin(jx)\sin(lx)\mathrm{d}x = \begin{cases} 0, & j \neq l \\ \pi, & j = l \neq 0. \end{cases}$$

Two functions $g(x)$ and $h(x)$ are called orthogonal on the interval $[a, b]$ if

$$\int_{a}^{b} g(x)h(x)\mathrm{d}x = 0.$$

With this definition, the functions of the system Eq. 1 are pairwise orthogonal on $[-\pi, \pi]$ or more briefly, the system Eq. 1 is orthogonal on $[-\pi, \pi]$.

Fourier representation can be expanded based on orthogonal systems. If the functions of the orthogonal system are continuous and the series expansion of $f(x)$ is uniformly convergent, it is defined as the Fourier series of $f(x)$.

If the function $y = f(x)$ is known, then the coefficients can be obtained from Eq. 2 for all j. If the function $y = f(x)$ is unknown, a common occurrence, this precludes such an analytic solution and recourse must be made to numerical integration methods such as the trapezoidal rule.

The smoothness of f is directly related to the decay of the Fourier coefficients, and in general, the smoother the function, the faster this decay. We can expect that relatively smooth functions equal their Fourier series. If Fourier coefficients of an integrable function f are provided, then the series of sum of Fourier coefficients converges, and in fact, Parseval's identity

$$\sum_{j=0}^{\infty} a_j^2 + b_j^2 = \int_{-\pi}^{\pi} f(x)^2\mathrm{d}x \tag{5}$$

holds. If the Fourier series of functions f converges to f in an appropriate sense, then a function is uniquely determined by its Fourier coefficients. This would lead to the following statement: if f and g have the same Fourier coefficients, then f and g are necessarily equal. This uniqueness of Fourier representation also enhances the essential use of numerical description in physics, geophysics, acoustics, and climatology and recently in the fields of pattern recognition, biology, and medicine (see Note 1).

3. Methods

The proposed method consists of four main steps. The first and second steps consist of modeling and representing a gene profile with sample Fourier coefficients, and then the calculation of derivatives from the Fourier coefficients. The third step is to cluster the derivative Fourier coefficients using model-based clustering. In the final step, genes with the same change pattern are clustered and the underlying change pattern is automatically estimated using the Fourier representation (see Note 2).

3.1. Model

Consider the data Y_{iu}, uth observation on the ith curve, of the form

$$Y_{iu} = f_i(t_{iu}) + \varepsilon_{iu} \quad i = 1, 2, \cdots n, \ u = 1, 2, \ldots, m \tag{6}$$

where $E(\varepsilon_{iu}) = 0$ and $\mathrm{Var}(\varepsilon_{iu}) = \sigma^2$. In the microarray experiment, Y_{iu} is the log gene expression of gene i at time t_{iu}.

We assume that the curve f_i belongs to a class of smooth functions F as defined below:

$$f_i(t) = \sum_{j=0}^{\infty} \phi_{ij} b_j(t), \tag{7}$$

where $\{b_j\}$ is an orthonormal basis system and

$$\phi_{ij} = \int f_i(\tau)\, b_j(\tau)\, \mathrm{d}\tau. \tag{8}$$

We can estimate f_i using Fourier coefficients by

$$\hat{f}_i(t) = \sum_{j=0}^{J} \hat{\phi}_{ij} b_j(t), \tag{9}$$

which is the projection onto the first J basis functions where J, $1 \leqslant J \leqslant m$, is a smoothing parameter to be chosen based on the data.

The sample Fourier estimate can be estimated as

$$\hat{\phi}_{ij} = \frac{1}{m} \sum_{r=1}^{m} Y_{ij} b_j(t_r), \tag{10}$$

with $t_r = r/m$ and $t \in [0, 1]$.

With regard to changes, the difference

$$D_{iu} = Y_{iu} - Y_{i,u-1} \tag{11}$$

can be approximated by $f_i^{'}(t_{iu})$, the derivative of f_i at t_{iu}, and $t_{iu} - t_{i,u-1}$, assuming that the first order derivative exists. Therefore, the following model can be considered:

$$D_{iu} = \frac{1}{m} f_i'(t_{iu}) + \eta_{iu}, \quad i = 1, 2, \ldots, n, \quad u = 2, \ldots, m \tag{12}$$

where $\eta_{iu} = \varepsilon_{iu} - \varepsilon_{i,u-1}$.

This setup can be extended to the cases where the design or time points are not the same for all curves. We want to classify the same patterns with differences or derivatives that give information about the underlying change pattern.

3.2. Trigonometric Fourier Series Estimators

The function represented with a Fourier series with the cosine bases is given as

$$f_i(t) = \phi_{i0} + \sum_{j=1}^{\infty} \phi_{ij}\sqrt{2}\cos \pi jt. \tag{13}$$

We can estimate f_i with J terms of Fourier coefficients as

$$\hat{f}_i(t) = \hat{\phi}_{i0} + \sum_{j=1}^{J} \hat{\phi}_{ij}\sqrt{2}\cos \pi jt \tag{14}$$

where Fourier coefficients are estimated as

$$\hat{\phi}_{ij} = \frac{1}{m}\sum_{r=1}^{m} Y_{ir}\sqrt{2}\cos \pi jt_r, j = 0, 1, 2, \ldots, m. \tag{15}$$

We also estimate the derivative of f_i as

$$\hat{f}'_i(t) = \frac{d\hat{f}_i(t)}{dt} = \pi \sum_{j=1}^{J} j\hat{\phi}_{ij}\sqrt{2}\sin \pi jt. \tag{16}$$

Note that the Fourier coefficients of the derivatives are calculated by weighting the coefficients from the original functions. The coefficients of the derivative have more weight j on the latter terms of the Fourier coefficients. This means that the higher frequency terms have more information about the derivative pattern of ups and downs.

The model in Eq. 12 can be expressed as

$$D_{iu} = \pi \sum_{j=1}^{J} \psi_{ij}\sqrt{2}\sin \pi jt + \eta_{iu} \tag{17}$$

where $\psi_{ij} = \frac{j}{m}\phi_{ij}$. Therefore, the Fourier coefficients of change can be estimated by $\hat{\psi}_{ij} = \frac{j}{m}\hat{\phi}_{ij}$. Since ψ_{ij} is a Fourier coefficient of the derivative function, we call ψ_{ij}, the derivative Fourier coefficient and $\hat{\psi}_{ij}$, the estimated derivative Fourier coefficient from the sample.

3.3. Selection of Smoothing Parameter

The parameter J controls the amount of smoothing and should be determined based on the data. Even though the optimal choice for J varies from function to function, we choose to use a single smoothing parameter that operates reasonably well for all of the curves. There has been some research on optimal choices for J.

For example, to find global smoothing parameter, J was calculated as the minimizer of the total regret (3). Eubank and Hart (24) suggested to choose the smoothing parameter J minimizing the risk.

With a large number of gene curves and various functional shapes, a universal rule for an optimal choice for J does not exist. Therefore, instead, we capitalize on the convergence property of Fourier transforms. Since the Fourier estimator converges to the true function, usually the first few Fourier coefficients contribute to the estimation of the whole function. In practice, we can select a smaller J for linear or smooth functions and a larger J for wigglier functions.

3.4. Clustering Gene Curves of the Same Change

The purpose of cluster analysis is to classify data of previously unknown structure into meaningful groupings. Discussed are methods of identifying groups of expressed genes useful for discovering patterns in microarray data when there is no predefined class variable to supervise the analysis (25). Cluster analysis techniques can be applied to construct classifications of genes.

Model-based clustering is a statistically based method involving the use of mixture models to determine clusters. The Gaussian mixture model approach assumes that the data have arisen from a mixture of multivariate Gaussian distributions. Yeung et al. (26) showed the performance of model-based clustering on several simulated and real gene expression data sets. The model-based approach has consistently selected the correct model and the number of clusters over other approaches. Fraley and Raftery (20) suggested model-based hierarchical agglomerative clustering based on computing an approximate maximum for the classification likelihood. Such clustering proceeds by successively merging pairs of clusters corresponding to the greatest increase in the classification likelihood among all possible pairs. Their strategy comprises three core elements: initialization via model-based hierarchical agglomerative clustering, maximum likelihood estimation via the EM algorithm, and selection of Bayes factors with Bayesian information criterion (BIC) approximation. Model-based agglomerative hierarchical clustering is successfully applied to problems in character recognition using a multivariate normal model (19) and is generalized to other models (27).

The similarity of cluster Fourier profiles of observed data $\phi_i = (\phi_{i1}, \phi_{i2}, \ldots, \phi_{iJ})$ and $\phi_j = (\phi_{j1}, \phi_{j2}, \ldots, \phi_{jJ})$, or derivatives $\psi_i = (\psi_{i1}, \psi_{i2}, \ldots, \psi_{iJ})$ and $\psi_j = (\psi_{j1}, \psi_{j2}, \ldots, \psi_{jJ})$ can be measured with Euclidean distance. It may be of interest to check the equivalence of the similarity of the estimated Fourier coefficients with the similarity of the estimated functions. A reasonable coordinate system via a Fourier transform of data has as much correct asymptotic coverage probability as the untransformed data (28).

As such, the sample Fourier coefficients can be used instead of the underlying functions (see Note 3).

After clustering with the estimated Fourier coefficients $\hat{\phi}_{ij}$s for the original function, we can estimate the function of each gene with these estimated Fourier coefficients using Eq. 9. The change pattern can also be estimated with derivative Fourier coefficients using Eq. 16. This automatic estimation is another capability of Fourier representation. These estimated periodic functions show the functional shape and periodicity.

3.5. Mixture Model of Fourier Coefficients

Clustering using a mixture model assumes that each group of the data is generated by an underlying probability distribution. Let us assume that data $X_1, \ldots, X_n$ are multivariate observations. In a Gaussian mixture model, each group k is modeled by the multivariate normal distribution with parameters μ_k(mean vector) and Σ_k(covariance matrix):

$$f_k(x_i|\mu_k, \Sigma_k) = \frac{1}{|2\pi\Sigma_k|^{1/2}} \exp\left\{-\frac{1}{2}(x_i - \mu_k)^t \Sigma_k^{-1} (x_i - \mu_k)\right\}. \quad (18)$$

Geometric features (shape, volume, and orientation) of each group k are determined by its covariance matrix Σ_k.

A general framework for exploiting the representation of the covariance matrix in terms of its eigenvalue decomposition as

$$\Sigma_k = \lambda_k \mathbf{D}_k \mathbf{A}_k \mathbf{D}_k^t$$

where $\mathbf{D}_k$ is the orthogonal matrix of eigenvectors, $\mathbf{A}_k$ is a diagonal matrix, and λ_k is an eigenvalue. $\mathbf{D}_k$ determines the orientation of the group, $\mathbf{A}_k$determines its shape, and λ_k determines its volume in (27). The equal volume spherical model is parameterized by $\Sigma_k = \lambda\mathbf{I}$ and the unequal volume spherical model by $\Sigma_k = \lambda_k\mathbf{I}$. The unconstrained model allows all variability in Σ_k. Each elliptical model is implemented in Mclust (29).

We consider model-based clustering with the estimated Fourier coefficients of change $\hat{\psi}_i = (\hat{\psi}_{i1}, \hat{\psi}_{i2}, \ldots, \hat{\psi}_{iJ})$. The sample Fourier coefficient $\hat{\phi}_{ij}$ in Eq. 10 is a form of weighted average of random variables with variance $O(m^{-1})$. The empirical distribution of Fourier coefficients is normal (30). By Central Limit Theorem for independently and identically distributed samples, the sample Fourier coefficient $\hat{\phi}_{ij}$ is asymptotically normally distributed as $m \to \infty$. As $m \to \infty$ and for a fixed $J << m$, $\hat{\phi}_i = (\hat{\phi}_{i1}, \hat{\phi}_{i2}, \cdots, \hat{\phi}_{iJ})$ follows an asymptotically J-dimensional multivariate normal distribution. Therefore, $\hat{\psi}_i = (\hat{\psi}_{i1}, \hat{\psi}_{i2}, \cdots, \hat{\psi}_{iJ})$ has an asymptotically multivariate normal distribution as a linear function of $\hat{\phi}_i = (\hat{\phi}_{i1}, \hat{\phi}_{i2}, \ldots, \hat{\phi}_{iJ})$. With this asymptotic property, we can use the Gaussian mixture model for clustering.

Model-based hierarchical agglomerative clustering (see Note 4) is an approach to compute an approximate maximum of the classification likelihood,

$$L_{cl}(\theta_1,\ldots,\theta_G;\, g_1,\ldots,g_n|\psi) = \prod_{i=1}^{n} l_{g_i}(\psi_{i1},\ldots,\psi_{iJ}|\theta_{g_i})$$

where g_is are labels indicating a unique classification for each observation and l_{g_i} is the probability function of the estimated Fourier coefficients of the *i*th gene. In the above Gaussian mixture likelihood, each component is weighted by the probability that a sample Fourier coefficient belongs to that component. Our clustering strategy is model-based agglomerative hierarchical clustering; and selection of the model and the number of clusters use approximate Bayes factors with the BIC approximation.

In the simulation study (11), overall the clustering estimation error is smaller in the model-based method with the Fourier coefficients than in the *K*-means method (see Note 5).

3.6. Cluster Validity

A major challenge in cluster analysis is the estimation of the optimal number of clusters. To identify the partition of clusters for which a measure of quality is optimal, as a cluster validity technique, silhouette method was proposed by Rousseeuw (31).

The silhouette width for the *i*th sample in the *j*th cluster is defined as:

$$s(i) = \frac{b(i) - a(i)}{\max\{a(i), b(i)\}} \tag{19}$$

where $a(i)$ is the average distance between the *i*th sample and all other samples included in the *j*th cluster, $b(i)$ is the minimum average distance between the *i*th sample and all of the samples clustered in *k*th cluster for $k \neq j$. A point is regarded as well clustered if $s(i)$ is large.

For a given cluster, a cluster silhouette S_j characterizes the heterogeneity:

$$S_j = \frac{1}{m_j}\sum_{i=1}^{m_j} s(i)$$

where m_j is number of samples in S_j.

The optimal number of clusters can be chosen as the value maximizing the average $s(i)$ over the data set (32). Silhouettes offer the advantage that they depend only on the actual partition of the objects, and not on the clustering algorithm that was used to obtain it. As a consequence, silhouettes could be used to improve the results of cluster analysis. The overall average silhouette value can be used as an effective validity index for any partition. We can consider the overall average silhouette in selecting the number of Fourier coefficients and the optimal number of clusters. A silhouette is generally known to work best with roughly spherical clusters. If the clustering algorithm does not result in this shape of cluster, the overall average silhouette width tends to become very low. The assessment of expression cluster validity with 18 measures showed that there is no universal validity paradigm to predict consistent results across different clustering techniques (33). Evaluation of biologically relevant results may support the cluster validity (see Note 6).

4. An Example with Yeast Cell Cycle Microarray Expression Data

4.1. Yeast Cell Cycle Data

Cell cycle is important in understanding cell replication, malignancy, and reproductive disease that are associated with genomicinstability and abnormal cell division. Biologists have been studying the cell cycle with budding yeast *Saccharomyces cerevisiae*, which is a free living, eukaryotic, and single cell but highly complex organism.

We applied our method of clustering to yeast cell cycle data (2) downloaded from http://genome-www.stanford.edu/cellcycle/. DNA microarrays and samples from yeast cultures were synchronized by three independent methods: α-factor arrest, elutriation, and the arrest of a cdc15 temperature-sensitive mutant. We used yeast alpha data collected at 18 time points for 120 min during two full cell cycles. After removing genes with the missing values, there were 4,489 genes remaining.

4.2. Choice of Fourier Coefficients and Clusters

To determine the J value and the number of clusters, we considered several J values and BIC with the assumption that each cluster covariance has the same elliptical volume and shape. Since we found that the optimal J value varied for each function, we surmised that a true optimal J value for all the functions may not exist (see Note 7).

For the comparison of the proposed method with the model-based clustering, *K*-means (see Note 8) clustering is adopted. Table 1 shows the median and average silhouette values with Euclidean distance between samples by model-based and *K*-means clustering methods for various J values in five clusters. Although $J = 1$ yields higher overall silhouette widths using both *K*-means and model-based clustering, we think a larger number than 1 is appropriate to extract enough information about the underlying change patterns. Judging from the highest overall silhouette value, the model-based with four Fourier coefficients and five clusters was considered most appropriate. With *K*-means, silhouette value with $J = 1$ is the largest. Silhouette value of *K*-means with $J = 3$ is larger than that of the model-based clustering with $J = 4$. Therefore, it should be noted that silhouette values of Euclidean distance between two clustering models may not be the only criterion for model comparison. Rather as in the

Table 1
Median and average silhouette values for five clusters with derivative Fourier coefficients of yeast data using Euclidean distance

	Model-based		*K*-means	
Number of FC	Median *S*	Average *S*	Median *S*	Average *S*
$J = 1$	0.555	0.339	0.594	0.520
$J = 2$	0.183	0.120	0.325	0.298
$J = 3$	0.103	0.007	0.253	0.239
$J = 4$	0.248	0.176	0.201	0.192
$J = 5$	0.043	0.001	0.174	0.165
$J = 6$	0.182	0.093	0.178	0.156
$J = 7$	0.138	0.065	0.135	0.140

following gene ontology analysis, biological interpretation should be done to validate clustering. However, the model-based method including density connects the probability-neighboring data, while *K*-means method measures intracluster homogeneity as cluster compactness.

Using the model-based and $J = 4$, each partition of five clusters has the following number of genes 3,032, 401, 164, 400, and 492. Figure 1 shows means, 5 and 95% percentiles of Fourier estimated gene scores in five clusters with sample derivative Fourier coefficients. The graph in the bottom right hand corner of Fig. 1 shows the estimated change patterns of the five clusters altogether. Figure 2 shows the means of four derivative Fourier coefficients as cluster profiles and gives the variation between clusters. Figure 3 shows chi-square plot of each cluster for multivariate normality with a dimension of 4. If the four derivative Fourier coefficients follow a multivariate normal distribution, they would scatter around the line with a slope of 1. Even though they satisfy asymptotic multivariate normality, this assumption can also be checked with chi-square plots. Except for cluster 4, they appear to have a slightly heavier tail than a normal distribution.

Gaussian mixture model clustering allows clusters to have different orientation or sizes while preserving some common features, such as an ellipsoidal shape. Cluster 5 in particular has a wide elliptical shape incorporated with the probability distribution. Owing to noise and the high dimensionality of data, careful consideration of statistical and biological validity is needed when analyzing the real microarray data.

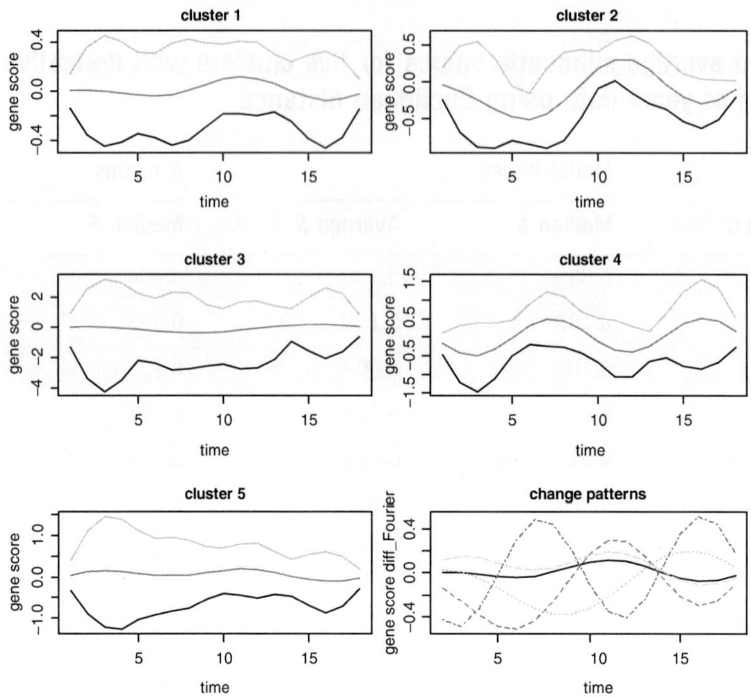

Fig. 1. Fourier estimated gene score mean, 5 and 95% percentiles in five clusters based on derivative Fourier coefficients with $J = 4$ of yeast data and Fourier estimated change patterns.

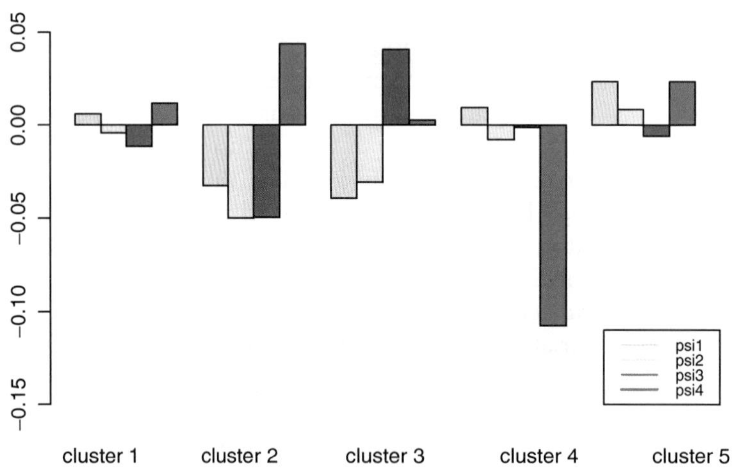

Fig. 2. Means of derivative Fourier coefficients ψ_1, ψ_2, ψ_3, ψ_4, in each cluster with $J = 4$ of yeast data.

4.3. Gene Ontology Analysis

In order to evaluate the result of the clustering analysis, we obtained Gene Ontology (GO) information for the clustered genes' biological processes, molecular functions, and cellular components. The GO database provides a useful tool to annotate and analyze the functions of a large number of genes. We searched

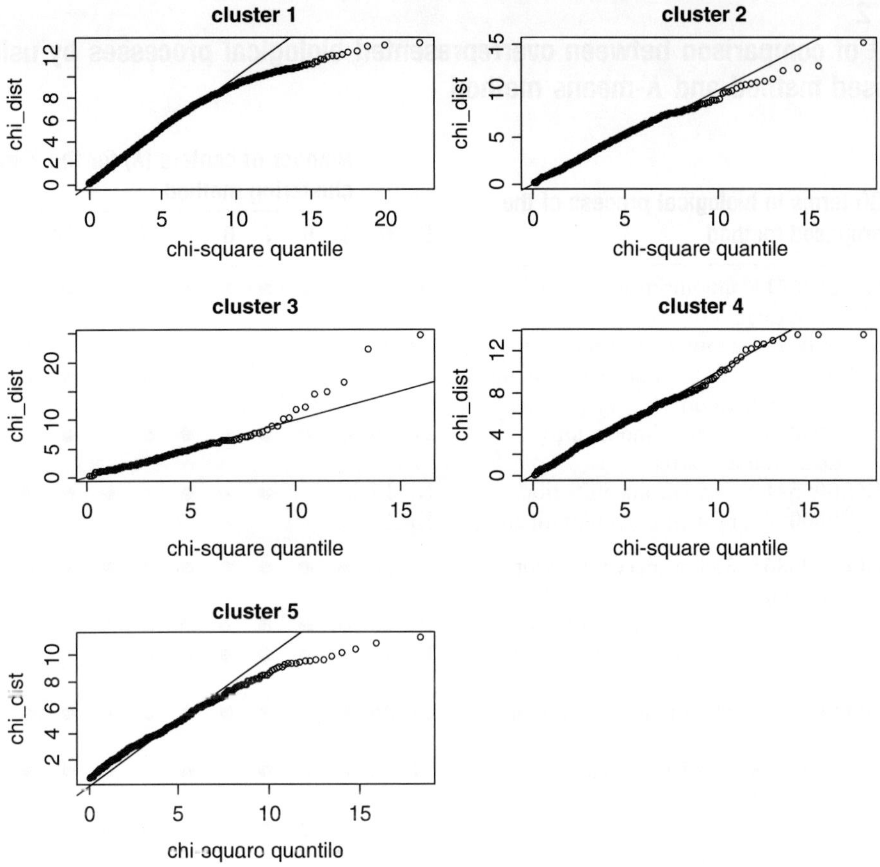

Fig. 3. Chi-square plots for multivariate normality in each cluster.

statistically overrepresented GO annotations using GOstat for evaluating statistical significance of overrepresented functional and molecular mechanisms (34). GOstat allows us to identify which annotations are typical for the group of genes. GOstat simply derives the statistical significance between expected and observed functional categories based on the Fisher's exact test.

In order to compare our method with other clustering methods, we also applied *K*-means clustering (35) to yeast cell cycle data. Table 2 shows some results of the overrepresented biological processes from the proposed method and the *K*-means clustering method for various values of k from 5 to 15.

In Table 2, the first column shows the cluster number of the proposed method. The second column summarizes the list of the selected overrepresented biological processes which had their children GO terms in the same cluster. For example, we first selected total 81 GO terms in cluster 1 by using GOstat and then selected six GO terms that had as many children nodes as possible in cluster 1. In the same way, *K*-means clustering results were obtained. We compared the list of the overrepresented GO

Table 2
Result of comparison between overrepresented biological processes by using proposed method and *K*-means method

No.	GO terms in biological process of the proposed method	*S*	*C*	Number of centers (*k*) for the *K*-means clustering method										
				5	6	7	8	9	10	11	12	13	14	15
1	GO:0006511, ubiquitin-dependent protein catabolism	26	27	●	●	●	●	●	●	●	●	●	●	●
	GO:0006445, regulation of translation	20	27								●	●	●	●
	GO:0006418, tRNA aminoacylation for protein translation	19	29					●	●	●	●	●	●	●
	GO:0030150, protein import into mitochondrial matrix	17	19	●	●	●	●	●	●		●		●	
	GO:0006413, translational initiation	16	17			●	●	●	●	●	●	●	●	●
	GO:0000209, protein polyubiquitination	16	17											
2	GO:0006365, 35S primary transcript processing	19	20	●	●	●	●	●	●	●	●	●	●	●
	GO:0030490, processing of 20S pre-rRNA	19	20	●	●	●	●	●	●	●	●	●	●	●
	GO:0000027, ribosomal large subunit assembly and maintenance	14	16	●	●	●	●	●	●	●	●	●	●	●
	GO:0000055, ribosomal large subunit export from nucleus	14	26	●	●	●	●	●	●	●	●	●	●	●
	GO:0030488, tRNA methylation	14	18	●	●	●		●		●	●	●	●	●
	GO:0000154, rRNA modification	13	15									●	●	●
	GO:0006325, establishment and/or maintenance of chromatin arch.	13	17	●	●	●	●	●	●	●	●	●	●	●
3	GO:0000750, pheromone-dependent signal transduction during conj.	21	28		●	●	●	●	●	●	●	●	●	●
	GO:0015892, siderophore-iron transport	11	15				●	●	●	●	●	●	●	●
	GO:0006827, high affinity iron ion transport	10	14					●	●					
	GO:0000079, regulation of cyclin-dependent protein kinase activity	9	16											
4	GO:0000710, meiotic mismatch repair	14	30	●	●	●	●	●	●	●	●	●	●	●
	GO:0007064, mitotic sister chromatid cohesion	10	14	●									●	●
	GO:0000731, DNA synthesis during DNA repair	8	18	●	●	●	●	●	●	●	●	●	●	●
	GO:0006301, postreplication repair	7	17	●		●	●	●	●	●	●	●	●	●
	GO:0006284, base-excision repair	7	17	●	●	●	●	●	●		●	●	●	●
	GO:0006289, nucleotide-excision repair	7	17	●						●	●	●	●	●
	GO:0006269, DNA replication, synthesis of RNA primer	6	16	●						●	●	●	●	●
5	*GO:0009086, methionine biosynthesis*	15	26											
	GO:0006526, arginine biosynthesis	14	23	●	●	●	●				●	●	●	●
	GO:0006537, glutamate biosynthesis	13	22			●							●	
	GO:0005978, glycogen biosynthesis	9	27											

terms from the proposed method (second column) with that from the *K*-means clustering method. The black dots in Table 2 represented the GO terms that were selected by both methods. In summary, there are some GO terms that can only be detected by the proposed method such as GO:0000209, GO:0000079, GO:0009086, and GO:0005978. In particular, all GO terms in cluster 5 of our proposed method are closely related to biosynthesis. The three GO terms in cluster 5, GO:0009086, GO:0006537, and GO:0005978, are rarely overrepresented by the *K*-means clustering method. Our proposed method not only found the GO terms that were not identified by the *K*-means method but also grouped them in the same cluster.

Furthermore, the genes in cluster 5 are closely related to the glucose metabolic pathway. For example, GLC3 (GO:0005978) encodes 1,4-glucan-6(1,4-glucano)-transferase, involved in glycogen accumulation. Glycogen in turn serves as a major storage carbohydrate (glucose) (36). Free glucose is oxidized to pyruvate. The other genes from GO:0006537, GO:0006526, and GO:0009086 are related to the synthesis of amino acids in the citric acid cycle, 15 ATPs and $3CO_2$ are produced from one pyruvate molecule. IDP1 (GO:0006526) catalyzes the oxidation of isocitrate to alpha-ketoglutarate (37). GLT1 (GO:0006537) synthesizes glutamate from glutamine and alpha-ketoglutarate (38). ARG1, ARG3, and ARG4 in GO:0006526 are involved in the synthesis of alginine from the glutamate (39, 40). Oxaloacetate, an intermediary in the citric acid cycle, is the entry point for the metabolism of the underlying carbon structure of the amino acids aspartate and asparagine. MET2 (GO:0009086) is involved in the synthesis of methionine from the aspartate (41). It catalyzes the conversion of homoserine to O-acetyl homoserine using one molecule of acetyl coenzyme A (acetyl-CoA) (42). These findings illustrate that our proposed methodology can identify genes which are biologically interpretable (see Note 9).

4.4. Using R Codes

Yeast alpha factor data is used to demonstrate the proposed method. The program derFcl.R contains main program part and several user-defined R functions. After running each function defined in the program, main part should be run.

1. The main part provides clustering results including a text outfile with identifying genes and their cluster numbers, silhouette plots, and chi-square plots. *J* indicates the number of derivative Fourier terms. Clustering results can be obtained according to *J* values. To use the program, modify the location of files downloaded at your PC. There are several functions in the program as follows.
2. The Fourier.est( ) function calculates the Fourier coefficients and derivative Fourier coefficients.

3. For model-based clustering, use library(mclust) and Mclust( ). For *K*-means clustering, use the kmeans( ) function.
4. The see_cluster_yeast( ) gives the profiles of each cluster.
5. To get the silhouette values after clustering, use library(cluster) and the silhouette( ) function.
6. chisq.plot.cl( ) provides the chi-square plot for each cluster.

5. Materials

1. The free software R can be downloaded from http://www.r-project.org/ and installed in the PC.
2. The yeast alpha factor data is arranged in trt.alpha.cdc.elu.txt file for compatibility with R codes in yeast folder at http://webhard.duksung.ac.kr/jaehee. This data file includes DNA microarrays and samples from yeast cultures synchronized by three independent methods: a factor arrest, elutriation, and arrest of a cdc15 temperature-sensitive mutant.
3. The R codes of our method are available in yeast folder at http://webhard.duksung.ac.kr/jaehee.

6. Notes

1. The Fourier coefficient is a transformation of observations based on the orthogonal system. Other transformation can be considered to represent the profile function. There are useful and meaningful gene profile functions such as wavelet transformation, Legendre polynomials, and Fourier transformation based on the orthogonal systems, etc.
2. Time course gene expression data can have temporal correlation according to time. The correlation structure should be incorporated in the statistical methods. We can screen out gene curves that are nearly flat, and then do clustering the transformed curves. Even if the observations are correlated or not, the limiting distribution of Fourier coefficients is multivariate normal. This fact is another advantage to utilize Fourier transformation.
3. Instead of the derivative Fourier coefficients, the original Fourier coefficients can be used for clustering. The choice depends on change patterns of interest.
4. One widely used class of methods involves hierarchical agglomerative clustering, in which two groups chosen to optimize some criterion are merged at each stage of the algorithm.

5. Partitional methods aim to produce a single partition of the gene, whereas hierarchical methods aim to find a nested series of partitions. *K*-means method is a popular partitional clustering procedure. Given some specified number of clusters ***k*** the goal is to segregate objects into ***k*** subgroups. The basic algorithm begins with either an initial partition of the object into ***k*** subgroups or an initial specification of ***k*** cluster centroids. The algorithm proceeds by considering each object, determining to which of the current cluster centroids the object is closest and assigning it to the cluster. The centroid of the recipient cluster is updated to reflect its new member, and the centroid of the donor cluster is updated. The algorithm cycles through the data again, relocating the whole objects among the clusters and updating cluster centroids. A major advantage of *K*-means clustering is computational feasibility. In the context of microarray data, this property is important for clustering genes. Two major disadvantages of *K*-means are that it requires specification of a number of clusters and an initial partitioning.

6. For clustering estimation error rate, the simulation study was done.

   Let T be a clustering map defined as

   $$T(f,g) = \begin{cases} 1, & \text{if } f \text{ and } g \text{ are in the same cluster} \\ 0, & \text{otherwise.} \end{cases}$$

   The clustering estimation error rate $\eta(K)$ is defined as

   $$\eta(K) = \frac{1}{{}_N C_2} \sum_{r<s} I(T_K(f_r, f_s) \neq \hat{T}_K(\hat{f}_r, \hat{f}_s))$$

   where $C = \{f_1, \ldots, f_N\}$ denotes the true curves and $\hat{C} = \{\hat{f}_1, \ldots, \hat{f}_N\}$ denotes the estimated curves. Let T and $\hat{T}$ represent the corresponding cluster maps, and K denotes the number of clusters. $\eta(K)$ then is the fraction of all pairs that are incorrectly put in separate clusters depending on K clusters. The clustering estimation error rates for the model-based method and *K*-means methods with Fourier coefficients and also with difference data for the number of Fourier terms and the number of repeated design points. Overall, the clustering estimation error is smaller in the model-based method with the Fourier coefficients than in the *K*-means with the Fourier coefficients since the Fourier coefficients follow the multivariate normal distribution. With Fourier coefficients, the clustering estimation error becomes smaller as the number of time points ***m*** becomes larger.

7. Selection of the number of Fourier coefficients J is another issue to be determined. The globally optimal estimator is not

optimal for each gene curve. A simulation study shows that once J exceeds 5, the clustering estimation error rate does not change appreciably. Therefore, we suggest using J around 5 firstly for dimension reduction and to perform the biological interpretation.

8. Each cluster can be represented by the cluster validity measures, which is based on the comparison of its tightness and separation. The silhouette measure show which objects lie well within their cluster, and which ones are merely somewhere in between clusters. The average silhouette width provides an evaluation of clustering validity, and might be used to select an appropriate number of clusters. There are other cluster validity measures such as Adjusted Rand Index, Dunn's index, and Gap.
9. From our analysis we observed that Fourier coefficients and derivative Fourier coefficients give efficient gene profiles for time course gene expression data and provide substantial dimension reduction especially clustering a large number of curves. We identified that some functional genes are not detected by K-means. Even though different clustering methods may result in slightly different clusters, each cluster has feature genes which give biological information to be discovered.

Acknowledgments

We thank Dr. Carroll for giving motivation and Haseong Kim for providing gene ontology results.

References

1. Cho, R. J., Campbell, M. J., Winzeler, E. A., Steinmetz, L., Conway, A., Wodicka, L., Wolfsberg, T. G., Gabrielian, A. E., Landsman, D., Lockhart, D. J. and Davi, R. W. (1998) A genome-wide transcriptional analysis of the mitotic cell cycle. *Mol. Cell* **2**, 65–73.
2. Spellman, P. T., Sherlock, G., Zhang, M. Q., Iyer, V. R., Anders, K., Eisen, M. B., Brown, P. O., Botstein, D. and Futcher, B. (1998) Comprehensive identification of cell cycle-regulated genes of the yeast Saccaromyces cerevisiae by microarray hybridization. *Mol. Biology of the Cell* **9**, 3273–3297.
3. Serban, N. and Wasserman, L. (2005) CATS: Clustering after transformation and smoothing. *J. Amer. Statist. Assoc.* **471**, 990–999.
4. Ernst, J., Nau, G. J. and Bar-Joseph, Z. (2005) Clustering short time series gene expression data. *Bioinformatics* **21**, 195–168.
5. Li, J. and Wong, L. (2002) Identifying good diagnostic gene groups from gene expression profiles using the concept of emerging patterns. *Bioinformatics* **18**, 725–34.
6. Park, T., Yi, S. G., Lee, S., Lee, S. Y., Yoo, D., Ahn, J., and Lee, Y. (2003) Statistical tests for identifying differentially expressed gene in time-course microarray experiments. *Bioinformatics* **19**, 694–703.
7. Lai, Y., Wu, B., Chen, L. and Zhao, H. (2004) A statistical method for identifying differential gene-gene co-expression patterns. *Bioinformatics* **20**, 3146–55.

8. Zhang, L., Zhang, A. and Ramanathan, M. (2003) Fourier harmonic approach for visualizing temporal patterns of gene expression data. *Proc. IEEE Comp. Sys. Bioinformatics Conf.* **2**, 137–147.
9. Murthy, K. R. K. and Hua, L. J. (2004) Improved Fourier transform method for unsupervised cell-cycle regulated gene prediction. *Proc. IEEE Comp. Sys. Bioinformatics Conf.* 194–203.
10. Kim, B., Littell, R. C. and Wu, R. (2006) Clustering periodic patterns of gene expression based on Fourier approximations. *Current Genomics* 7, 197–203.
11. Kim, J. and Kim, H. (2008) Clustering of change patterns using Fourier coefficients. *Bioinformatics* **24**, 184–191.
12. Peddada,S., Lobenhofer, E., Li L., Afshari C., Weinberg C. and Umbach D. M. (2003) Gene selection and clustering for time-course and dose response microarray experiments using order-restricted inference. *Bioinformatics* **19**, 834–841.
13. Johansson, D., Lindgren, P., Berglund, A., (2003) A multivariate approach applied to microarray data for identification of genes with cell cycle-coupled transcription. *Bioinformatics* **19**, 467–473.
14. Schliep, A., Schōnhuth, A., Steinhoff, C., (2003) Using hidden Markov models to analyze gene expression time course data. *Bioinformatics* **19** (Suppl.), i255-i263.
15. Luan and Li (2003) Clustering of time-course gene expression data using a mixed-effects models with B-splines. *Bioinformatics* **19**, 474–482.
16. Song J. J., Lee, H. J., Morris, J. S. and Kang, S. (2007) Clustering of time-course gene expression data using functional data analysis. *Comp. Biol. and Chem.* **31**, 4, 265–274.
17. Bar-Joseph, Z. (2004) Analyzing time series gene expression data. *Bioinformatics* **20**, 2493–2503.
18. Yeung, K. Y., Fraley, C., Murua, A., Raftery, A. E. and Ruzzo, W. L. (2001) Model based clustering and data transformations for gene expression data. *Bioinformatics* **17**, 977–998.
19. Murtage, C. and Raftery, A. E. (1984) Fitting straight lines to point patterns. *Pattern Recognition* **17**, 479–483.
20. Fraley, C. and Raftery, A. E. (2002) Model-based clustering, discriminant analysis and Density Estimation. *J. Amer. Statist. Assoc.* **97**, 611–631.
21. Tolstov, G. P. (1962) *Fourier analysis.* McGraw-Hill, New York.
22. Stein, E. M. and Shakarchi, R. (2003) *Fourier analysis.* Princeton University Press, Princeton.
23. Lestrel, P. E. (1997) *Fourier descriptors and their applications in biology.* Cambridge University Press, London.
24. Eubank, R. and Hart, J. D. (1992) Testing goodness-of-fit via order selection criteria. *Ann. Stat.* **20**, 3, 1412–1425.
25. Simon, R. M., Korn, E. L., McShane, L. M., Radmacher, M. D., Wright, G. W. and Zhao, Y. (2003) *Design and analysis of DNA microarray investigations.* Springer, New York.
26. Yeung, K. Y. and Ruzzo,W. L. (2001) An empirical study on principal component analysis for clustering gene expression data. *Bioinformatics* **17**, 763–774.
27. Banfield, J. D., and Raftery, A. E. (1993) Model-based Gaussian and non-Gaussian clustering *Biometrics* **49**, 803–821.
28. Beran, R. and Dumbgen, L. (1998) Modulation of estimators and confidence Sets. *Ann. Stat.* **26**, 1826–1856.
29. Fraley, C. and Raftery, A.E. (1999) MCLUST: software for Model-based cluster analysis. *J. Classif.* **16**, 297–306.
30. Freedman, D. and Lane, D. (1980) The Empirical distribution of Fourier coefficients. *Ann. Stat.* **8**, 1244–1251.
31. Rousseeuw, P. J. (1987) Silhouettes: graphical aid to the interpretation and validation of cluster analysis. *J. Comp. and Appl. Math* **20**, 53–65.
32. Kaufman, L. and Rousseeuw, P. J. (1990) Finding groups in data: An introduction to cluster analysis. Wiley, New York.
33. Ajuaje, F. (2002) A cluster validity framework for genome expression data. *Biometrics* **18**, 319–320.
34. Beissbarth, T. and Speed, T. P. (2004) GOstat: Find statistically overrepresented Gene Ontologies within a group of genes. *Bioinformatics* **6**, 20(9), 1464–1465.
35. MacQueen, J. B. (1967) Some Methods for classification and Analysis of Multivariate Observations, *Proceedings of 5-th Berkeley Symposium on Mathematical Statistics and Probability.* Berkeley. University of California Press, **1**, 281–297.
36. Rowen, D. W., Meinke, M. and LaPorte, D. C. (1992) GLC3 and GHA1 of Saccharomyces cerevisiae are allelic and encode the glycogen branching enzyme. *Mol. Cell. Biol.* Jan;**12**(1), 22–29.
37. Haselbeck, R. J. and McAlister-Henn, L. (1993) Function and expression of yeast mitochondrial NAD- and NADP-specific isocitrate dehydrogenases. *J. Biol. Chem.* **268**(16), 12116–12122.
38. Valenzuela, L., Ballario, P., Aranda, C., Filetici, P. and A. Gonzalez, A. (1998) Regulation of expression of GLT1, the gene encoding glutamate synthase in Saccharomyces cerevisiae. *J. Bacteriol.* **180**(14), 3533–3540.

39. Jauniaux, J. C., Urrestarazu, L. A., and Wiame, J. M. (1978) Arginine metabolism in Saccharomyces cerevisiae: subcellular localization of the enzymes. *J. Bacteriol.* **133**(3), 1096–1107.
40. Crabeel, M., Seneca, S., Devos, K. and Glansdorff, N. (1988) Arginine repression of the Saccharomyces cerevisiae ARG1 gene comparison of the ARG1 and ARG3 control regions. *Curr. Gen.* **3**(2), 113–124.
41. Masselot M. and De Robichon-Szulmajster, H. (1975) Methionine biosynthesis in Saccharomyces cerevisiae. I. Genetical analysis of auxotrophic mutants. *Mol. Gen. Genet.* **139** (2):121–132.
42. Thomas, D. and Surdin-Kerjan, Y. (1997) Metabolism of sulfur amino acids in Saccharomyces cerevisiae. *Microbiol. Mol. Biol. Rev.* **61** (4), 503–532.

Part III

Analysis of Network Behaviour by Quantitative Genetics

Chapter 11

Finding Modulators of Stochasticity Levels by Quantitative Genetics

Steffen Fehrmann and Gaël Yvert

Summary

Although bakers and wine makers constantly select, compare, and hunt for new wild strains of *Saccharomyces cerevisiae*, yeast geneticists have long focused on a few "standard" strains to ensure reproducibility and easiness of experimentation. And so far, the wonderful natural resource of wild genetic variation has been poorly exploited in most academic laboratories. We describe here how one can use this resource to investigate the molecular sources of stochasticity in a gene regulatory network. The approach is general enough to be applied to any network of interest, as long as the experimental read-out offers robust statistics. For a given network, a typical study first identifies two backgrounds A and B displaying different levels of stochasticity and then study the network in A × B progeny. Taking advantage of microarrays or resequencing technologies, genotyping of appropriate segregants can then lead to the genomic regions housing modulators of stochasticity. The powerful toolbox available to manipulate the yeast genome offers several ways to narrow these regions further and to unambiguously demonstrate the regulatory consequences of DNA polymorphisms.

Key words: Stochasticity, Noise in gene expression, Cell to cell variation, QTL, Genetic variation

1. Introduction

The concept here is to treat the stochasticity level of interest as a quantitative phenotype, and to map Quantitative Trait Loci (QTL) controlling this phenotype. This is very similar to the many QTL approaches that identified genetic variations responsible for disease in humans, or agronomical traits in plants and animals, except that the organism is yeast and the phenotype is derived from thousands of measurements in single cells. As for any QTL study, it is essential to properly design the phenotyping assay and the choice of strains to use and we provide guidelines to achieve this in Subheading 1.1. Subheading 1.2 is devoted to the identification of two differing backgrounds that can then be used for genetic mapping, which is described in Subheadings 1.3

Attila Becskei (ed.), *Yeast Genetic Networks: Methods and Protocols*, Methods in Molecular Biology, vol. 734,
DOI 10.1007/978-1-61779-086-7_11,

and 1.4. Subheading 1.5 describes powerful methods available in yeast to narrow the QTL locus down to the polymorphism(s) responsible for noise modulation. Although the precise methodology greatly differs from one network study to another, our experimental protocol to transform a reporter system and analyze it by flow-cytometry might provide useful guidelines to others and is presented in Subheading 2.

1.1. Design

1.1.1. Choice of Natural Backgrounds

In many cases, no prior knowledge or intuition drives the choice of the appropriate natural genetic backgrounds to use. A typical study thus collects many unrelated strains and studies the network of interest in those strains. Collecting starts with simple requests to colleagues and may continue by purchasing strains from repositories or even, if motivation justifies the effort, collecting directly from natural habitats or industrial plants. This section describes the sources of strains already documented. For direct isolation of *Saccharomyces cerevisiae* strains from field samples, please see for example Cappello et al. (1).

Some extent of genetic diversity is already available among the commonly used laboratory strains of *S. cerevisiae*. S288c is famous because it was chosen for the genome sequencing project of the 1990s (2). A large panel of useful strains derives from it, such as the BY series of strains described in Brachmann et al. (3) and panels of systematic gene deletion (4) or tagging (5, 6). Strains W303, A364A, CEN.PK, and FL100 are also frequently used in laboratories. They are closer to S288c than most natural isolates (7) but may contain enough genetic differences to harbor differential properties of the network of interest. The laboratory strain SK1 sets apart from this group (8, 9). Its origin is totally independent, and it was cherished by scientists studying sporulation because of its high level of ascus production (10).

An enormous resource of yeast natural genetic variation resides in public repositories of microorganisms. Large collections of wild yeast strains have been archived in various countries. We cite in Table 1 some of these resources which gather yeasts from all around the globe and isolated from diverse ecological environments. In addition to these programs, specific colleagues such as Robert Mortimer (11) and John Mc Cusker (12) have isolated and described many wild *S. cerevisiae* strains that are now studied in several laboratories (8, 9, 13). Noteworthy, "clinical" isolates found in immunocompromised patients were shown to be poorly related to one another and therefore offer an interesting panel of diversity (8).

Ideally, one would like to study his/her favorite genetic network in backgrounds for which a full genome sequence is available. Considering the recent collapse of sequencing costs, it is likely that this limitation is soon bypassed. As we write this paper, three genomes of *S. cerevisiae* have been fully sequenced at high coverage

Table 1
Collections of wild *S. cerevisiae* strains

Acronym	Full name	URL	Country
CLIB	Collection de Levures d'Intérêt Biotechnologique	http://www.inra.fr/clib/	France
CECT	The Spanish Type Culture Collection	http://www.cect.org/	Spain
DBVPG	Industrial Yeasts Collection	http://www.agr.unipg.it/dbvpg/	Italy
CBS	Fungal Biodiversity Center	http://www.cbs.knaw.nl/	The Netherlands
NCYC	National Collection of Yeast Cultures	http://www.ncyc.co.uk/index.html	The UK

and assembled: S288c (2), RM11-1a (14) and YJM789 (15). In addition, two recent surveys of yeast polymorphisms have investigated a very large panel of strains at lower coverage. Schacherer et al. (8) described population structure among 63 *S. cerevisiae* strains based on microarray-inferred polymorphic sites, and simultaneously, Liti et al. (9) described 38 *S. cerevisiae* and 35 *Saccharomyces paradoxus* strains by direct sequencing. These genomic resources are obviously extremely valuable when using natural genetic variation to study gene network properties.

1.1.2. Technical Considerations Regarding Wild Strains

While laboratory strains can be manipulated as haploids, the majority of wild *S. cerevisiae* strains are diploid, and many are polyploid or aneuploid (especially among industrial strains). Heterozygosity is frequent in these wild strains, and it is therefore important to clarify, as early as possible in the project, the level of ploidy and homozygosity of the organism investigated.

One major consideration then is homothallism. Most laboratory strains were selected for their heterothallic phenotype: haploid cells fail to switch mating type and therefore cannot self-cross to regenerate diploids. Most wild yeasts are homothallic, preventing the maintenance of stable haploids. This property can be useful to "purify" the genetic background of interest. The wild diploid strain potentially heterozygote is sporulated; a spore is isolated by microdissection and restreaked on plate for isolated colonies. This way, a diploid strain purely homozygous can be obtained because mating-type switching allowed two isogenic strains to mate. In many cases, it is then desirable to generate a haploid derivative of this "pure" strain. This can be done by disrupting the HO gene responsible for the mating-type switch. Several constructs are already available to target this gene, such as plasmid pHO-poly-KanMX4-HO (16).

When studying single-cell phenotypes, such as the level of a fluorescent reporter system, investigators often need to properly separate cells from one another before acquisition (e.g., during flow-cytometry). In this case, another important consideration is the propensity of the strain to generate isolated cells. Many wild strains display undesirable phenotypes, such as flocculation or clumpiness. Flocculation is driven by calcium-dependent aggregation of cells via agglutinins anchored at the cell wall (17). This trait is visible by eye when growing liquid cultures, with flocks of cells rapidly sedimenting and leaving a clear medium above them. If the context of the study allows it, treating cultures with chelators of calcium, such as EDTA, will efficiently disrupt the flocks and generate isolated cells. However, depending on the gene network of interest, the cellular response to such treatments may affect network properties. Clumpiness refers to incomplete separation of daughter cells from their mother (18). This attachment usually results from incomplete digestion of the cell wall at bud neck, and we do not know of any efficient treatment to achieve separation. However, several genes are known to be essential for this phenotype and, if the study allows it, disrupting such a target can generate single-cell suspensions. This was observed in the RM11-1a *amn1Δ* mutant for example (19).

Another frequent consideration is the spore viability observed when crossing the wild strain with a reference. This is why RM11-1a was chosen for genetical genomics studies: a large progeny was needed when crossing it with S288c (20). If viability is low, mapping will still be possible, but some loci will be biased because selection will prevent their proper randomization in the progeny. If network regulators are linked to determinants of viability, they will be missed.

Finally, gene network investigations largely require DNA transformations. Transformation efficiency varies greatly between backgrounds, and this may sometimes be a limitation. We observed that some strains were difficult to transform by the usual LiAc technique (21) but displayed high efficiencies by the electroporation method (22).

1.1.3. Useful Considerations Regarding Network Constructions

Yeast geneticists commonly use auxotrophic markers when manipulating gene networks. However, most wild strains are prototrophs and these markers are therefore useless unless one mutates the appropriate genes (such as *URA3, LEU2, HIS3,*…). This can be done using specific targeting plasmids (3) but is nonetheless time-consuming if numerous natural strains are considered. A better strategy is to use dominant markers. Several drug-resistance cassettes are now available and transformants can be selected on Nourseothricin, Hygromycin B, Kanamycin, or Phleomycin from most natural strains (23).

A reporter system of the network activity is generally essential for experimental readouts. For some studies, the same reporter system may be introduced in all strains, as in Ansel et al. (24), where the same plasmid was integrated in the genome of wild strains. For other studies, the reporter system may need to possess the polymorphisms of the wild strain considered. For example, if polymorphisms in a promoter region are of interest.

Most investigations of stochasticity in gene expression use integrated constructs to avoid plasmid copy number variations. Once a promising difference in stochasticity is seen between two wild backgrounds A and B, we recommend to confirm the result after targeting the insertion at a different locus of the genome. This rules out any locus-specific effect that may mislead further characterizations.

1.2. Identification of Genetic Backgrounds with Differential Levels of Stochasticity

Once reporter systems and desired network constructions are inserted into "cleaned" wild strains, these can be compared for their level of stochasticity. Usually, when single-cell data are acquired by flow-cytometry, noise values are derived from a sub-population of cells that have similar size and shape, to avoid sources of variability related to cellular morphology. This is usually achieved by gating the data, either during acquisition or in silico afterward. Gating is done by selecting cells that fall into a narrow window of Forward Scatter (FSC) and Side Scatter (SSC) values. This is sometimes not appropriate when dealing with wild natural backgrounds because the distribution of (FSC, SSC) values can greatly differ between strains, and a predefined gate does not select the same type of cells in all backgrounds. We therefore recommend to use full-sample data, and apply a size correction by linear regression: FL1 ~ log(FSC) + log(SSC), where FL1 are GFP values acquired on a log scale, and FSC and SSC are acquired on a linear scale. For each cell i, the fluorescent FL1 intensity y_i is then replaced by $\langle y \rangle + e_i$ where e_i is the residual of the regression.

After a difference is found, it is of course important to ensure reproducibility and to pay attention to confounding effects. In the simple case of the regulation of a GFP reporter, we observed that many wild backgrounds had differential noise values simply because their mean expression differed (24). Ideally, one would like to see a noise difference between two backgrounds having similar mean. If this is not obtained, the experimenters will need to decompose mean and noise effects, for example by varying induction levels if that is possible.

1.3. Genetic Mapping by QTL Scan

There are several designs to map QTLs in experimental crosses. We describe here a few ones before detailing the introgression strategy in the next section.

1.3.1. Single-Marker QTL Scan

A blind and usually efficient strategy is to scan for correlation between inheritance of a genetic marker and phenotypic values in an F1 population: haploid strain A is mated with haploid strain B,

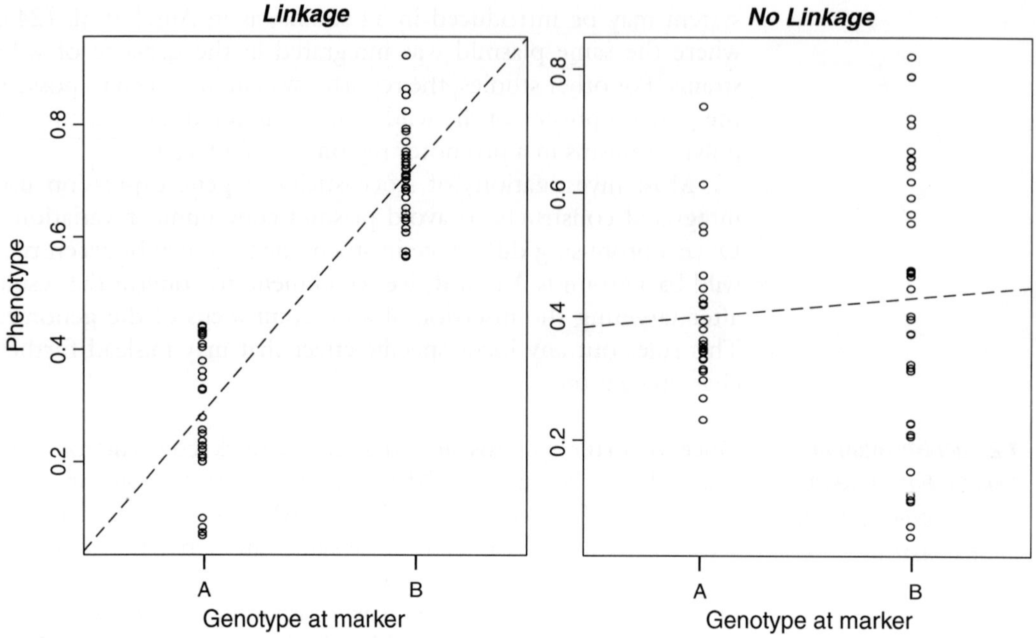

Fig. 1. Test for genetic linkage between a phenotype and a single genetic marker. Each *dot* represents one F1 spore from A × B, with its genotype on the *x*-axis and its phenotypic value on the *y*-axis. Either the genotypes at this marker significantly discriminate phenotypic values (*left*) and a QTL is found, or no correlation is seen between genotypes and phenotypes (*right*).

the hybrid is sporulated, and spores from tetrads are used. These spores are genotyped and their level of network stochasticity (the phenotype of interest) is monitored. Thus, for every haploid spore *i*, genotype $g_{i,m}$ at every marker *m* is either A or B, and a quantitative noise value y_i is available. The key principle of a test for genetic linkage between marker *m* and phenotype *y* is presented on Fig. 1. Typically, spores are split in two groups according to their genotype at *m* (either A or B), and a statistical test is performed to reject the null hypothesis that the two groups do not have a phenotypic difference. This test can be a simple Student's *t*-test if the criteria of normality and homogeneity of variance are met, a linear regression or a nonparametric Wilcoxon Mann–Whitney test. Although less sensitive, this latter test is usually very robust as it is based on ranks of values and does not assume any particular properties of the data. In any case, a *P*-value is generated to estimate the significance of genetic linkage between *m* and *y*. This single-marker test is iterated at every marker position *m* to generate a vector of *P*-values.

For this strategy to be fruitful, a marker must be available "near" the causative gene(s). Otherwise, many spores have recombined between this regulator and the closest marker, and no correlation is seen. Fortunately, dense genetic maps can be obtained for *S. cerevisiae* using whole genome tiling arrays (25, 26) which makes single-marker based tests usually powerful enough. When using fewer markers (such as in animals, plants, or an uncommon yeast species), several improvements of the basic test are available, such as interval mapping (27) or multipoint mapping (28) which allow to map QTLs within gaps (a region where no marker is available) by using observations from neighboring markers.

Two important things must be kept in mind. First, for the test to be valid, segregants must be independent from one another which means that every spore chosen must come from a distinct tetrad. Secondly, a multiple-testing issue arises from the fact that the single-marker test is applied to all markers, which means several hundreds or thousands of times. It is therefore important to correct for this multiplicity.

If you are not familiar with this issue, imagine you would like to test if any page of this book has a significant difference in the occurrence of the letter *U* at its top ten lines vs. its bottom ten lines. At every page, you will count *U*s in the top and bottom lines and apply a test to reject the null hypothesis H0 = "AT THIS PAGE, there is no difference in the number of *U*s between the two groups," and obtain a *P*-value. And you may want to reject H0 if $P < 0.01$. But H0 is not the null hypothesis you really want to test against. When scanning the entire book, you want to reject H0* = "THERE IS NO PAGE displaying a difference in the number of *U*s between the two groups." If the book has 200 pages, the single-page test is likely to give $P < 0.01$ at about two pages by chance only. So, to confidently reject H0*, you need to know how small *P* must really be. A simple way to do so is to divide your single-test significance level by the number of tests applied, which is called the Bonferroni correction. In other words, if one of the 200 pages shows $P < 0.00005$, then something happens at this page, at a book-wide significance level of 0.01. However, this correction is often too conservative, especially when data are correlated (for example, if a footnote containing many *U*s appears at all pages of a chapter).

We therefore recommend to perform empirical permutations, which is more powerful than the simple Bonferroni correction. To do so, one generates *N* randomized datasets, where the genotypes are unchanged but the phenotype vector is shuffled: phenotypic values are randomly reassigned to segregants, without replacement (Fig. 2). This can be done in R using the sample() function. On each dataset, rescan the genotypic map for linkage with the same method as used above, and store the ***lowest*** *P*-value obtained. Let P_{rd} be the resulting vector of *N P*-values. The genome-wide significance of a given P_o value is then the number of P_{rd} values lower than P_o, divided by *N*. In other words, if P_o is a linkage obtained in the actual dataset, and if less than 5% of P_{rd} values are lower than P_o, then genetic linkage is significant at the 0.05 level genome-wide.

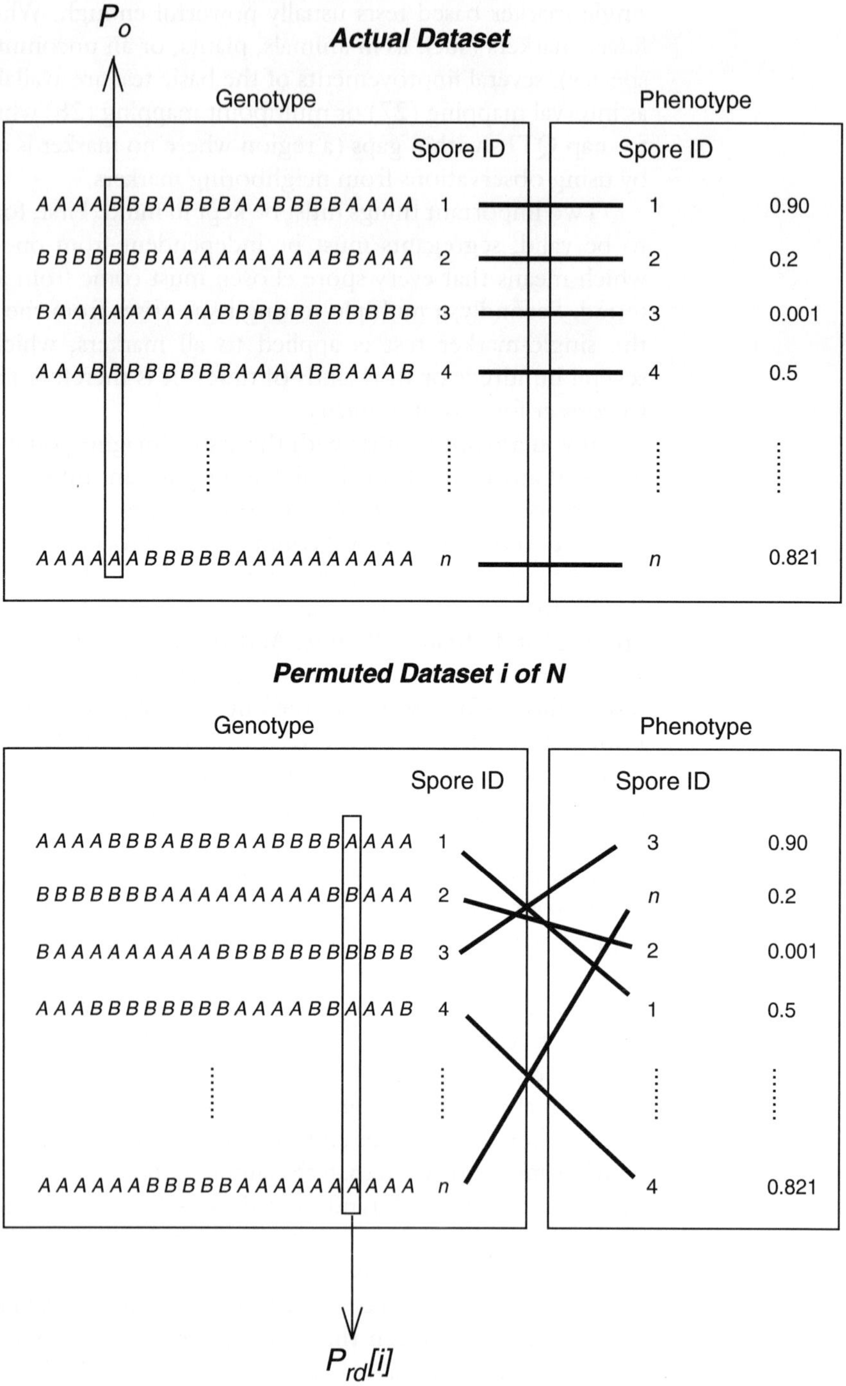

Fig. 2. Permutation test to determine genome-wide significance. The actual dataset is presented on *top*, with genotypes A or B at consecutive markers and a set of phenotypic values for each F1 spores of A $\times$ B. A marker showing the lowest *P*-value (P_0) is *boxed*. *N* permuted datasets are generated by randomly reassigning phenotypic values to spores. One such dataset is represented, where the marker with lowest *P*-value (P_{rd}) is framed. The genome-wide significance of P_0 is then derived from the vector of *N* P_{rd} values as described in text.

As shown in Ansel et al. (24), such a blind QTL-scan is poorly appropriate when considering network stochasticity values, as it is very sensible to the confounding effect of mean expression variation. We therefore recommend to select the spores showing suitable phenotypic values and apply linkage tests on this selected set.

1.3.2. Selective Genotyping

When phenotyping is cheap enough (which is usually the case in yeast experiments), mapping studies can be greatly improved by choosing judicious spores to genotype. In the case of network stochasticity, a powerful approach may be to analyze a large amount of spores (>200) and select one set of about 30 spores all having low noise and a rather similar mean value, and one set of about 30 spores all having high noise and a mean value similar to the first set. This way, genetic factors influencing mean expression are ignored, and the major contributors of noise are focused on. These ~60 spores are then genotyped and used to scan the genetic map as explained above.

1.3.3. Bulk Genotyping

If the phenotyping throughput is high enough, and if genotyping cost is an issue, a nice alternative to selective genotyping is the direct genotyping of many segregants pooled together. This bulk genotyping was efficient in some yeast studies using microarrays and is likely facilitated by direct resequencing of pooled individuals. In this case, a large pool of F1 individuals showing high phenotypic values is obtained (either one by one, or from a population on which the desired phenotypic value has been selected), and the DNA of all individuals is pooled (for example, by pooling equal amounts of cells of each segregant and extracting DNA from this heterogeneous population). The same procedure is applied to a population of F1 strains displaying low phenotypic values. The two DNA samples are then compared on a microarray or by sequencing to search for allele frequencies that best discriminate the phenotypes of two sets (29).

1.4. Genetic Mapping by Introgression

One major advantage of yeast over other eukaryotes is its very short life cycle, allowing introgression to be made very rapidly. This can be used to reduce the size and number of possibly implicated genomic intervals. Mapping by introgression has two advantages over a standard QTL scan: if several phenotypes with confounding effects segregate (e.g., both mean and noise values), one can desire to introgress only the one phenotype of interest. In addition, fewer genotyping is required, which may considerably reduce the overall financial cost of the study. We thus advice to choose this strategy if phenotyping of the network properties of interest is straightforward enough, such as a "noise" estimate in a flow cytometry measurement.

As described in Ansel et al. (24), critical genes of a high noise strain B are introduced into the genome of a low noise strain A

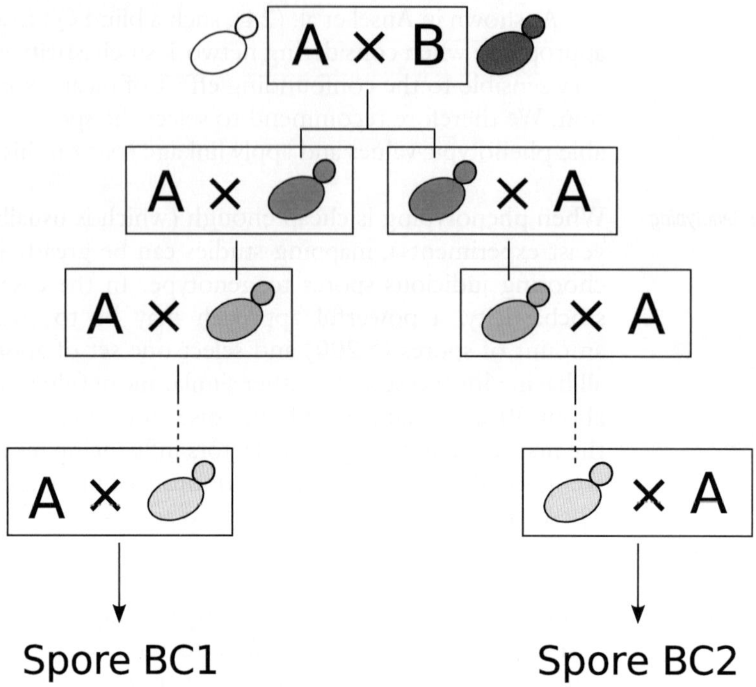

Fig. 3. Introgression design. Two independent introgressions are performed. For each one, every step consists of selecting a spore displaying a stochasticity level similar to the one of strain B and backcrossing it with strain A.

by consecutive backcrosses. First, an F1 progeny from A × B is generated and scored for noise (Fig. 3). Then, one segregant showing the most extreme noise value is backcrossed with strain A and the progeny reanalyzed. Again a segregant with extreme noise is chosen for backcross and so on to gradually reduce the fraction of B's original genome while maintaining as much as possible of B's phenotypic characteristics. The aim is to obtain a strain (BC1) having a minimal amount of the B genome (typically less than 1%) but still its high noise properties. If every intermediate spore is chosen so that its mean GFP expression is similar to the parental values, then one may reasonably assume that the introgressed alleles are selected on the basis of their contribution to noise and not to mean expression, uncoupling the two confounded phenotypes. This entire procedure is then repeated totally independently to obtain a second strain (BC2) having also a minimum amount of the B genome, high noise, and similar mean expression. The two strains are then genotyped to identify their regions inherited from B. For example, if backcross was done at 99%, for any remaining B locus of BC1, the probability that it is also of B genotype in BC2 by chance only is 1% (this assumes total independence, which may be violated if deleterious alleles have been counter-selected in the process). Any region sharing B alleles in BC1 and BC2 is therefore a candidate QTL for noise.

Mapping by introgression is feasible because current technologies offer very dense genotyping capabilities, especially for yeast. For example, BC1 and BC2 spores can be characterized using only two DNA microarrays (one per spore). An algorithm was developed by Gresham et al. (25) to infer the presence of polymorphisms from the hybridization values of genomic DNA to Affymetrix Yeast Tiling Microarrays. These microarrays cover the entire genome in overlapping 25 oligomers that exactly match the reference sequence, the so called perfect match probes (PM). They are spaced every 4 bp on the reference genome (S288c) so that each nucleotide is covered multiple times with a different position relative to the probe. The SNPScanner algorithm (25) then infers sequence differences at every nucleotide position as a log-likelihood ratio comparing two linear models: one assuming the presence of a polymorphism that destabilizes hybridization, and one assuming no polymorphism. High values of this ratio (called prediction signal) are indicative of a polymorphism. At a prediction signal threshold of 5 and in combination with a heuristic filter, the authors could exclude all false positives of a validation set while still retrieving 77.5% of the SNPs.

However, SNPScanner genotyping presents two limitations. First, although very high, genotyping resolution is not single-base and the identity of polymorphisms is not known. This can be frustrating when one wants to know what polymorphisms are likely to be causal. For example, a nonsynonymous SNP or a frame-shift may be favored in comparison to synonymous SNPs. Secondly, any sequence not present in the reference S288c genome is not analyzed because it is not represented on the microarray. This limitation can now be overcome by direct genome resequencing which is becoming affordable for a growing community of researchers.

Once a candidate locus is found by genotyping introgressed strains, more data from further spores is needed to validate the implication of the candidate locus in noise modulation. If DNA polymorphisms are known, restriction fragment length polymorphisms (RFLP) can be used to genotype any additional progeny. To do so, the candidate locus is searched for a sequence polymorphism that changes a restriction site, and primers are designed to amplify this site. Digestion of PCR amplicons then reveals the genotype of the locus in any additional spore. Strain BC1 is backcrossed again with A and 50–100 segregants of this cross is phenotyped for noise and genotyped at the locus using the RFLP marker. The correlation between noise values and genotype at the candidate locus is tested as for single-marker QTL scan, except that no correction for multiple-testing is required since a single test is applied.

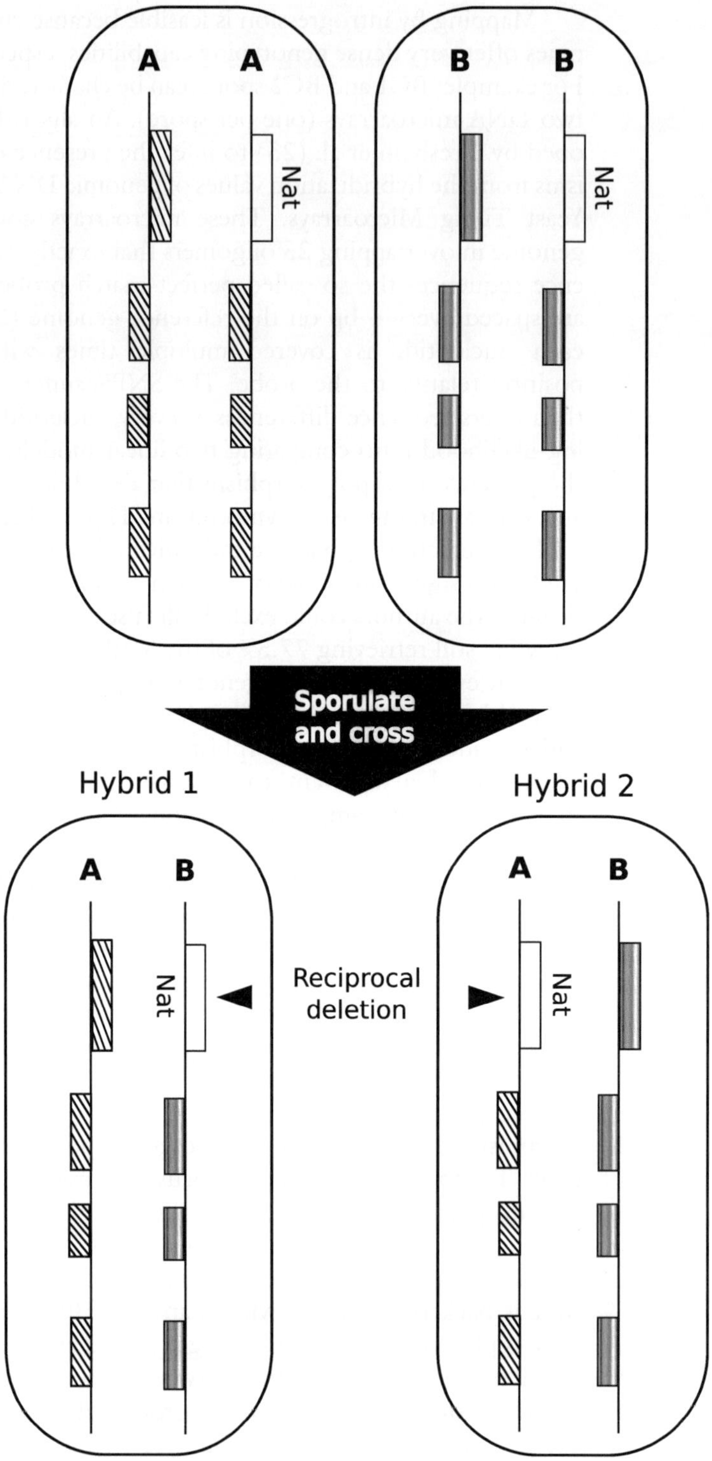

Fig. 4. Reciprocal hemizygous strains. *Boxes* represent genes. Nat, a deletion cassette such as *NatMX4* that disrupts the corresponding gene.

1.5. From a Genetic Locus to the Modulator(s)

The identified locus usually harbors far more than only one gene or predicted ORF. Thus, the locus needs to be dissected into each possible gene products and a technique allowing such an accuracy is reciprocal-hemizygosity analysis (RHA) (30). It is based on hybrids of the two genetic backgrounds hemizygotic for exactly one gene (functional vs. deleted) but otherwise completely isogenic. Reciprocal crosses are performed with the isogenic strains deleted for either the B allele of the gene of interest or for the A form (Fig. 4). Thanks to near-isogenicity, the exact contribution of each allele to the phenotype can be concluded. Since a functional copy of the gene is kept, the effect of each allele can be observed even for essential genes. The main advantage of this method is the ability to test every gene of the locus systematically. One limitation is that some phenotypes are sensible to ploidy, and recourse to diploid hybrids is not always possible.

A nice alternative to test the involvement of a given subregion is allele swapping, which results in a haploid strain isogenic to A except from a small region harboring the B allele. One powerful two-step method to achieve this without cloning efforts is described in Gray et al. (31). First, the URA3 gene is inserted at the locus of interest in a *leu2* strain of background A, and then a PCR fragment carrying the B allele is cotransformed with a LEU2 centromeric plasmid, followed by replica-plating on 5FoA to select recombinational events that excised URA3. Similarly, Storici et al. described the "delitto perfetto" approach to introduce specific mutations or polymorphisms at a precise locus in the genome. They elegantly improved local recombination rate by conditionally forcing double-strand cleavage within the insert (32).

2. Materials

2.1. YPD

1% w/v Bacto Peptone, 1% w/v Bacto Yeast Extract, 2% w/v D(+)-Glucose monohydrate, for agar plates add 2% w/v agar. Autoclave the medium. Nourseothricin-sulfate (Jena Bioscience, Jena, Germany) was added to YPD agar at a temperature of roughly 50°C to a final concentration of 60 mg/l to make YPD + NAT plates.

2.2. Synthetic Medium Without Methionine

0.67% w/v Difco Yeast Nitrogen Base w/o amino acids, 2% w/v D(+)-Glucose monohydrate, 0.2% amino acid mix without methionine (made of 2 g uracil , 4 g leucine, 1 g adenine, and 2 g of each of the following amino acids: A, R, D, N, C, E, Q, G, H, I, K, F, P, S, T, W, Y, V). Autoclave the medium.

2.3. Transformation Reagents

2.3.1. Lithium Acetate Solution

Lithium acetate solution is needed in two concentrations. For the first solution, dissolve 0.1 mol/l LiAc and for the second 1 mol/l LiAc in deionized water, autoclave.

2.3.2. Polyethylene Glycol Solution

Dissolve Polyethylene Glycol MW 3350 (PEG) 50% w/v in deionized water, autoclave. PEG may not entirely dissolve before autoclaving.

2.3.3. Salmon Sperm DNA Solution

Add salmon sperm DNA (SS-DNA) 10 mg/ml to sterile deionized water, boil for 10 min in a water bath to denature the DNA, store in aliquots at −20°C and thaw according to requirements.

2.4. Flow Cytometry

2.4.1. Methionine Stock Solution

Prepare methionine solution 0.1 M in deionized water (see Note 1) and filter sterile with 0.22-μm filter membranes.

2.4.2. Tris Buffered Saline

Tris buffered saline (TBS) consists of 0.3% w/v Tris, 0.8% w/v NaCl, 0.02% w/v KCl, adjust pH to 7.4 with fuming HCl, autoclave.

2.4.3. Flow Cytometer

We are using for all our acquisitions a FACSCalibur flow cytometer (Becton Dickinson, Sparks, MD, USA) and 5-ml Falcon round bottom FACS tubes.

3. Methods

All liquid cultures are grown at 30°C with 220 rpm shaking.

3.1. Transformation of an Nat^R Reporter Construct

Transformation is known to be sometimes mutagenic, and it is therefore important to generate independent transformants of the construct of interest. This way, unknown side-effect mutations differ between the replicates, and if one particular network property (such as high noise) is seen in all replicates, one can attribute it to the original genetic background of the recipient strain. If transformants are isolated on the basis of auxotrophy complementation, several colonies can be picked. But in the case of drug resistance as presented here, the culture is split into three different tubes right after heat-shock (step 11) and one transformant from each subpopulation is selected.

1. Inoculate 5 ml YPD with an isolated colony and grow overnight.
2. The next morning, prepare a tube with 5 ml fresh YPD, add 300 μl of the overnight culture and incubate for 3–5 h (2–4 divisions).

3. Prepare two sterile 1.5-ml microfuge tubes per transformation.
4. Spin the cells down at 1,000 × *g* for 5 min, pour off the medium and resuspend the pellet in 5 ml of sterile deionized water.
5. Prepare the transformation mix described below. In case, one DNA product is transformed into multiple strains the mix can be scaled up.
6. Collect cells at 1,000 × *g* for 5 min, resuspend them in 1 ml lithium acetate 0.1 M and transfer the suspension to one of the prepared 1.5-ml microfuge tube.
7. Cells are pelleted down at maximum speed in a tabletop centrifuge for 15 s. Aspirate supernatant and resuspend the pellet in 200 μl fresh lithium acetate 0.1 M. Keep tubes on ice in case they cannot immediately be processed further.
8. Transfer 50 μl of the LiAc cell suspension to the second prepared microfuge tube and add either 360 μl of a transformation mix consisting of
   (a) 240 μl PEG 50% w/v
   (b) 36 μl Lithium acetate 1.0 M
   (c) 50 μl boiled single-stranded salmon sperm DNA (10 mg/ml)
   (d) 34 μl H_2O + DNA to transform (usually 0.1–10 μg)
   or add the reagents respecting this order. PEG protects cells from the high concentration of LiAc and shall come in first.
9. Vortex well and heat shock the mix in a water bath at 42°C for 40 min.
10. Centrifuge in a tabletop centrifuge for 15 s at 6,500 rpm and remove the transformation mix by aspiration.
11. Gently resuspend the pellet in 300 μl YPD and split it into three independent 15-ml tubes with 2 ml fresh YPD.
12. Incubate the three tubes for 4 h to allow expression of the drug resistance protein.
13. After 4 h, spin the cultures down at 1,000 × *g*, remove the supernatant, resuspend the cells in 100 μl YPD, and plate the cells of each tube on a separate YPD-NAT plate.
14. Keep one colony per plate for your analysis.

3.2. FACS Acquisition

In our case, the MET17 promoter was used to drive GFP expression and our protocol is adapted to the repressive effect of methionine. While readers likely use different reporter constructs, our procedure may give guidelines for other designs.

We detected maximum noise differences when we applied a full repression of the promoter followed by a moderate activation. Methionine concentrations used were determined empirically.

1. Start a culture from an isolated colony in 3 ml YPD and incubate over night.
2. The next day, prepare repression medium [synthetic medium without methionine (SD-M) + methionine stock solution to a final concentration of 1 mM] and activation medium (SD-M + methionine stock solution to a final concentration of 50 μM) (see Note 1).
3. Prepare cell culture tube using 4 ml repression medium.
4. Take the OD_{600} from a dilution of 100 μl of the overnight cell suspension in 900 μl H_2O.
5. Adjust the tube prepared in step 3 to an OD_{600} of 0.1.
6. The culture is incubated for precisely 3 h. Then, harvest the culture at 3,400 × *g* for 5 min, pour the medium off and resuspend the pellet in 4 ml activation medium.
7. Incubate cells for 2 h.
8. In the meanwhile, prepare FACS tubes with 1 ml TBS.
9. For acquisition, dilute the cultures in order to record less than 200 events per second. Collect linear data for the FSC and the Sideward Scatter channels (SSC) and logarithmic data for FL1.
10. Listmode data files can then be imported into the R statistical software environment (http://www.r-project.org/) using the flowCore package from Bioconductor (http://www.bioconductor.org/).

4. Note

1. The methionine solution should not be kept longer than 2 weeks at 4°C and we recommend to add it fresh in SD-M medium right before starting the experiment (possible degradation).

Acknowledgments

Supported by grants ATIP and ANR-07-BLAN-0070 from CNRS and Agence Nationale de la Recherche, France, respectively. S.F. was supported by the Deutscher Akademischer Austausch Dienst (DAAD) foundation, Germany.

References

1. Cappello, M. S., Bleve, G., Grieco, F., Dellaglio, F., and Zacheo, G. (2004) Characterization of Saccharomyces cerevisiae strains isolated from must of grape grown in experimental vineyard, *Journal of applied microbiology* **97**, 1274–1280.
2. Mewes, H. W., Albermann, K., Bahr, M., Frishman, D., Gleissner, A., Hani, J., Heumann, K., Kleine, K., Maierl, A., Oliver, S. G., Pfeiffer, F., and Zollner, A. (1997) Overview of the yeast genome, *Nature* **387**, 7–65.
3. Brachmann, C. B., Davies, A., Cost, G. J., Caputo, E., Li, J., Hieter, P., and Boeke, J. D. (1998) Designer deletion strains derived from Saccharomyces cerevisiae S288C: a useful set of strains and plasmids for PCR–mediated gene disruption and other applications, *Yeast* **14**, 115–132.
4. Winzeler, E. A., Shoemaker, D. D., Astromoff, A., Liang, H., Anderson, K., Andre, B., Bangham, R., Benito, R., Boeke, J. D., Bussey, H., Chu, A. M., Connelly, C., Davis, K., Dietrich, F., Dow, S. W., El Bakkoury, M., Foury, F., Friend, S. H., Gentalen, E., Giaever, G., Hegemann, J. H., Jones, T., Laub, M., Liao, H., Davis, R. W., and et al. (1999) Functional characterization of the S. cerevisiae genome by gene deletion and parallel analysis, *Science* **285**, 901–906.
5. Ghaemmaghami, S., Huh, W. K., Bower, K., Howson, R. W., Belle, A., Dephoure, N., O'Shea, E. K., and Weissman, J. S. (2003) Global analysis of protein expression in yeast, *Nature* **425**, 737–741.
6. Newman, J. R., Ghaemmaghami, S., Ihmels, J., Breslow, D. K., Noble, M., DeRisi, J. L., and Weissman, J. S. (2006) Single–cell proteomic analysis of S. cerevisiae reveals the architecture of biological noise, *Nature* **441**, 840–846.
7. Schacherer, J., Ruderfer, D. M., Gresham, D., Dolinski, K., Botstein, D., and Kruglyak, L. (2007) Genome–wide analysis of nucleotide–level variation in commonly used Saccharomyces cerevisiae strains, *PLoS ONE* **2**, e322.
8. Schacherer, J., Shapiro, J. A., Ruderfer, D. M., and Kruglyak, L. (2009) Comprehensive polymorphism survey elucidates population structure of Saccharomyces cerevisiae, *Nature* **458**, 342–345.
9. Liti, G., Carter, D. M., Moses, A. M., Warringer, J., Parts, L., James, S. A., Davey, R. P., Roberts, I. N., Burt, A., Koufopanou, V., Tsai, I. J., Bergman, C. M., Bensasson, D., O'Kelly, M. J., van Oudenaarden, A., Barton, D. B., Bailes, E., Nguyen, A. N., Jones, M., Quail, M. A., Goodhead, I., Sims, S., Smith, F., Blomberg, A., Durbin, R., and Louis, E. J. (2009) Population genomics of domestic and wild yeasts, *Nature*.
10. Deutschbauer, A. M., and Davis, R. W. (2005) Quantitative trait loci mapped to single–nucleotide resolution in yeast, *Nat Genet* **37**, 1333–1340.
11. Mortimer, R. K., Romano, P., Suzzi, G., and Polsinelli, M. (1994) Genome renewal: a new phenomenon revealed from a genetic study of 43 strains of Saccharomyces cerevisiae derived from natural fermentation of grape musts, *Yeast* **10**, 1543–1552.
12. McCusker, J. H., Clemons, K. V., Stevens, D. A., and Davis, R. W. (1994) Genetic characterization of pathogenic Saccharomyces cerevisiae isolates, *Genetics* **136**, 1261–1269.
13. Fay, J. C., and Benavides, J. A. (2005) Evidence for domesticated and wild populations of Saccharomyces cerevisiae, *PLoS Genet* **1**, 66–71.
14. Ruderfer, D. M., Pratt, S. C., Seidel, H. S., and Kruglyak, L. (2006) Population genomic analysis of outcrossing and recombination in yeast, *Nat Genet* **38**, 1077–1081.
15. Gu, Z., David, L., Petrov, D., Jones, T., Davis, R. W., and Steinmetz, L. M. (2005) Elevated evolutionary rates in the laboratory strain of Saccharomyces cerevisiae, *Proc Natl Acad Sci U S A* **102**, 1092–1097.
16. Voth, W. P., Richards, J. D., Shaw, J. M., and Stillman, D. J. (2001) Yeast vectors for integration at the HO locus, *Nucleic Acids Res* **29**, E59–59.
17. Lo, W. S., and Dranginis, A. M. (1996) FLO11, a yeast gene related to the STA genes, encodes a novel cell surface flocculin, *J Bacteriol* **178**, 7144–7151.
18. Colman–Lerner, A., Chin, T. E., and Brent, R. (2001) Yeast Cbk1 and Mob2 activate daughter–specific genetic programs to induce asymmetric cell fates, *Cell* **107**, 739–750.
19. Yvert, G., Brem, R. B., Whittle, J., Akey, J. M., Foss, E., Smith, E. N., Mackelprang, R., and Kruglyak, L. (2003) Trans–acting regulatory variation in Saccharomyces cerevisiae and the role of transcription factors, *Nat Genet* **35**, 57–64.
20. Brem, R. B., Yvert, G., Clinton, R., and Kruglyak, L. (2002) Genetic dissection of transcriptional regulation in budding yeast, *Science* **296**, 752–755.

21. Ito, H., Fukuda, Y., Murata, K., and Kimura, A. (1983) Transformation of intact yeast cells treated with alkali cations, *J Bacteriol* **153**, 163–168.
22. Thompson, J. R., Register, E., Curotto, J., Kurtz, M., and Kelly, R. (1998) An improved protocol for the preparation of yeast cells for transformation by electroporation, *Yeast* **14**, 565–571.
23. Goldstein, A. L., and McCusker, J. H. (1999) Three new dominant drug resistance cassettes for gene disruption in Saccharomyces cerevisiae, *Yeast* **15**, 1541–1553.
24. Ansel, J., Bottin, H., Rodriguez-Beltran, C., Damon, C., Nagarajan, M., Fehrmann, S., Francois, J., and Yvert, G. (2008) Cell-to-cell stochastic variation in gene expression is a complex genetic trait, *PLoS Genet* **4**, e1000049.
25. Gresham, D., Ruderfer, D. M., Pratt, S. C., Schacherer, J., Dunham, M. J., Botstein, D., and Kruglyak, L. (2006) Genome-wide detection of polymorphisms at nucleotide resolution with a single DNA microarray, *Science* **311**, 1932–1936.
26. Mancera, E., Bourgon, R., Brozzi, A., Huber, W., and Steinmetz, L. M. (2008) High-resolution mapping of meiotic crossovers and non-crossovers in yeast, *Nature* **454**, 479–485.
27. Lander, E. S., and Botstein, D. (1989) Mapping mendelian factors underlying quantitative traits using RFLP linkage maps, *Genetics* **121**, 185–199.
28. Liang, K. Y., Hsu, F. C., Beaty, T. H., and Barnes, K. C. (2001) Multipoint linkage-disequilibrium-mapping approach based on the case-parent trio design, *Am J Hum Genet* **68**, 937–950.
29. Brauer, M. J., Christianson, C. M., Pai, D. A., and Dunham, M. J. (2006) Mapping novel traits by array-assisted bulk segregant analysis in Saccharomyces cerevisiae, *Genetics* **173**, 1813–1816.
30. Steinmetz, L. M., Sinha, H., Richards, D. R., Spiegelman, J. I., Oefner, P. J., McCusker, J. H., and Davis, R. W. (2002) Dissecting the architecture of a quantitative trait locus in yeast, *Nature* **416**, 326–330.
31. Gray, M., Kupiec, M., and Honigberg, S. M. (2004) Site-specific genomic (SSG) and random domain-localized (RDL) mutagenesis in yeast, *BMC Biotechnol* **4**, 7.
32. Storici, F., and Resnick, M. A. (2006) The delitto perfetto approach to in vivo site-directed mutagenesis and chromosome rearrangements with synthetic oligonucleotides in yeast, *Methods Enzymol* **409**, 329–345.

Chapter 12

Functional Mapping of Expression Quantitative Trait Loci that Regulate Oscillatory Gene Expression

Arthur Berg, Ning Li, Chunfa Tong, Zhong Wang, Scott A. Berceli, and Rongling Wu

Abstract

Genetic networks underlying many biological processes, such as vertebrate somitogenesis, cell cycle, hormonal signaling, and circadian rhythms, are characterized by oscillations in gene expression. It has been recognized that the frequency and amplitude of gene expression oscillations vary among individuals and can be controlled by specific expression quantitative trait loci (eQTLs). In this chapter, we develop a dynamic model for mapping and identifying such eQTLs by integrating mathematical aspects of oscillatory dynamics into the functional mapping framework. The model can determine whether and how eQTLs regulate individual genes' activation kinetics and expression dynamics by estimating and testing Fourier series parameters for different eQTL genotypes. We incorporate a general autoregressive moving-average process of order (r,s), the so-called ARMA(r,s), to model the covariance structure for gene expression profiles measured in time course, broadening the applicability of the new dynamic model to mapping eQTLs in practice. The expectation-maximization algorithm (EM algorithm) was derived to estimate all parameters modeling the mean–covariance structures within a mixture model setting. Simulation studies were performed to investigate the statistical behavior of the model. The model will provide a powerful statistical tool for mapping eQTLs and their epistatic interactions that regulate oscillations in gene expression, helping to construct a regulatory genetic network for those periodic biological phenomena.

Key words: Akaike information criterion, Bayesian information criterion, Cell cycle, Fourier series, Quantitative trait locus, Single-nucleotide polymorphism

1. Introduction

All complex traits or diseases are regulated by gene expression. To understand this process, many studies have been initiated to identify regulators, such as transcription factors and their regulatory mechanisms. These studies have been instrumental in improving our understanding of how gene expression is regulated in cells and how its disruption can lead to phenotypic alterations, but they did not attempt to identify genetic variation in gene expression.

Attila Becskei (ed.), *Yeast Genetic Networks: Methods and Protocols*, Methods in Molecular Biology, Vol. 734,
DOI 10.1007/978-1-61779-086-7_12, © Springer Science+Business Media, LLC 2011

In fact, a growing body of evidence has indicated that there is great variability in gene expression level among individuals (1–5). The incorporation of genetic information into gene expression studies has demonstrated tremendous potential to study the genetic basis of variation in gene expression by identifying so-called expression quantitative trait loci (eQTLs) (6). The characterization of eQTLs in the past years has stimulated a fruitful merger of genetics and genomics (7), allowing a deeper understanding of genetic regulatory mechanisms involved in phenotypic formation and development (3).

As transcriptomic technologies become both faster and cheaper in the recent years, it has been increasingly possible to measure time-related fluctuations in gene expression. The studies of time-series gene expression profiles can particularly facilitate the construction of detailed genetic networks underlying many fundamental biological processes, such as vertebrate somitogenesis, cell cycle, hormonal signaling, and circadian rhythms (8–10). With these dynamic gene expression data, we can characterize the behavior of genetic networks using the geometric shape of the time response rather than just its magnitude. There has been a wealth of literature that describes statistical and computational models for clustering the temporal patterns of gene expression profiles based on their physiological function (11–21). More recently, Kim et al. (22) integrated the Fourier series approximation into a mixture model framework for functional clustering of gene expression according to its periodic pattern, allowing a quantitative test of many biologically meaningful characteristics about gene expression trajectories and the duration of biological rhythms. Although it is clear that eQTLs play a ubiquitous role in regulating biological networks, a method for mapping eQTLs involved in rhythmic patterns of gene expression has not been developed thus far.

In this chapter, we develop a new statistical model for functional mapping of rhythmic eQTLs that regulate dynamic gene expression. The central idea of this model lies in the incorporation of Fourier series functions into a mapping framework constructed by a mixture model and the implementation of a general autoregressive moving-average process of order (r,s), known as ARMA (p,q), to model the covariance structure. A sophisticated EM algorithm was derived to estimate the Fourier parameters that define the periodic pattern of gene expression and the ARMA (r,s) parameters. The model has a capacity to test whether a specific eQTL govern transcriptional trajectories and how it affects the temporal pattern of gene expression by testing individual Fourier parameters or their combinations. We show how an eQTL affects quantitative aspects of dynamic expression through numerical simulation. Furthermore, we test the statistical behavior of the model through statistical simulation with parameters based on a real-time course gene expression clustering analysis.

These results are of help to guide the use of the new model to understand developmentally and genetically regulated networks where transcriptional expression is altered during patterning such as those underlying vertebrate somitogenesis.

2. Model

2.1. Likelihood

Suppose there is a natural population from which a set of subjects is sampled randomly for eQTL mapping. All the sampled subjects are genotyped genome wide by single-nucleotide polymorphism (SNP) high-throughput technologies. Also, microarray gene expression is measured at multiple equally spaced time points, $t_1, \ldots, t_M$, during an oscillatory period of interest. Assume that there are n such genes measured for one of the subjects. Kim et al. (22) used the Fourier series approximation to catalog gene expression profiles into distinct groups based on the physiological function of genes. Let us consider one of such gene groups. For subject i, time-dependent expression of a gene from this group is denoted as $y_i = (y_i(t_1), \ldots, y_i(t_M))$. The question we are interested to address is whether there is a specific eQTL that regulate the rhythmic pattern of this gene.

To address this question, we will perform genome-wide association studies of individual SNP markers with gene expression profiles. Assume that there is an eQTL, segregating with two alleles A (in a frequency of q) and a (in a frequency of $1 - q$), responsible for such gene expression profiles. This eQTL, with three genotypes AA (labeled as 2), Aa (labeled as 1), and aa (labeled as 0), is associated with a marker with two alleles M (in a frequency of p) and m (in a frequency of $1 - p$). The linkage disequilibrium between the eQTL and marker is denoted as D. The expression dynamics of the gene is determined by the eQTL with known genotypes but associated with the marker (**M**). Thus, the likelihood of gene expression dynamics is formulated by a mixture model, expressed as:

$$L(\omega, \Theta, \Psi | y, M) = \sum_{i=1}^{n} \sum_{j=0}^{2} \omega_{j|i} f_j(y_i; \Theta_j, \Psi), \tag{1}$$

where $\omega = (\{\omega_j\}_{j=0}^{2})$ is a vector of nonnegative frequencies for three eQTL genotypes that sum to unity, $\Theta = (\{\Theta_j\}_{j=0}^{2})$ contains the parameters that specifically describe the characteristics of the jth eQTL genotype, Ψ is the parameter that common to all the genotypes, and $f_j(y_i; \Theta_j, \Psi)$ denotes the density function for the jth eQTL genotype that is usually assumed to be a multivariate normal with mean vector:

$$\mu_j = (\mu_j(t_1), \ldots, \mu_j(t_M)) \tag{2}$$

and common $M \times M$ covariance matrix

$$\Sigma = \begin{bmatrix} \sigma_1^2 & \sigma_{12} & \ldots & \sigma_{1M} \\ \sigma_{21} & \sigma_2^2 & \ldots & \sigma_{2M} \\ \vdots & \vdots & \ddots & \vdots \\ \sigma_{M1} & \sigma_{M2} & \ldots & \sigma_M^2 \end{bmatrix}. \tag{3}$$

Wang and Wu (23) showed that the frequencies of eQTL genotypes, although unknown, can be inferred from marker genotypes. The eQTL and marker form four haplotypes *MA*, *Ma*, *mA*, and *ma* whose frequencies are $p_{11} = pq + D$, $p_{10} = p(1-q) - D$, $p_{01} = (1-p)q - D$, and $p_{00} = (1-p)\ (1-q) + D$, respectively. Thus, conditional on the marker genotype (*MM*, *Mm*, or *mm*) of the *i*th subject, the probability of eQTL genotype *j* is expressed as $\omega_{j|i}$, which is a function of p_{11}, p_{10}, p_{01}, and p_{00} (see Table 1).

2.2. Modeling the Mean–Covariance Structures

The traditional approach for mapping a QTL with likelihood (Eq. 1) is to estimate each genotypic value at individual time points in mean vector (Eq. 2) and all possible variances and covariances in matrix (Eq. 3). But this approach is statistically not parsimonious because there exists some structure for the mean vector and matrix. Also, in biology, this approach does not make use of mechanistic rules behind in the regulation of temporal gene expression. Here, we will incorporate the idea of functional mapping pioneered by Ma et al. (24) to model the mean vector and covariance structure.

As shown in (22, 25), we use the Fourier series to model time-dependent patterns of gene expression (see also ref. 26). By decomposing the periodic expression level into a sum of the

Table 1
Joint genotype frequencies at the marker and QTL in terms of gametic haplotype frequencies, from which the conditional probabilities of QTL genotypes given marker genotypes can be calculated according to Bayes' theorem

Genotype	Diplotype	*AA* *A*|*A*	*Aa* *A*|*a*+*a*|*A*	*aa* *a*|*a*	Observations
MM	*M*|*M*	p_{11}^2	$2p_{11}p_{10}$	p_{10}^2	n_1
Mm	*M*|*m*	$2p_{11}p_{01}$	$2p_{11}p_{00} + 2p_{10}p_{01}$	$2p_{10}p_{00}$	n_2
mm	*m*|*m*	p_{01}^2	$2p_{01}p_{00}$	p_{00}^2	n_3

orthogonal sinusoidal terms, we have a general form of the Fourier signal as:

$$F(t) = \alpha_0 + \sum_{k=1}^{\infty} \left(\alpha_k \cos\left(\frac{2\pi kt}{\tau}\right) + \beta_k \sin\left(\frac{2\pi kt}{\tau}\right) \right). \quad (4)$$

The coefficients α_k and β_k determine the times at which the expression level achieves maximums and minimums, α_0 is the average expression level of the gene, and τ specifies the periodicity of the regulation. The gene expression value over time can be approximated by partial sum of the Fourier series decomposition where the sum in Eq. 4 only contains K terms. We denote this Fourier series approximation by $F_K(t)$, specifically,

$$F_K(t) = \alpha_0 + \sum_{k=1}^{K} \left(\alpha_k \cos\left(\frac{2\pi kt}{\tau}\right) + \beta_k \sin\left(\frac{2\pi kt}{\tau}\right) \right). \quad (5)$$

We will use this Fourier series approximation to model time-depedent changes of gene expression for each eQTL genotype shown in the mean vector (Eq. 2). The mean value of gene expression for eQTL genotype j at time t_ℓ ($\ell = 1, \ldots, M$) is expressed as $\mu_j(t_l)F_K(t_l; \Theta_j)$, where $\Theta_j = \{\alpha_{0j}, \alpha_{1j}, \ldots, \alpha_{Kj}, \beta_{1j}, \ldots, \beta_{Kj}, \tau_j\}$ denotes the vector of Fourier parameters of the first K orders.

In addition, the covariance structure (Eq. 3) is modeled by a statistical approach. The most convenient method for doing this to use the first-order autoregressive model [AR(1)]. Although the AR(1) covariance matrix has computational advantages due to the existence of closed form expressions of its inverse and determinant, there is a concern about its flexibility being parameterized by only two parameters (typically denoted by σ^2 and ρ). To accommodate more robust covariance structures, Li et al. (25) adopted a flexible approach using the autoregressive moving-average process, ARMA(r,s) (27). The zero-mean random error, ε_{ij}, for subject i with eQTL genotype j is generated according to the following process:

$$\varepsilon_{ij}(t) = \eta_t + \sum_{k=1}^{r} \varphi_k \varepsilon_{ij}(t-k) + \sum_{k=1}^{s} \theta_k \eta_{t-k} \quad (6)$$

where $\varphi_1, \ldots, \varphi_r$ and $\theta_1, \ldots, \theta_s$ are unknown parameters, and $\{\eta_t\}$ is a sequence of independent and identically distributed (iid) normal random variables with zero mean and variance σ^2. Certain restrictions are imposed on the parameters of the ARMA model to insure estimability; further details can be found in (27, 28). The ARMA(p,q) model parameters are listed in $\Psi = \{\varphi_1, \ldots, \varphi_r, \theta_1, \ldots, \theta_s, \sigma^2\}$. The total number of parameters to be estimated with three eQTL genotypes, an ARMA(*r*,*s*) covariance structure, and a Fourier series of degree K comes to $r+s+1+6(K+1)$.

2.3. The EM Algorithm

We propose to obtain the maximum likelihood estimates of ω, Θ, and Ψ in the mixture model (Eq. 1) through an EM algorithm. In the E-step, we calculate the posterior probability that subject i has eQTL genotype j

$$P_{ji} = \frac{\omega_{j|i} f_j(y_i; \Theta_j, \Psi)}{\sum_{j'=0}^{2} \omega_{j'|i} f_{j'}(y_i; \Theta_{j'}, \Psi)}. \tag{7}$$

In the M-step, the posterior probabilities are used to update the parameters. In particular, the haplotype frequencies are calculated as:

$$\hat{p}_{11} = \frac{\left[\sum_{i=1}^{n_1} (2P_{2i} + \hat{P}_{1i}) + \sum_{i=1}^{n_2} (P_{2i} + \phi P_{1i})\right]}{2n}, \tag{8}$$

$$p_{10} = \frac{\left[\sum_{i=1}^{n_1} (P_{1i} + P_{0i}) + \sum_{i=1}^{n_2} (P_{0i} + (1-\phi) P_{1i})\right]}{2n}, \tag{9}$$

$$p_{01} = \frac{\left[\sum_{i=1}^{n_3} (2P_{2i} + P_{1i}) + \sum_{i=1}^{n_2} (P_{2i} + (1-\phi) P_{1i})\right]}{2n}, \tag{10}$$

$$\hat{p}_{00} = \frac{\left[\sum_{i=1}^{n_1} (2P_{2i} + \hat{P}_{1i}) + \sum_{i=1}^{n_2} (P_{2i} + \phi P_{1i})\right]}{2n}, \tag{11}$$

where $\phi = p_{11}p_{00}/(p_{11}p_{00} + p_{10}p_{01})$, and n_2, n_1, and n_0 are the numbers of subjects with marker genotypes *MM*, *Mm*, and *mm*, respectively. We then update the parameters p, q, and D as follows:

$$\begin{aligned} p &= p_{11} + p_{10}, \\ q &= p_{11} + p_{01}, \\ D &= p_{11}p_{00} - p_{10}p_{01}. \end{aligned} \tag{12}$$

The parameters Θ and Ψ can be estimated using the procedure described in the Appendix.

2.4. Model Selection

In practice, we do not know the best order of K to approximate the Fourier series for genotypic mean vector (Eq. 1) and the best order of p and q in the ARMA covariance structure (Eq. 2). The best fit to the data in terms of K, p, and q can be identified using the Akaike information criterion (AIC) (29) and the Bayesian information criterion (BIC) (30), which are defined as follows:

$$\mathrm{AIC}(K, r, s) = -2 \log L_c(\hat{\Theta}_K, \hat{\Psi}_{r,s}|y) + 2N(K, r, s)$$

$$\mathrm{BIC}(K, r, s) = -2 \log L_c(\hat{\Theta}_K, \hat{\Psi}_{r,s}|y) + \log(n) N(K, r, s)$$

where $\hat{\Theta}_K$ and $\hat{\Psi}_{p,q}$ are the maximum likelihood estimate of Θ and Ψ under orders K and (r,s), respectively, and $N(K,r,s)$ is the number of parameters in the mixture model determined by K and (r,s). The selected model has the smallest AIC and BIC.

Under the framework of maximum likelihood estimation, it is possible that the likelihood increases when more parameters are added into the model, which could lead to overfitting. Both AIC and BIC resolve this problem by including a penalty term for the number of parameters, but BIC imposes a stronger penalty than AIC, and as a result, it tends to select models with smaller number of parameters than those chosen by AIC method.

The dimension of our model parameters can be viewed as growing in two directions, one determined by K and the other by (r,s). A one unit increase in J gives arise to the addition of $2K+2$ parameters, which is always larger than a one unit increase in $r+s$, we propose a three-step procedure to select the best model. First, we fit an ARMA covariance structure with relatively low orders (r,s) to $\sum$, i.e., ARMA(1,0) or ARMA(1,1), and calculate AIC or BIC values by varying K starting from 1. The model with the smallest AIC or BIC is identified. We denote the corresponding K as K_b. We then fit the Fourier series with K_s orders, but this time vary (r,s) to find the best combination (r_b,s_b). In the third step, we go back to step 1 and refit the model with ARMA(r_b,r_b) and select K again. The resulting model with the smallest AIC or BIC is our final choice. Alternative to the three-step procedure, if the amount of computation is not a limiting factor, one could simply calculate the AIC or BIC values for all models under consideration and select the model that minimizes the criterion of choice.

2.5. Hypothesis Tests

The existence of an eQTL that regulates transcriptional expression profiles can be tested by formulating the following hypothesis:

$$\begin{aligned} H_0 &: \Theta_j \equiv \Theta \quad j = 0,\ 1,\ 2 \\ H_1 &: \text{At least one of the equalities above does not hold,} \end{aligned} \tag{13}$$

where Θ is the vector of the Fourier series parameters when there is no eQTL for the given data. The likelihood ratio test statistic can be calculated under the null and alternative hypotheses; that is,

$$L_r = -2\left[\log L\left(\hat{\Theta}_{H_0}|y\right) - \log L\left(\hat{\Theta}|y\right)\right],$$

where $\hat{\Theta}_{H_0}$ and $\hat{\Theta}$ stand for the MLEs of the parameters under the null hypothesis and the alternative, respectively.

Since there is no closed-from distribution for L_r, the critical value for claiming the existence of at least two different expression patterns is determined by a parametric bootstrap method. We simulate n gene expression profiles at the observed time points under the multivariate normal model indicated by the null hypothesis. The true values of the parameters in the simulation are taken to be the MLEs under the null hypothesis, i.e., $\hat{\Theta}_{H_0}$. For each simulated dataset, the likelihood ratio test statistic L_r is calculated by fitting the models under the null and the alternative hypotheses. This procedure is repeated for a large number of times, say 1,000, and the 95th percentile of the empirical distribution of L_r is then regarded as the critical value of the test (Eq. 13).

3. Computer Simulation

Simulation studies were performed to test the statistical behavior of the model. We simulated a sample of subjects from a natural Hardy–Weinberg equilibrium population, in which the cosegregation between a marker and QTL was simulated by a given allele frequency for the marker (p) and QTL (q), respectively, and their linkage disequilibrium coefficient (D). Marker and QTL genotypes were then simulated according to the joint probabilities described in Table 1, from which the numbers of each genotype at the marker can be obtained for a given total size of samples.

We assume that the QTL (whose genotypes are unknown) controls the time-dependent pattern of gene expression profiles. These profiles were determined using the parameters estimated from a real-world time-course gene expression functional clustering. Rustici et al. (9) published a time-course microarray experiment, in which 407 periodically expressed genes regulating the genome-wide transcriptional program of the *Schizosaccharomyces pombe* cell cycle were clustered into several major waves of expression. Li et al. (24) applied Fourier clustering techniques to analyze this data, leading to the identification of nine distinct time-dependent clusters. In this study, we chose three of these clusters, each corresponding to a different QTL genotype (QQ, Qq, or qq), to simulate periodically expressed gene profiles using the Fourier parameters of each cluster.

Given a QTL genotype, time-course gene expression data with 21 time points were simulated by adding correlated residual noise to the mean expression profile that corresponds to the QTL genotype. The three mean curves and parameters for the correlated residual noise are taken from the real data analysis mentioned above. Specifically, the mean curves are parameterized by an order-two truncated Fourier series and the residual error is generated from an order-two autoregressive process with parameters $(\sigma, \phi_1, \phi_2) = (0.25, 0.515, 0.081)$.

To give a clearer picture of the simulated data, one instance of the simulated data with parameters $n = 200$, $p = 0.6$, $q = 0.6$, $D = 0.05$, $\sigma = 0.25$, $\phi_1 = 0.515$, and $\phi_2 = 0.081$ was generated and graphed in Fig. 1 together with the true mean curves.

Three levels of marker-QTL associations were considered in the simulations by varying LD $D = 0.1$, 0.05, and 0, which correspond to normalized linkage disequilibrium of 0.417, 0.208, and 0. Two sample sizes were considered, $n = 200$ and $n = 400$. The mean

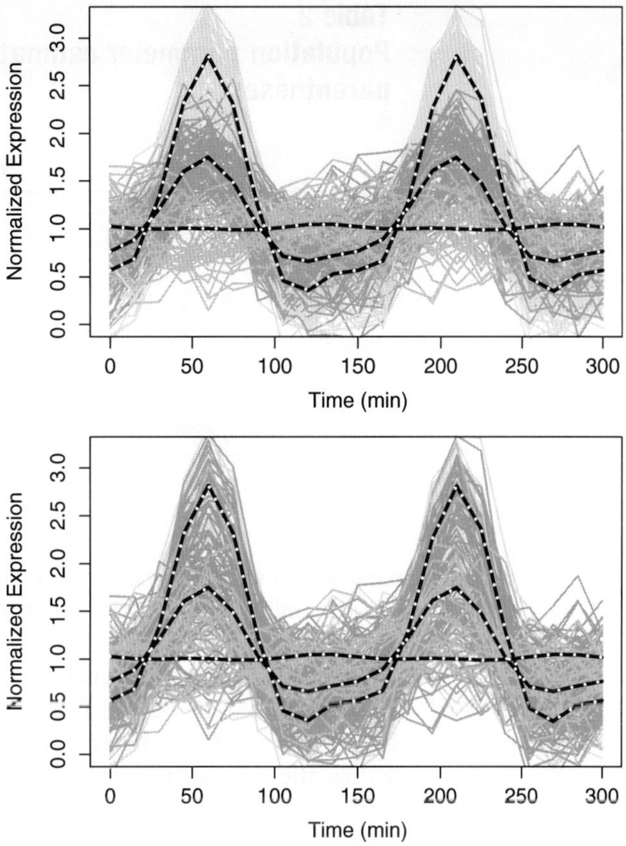

Fig. 1. Graph of simulated data with parameters $n = 100$, $p = 0.6$, $q = 0.6$, $D = 0.05$, $\sigma = 0.2$, $\phi_1 = 0.515$, $\phi_2 = 0.081$, and *gray-level coloring* is according to (*top*) unobserved QTL genotype and (*bottom*) observed marker genotype.

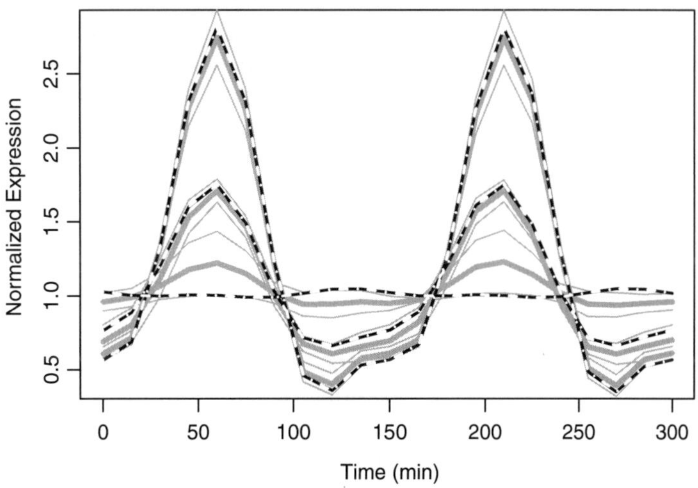

Fig. 2. Mean cuver estimates of clusters with sample size $n = 200$ and linkage disequilibrium $D = 0.05$. The true means are drawn in *black* with *white dashes*, the estimated means are drawn in *gray*, and the estimated standard errors are drawn with a *thinner gray line* above and below the estimated means.

Table 2
Population parameter estimates with standard errors in parentheses

	p (0.6)	q (0.6)	D
$n=200$			
$D=0.1$	0.603 (0.032)	0.581 (0.029)	0.084 (0.025)
$D=0.05$	0.604 (0.026)	0.590 (0.043)	0.051 (0.023)
$D=0$	0.592 (0.021)	0.596 (0.012)	0.006 (0.016)
$n=400$			
$D=0.1$	0.596 (0.017)	0.597 (0.018)	0.094 (0.005)
$D=0.05$	0.600 (0.024)	0.588 (0.029)	0.047 (0.017)
$D=0$	0.603 (0.006)	0.593 (0.038)	0.002 (0.010)

parameter estimates consistently converged to the correct mean curve profiles. There is little variation in the mean curves across the levels of linkage disequilibrium and two sample sizes, therefore a single graphic of the mean curves is provided in Fig. 2 corresponding to $n=200$ and $D=0.05$. The mean curves were computed over 20 realizations with the true means drawn in black with white dashes, the estimated means drawn in gray, and the estimated standard errors drawn with a thinner gray line above and below the estimated means. The results from the simulation studies suggest that different curves of QTL genotypes can be reasonably identified and estimated.

Table 2 gives mean estimates of the population parameters p, q, and D over the three levels of linkage disequilibria and two different sample sizes. Estimation of the allele frequency is naturally accurate independent of the linkage disequilibrium and for any modest sample size, but the estimation of the QTL allele frequency and linkage disequilibrium parameter considerably improve with the sample size.

4. Discussion

Microarray technologies allow a large-scale measure of gene expression for large numbers of individuals. This will not only provide a tool to construct a detailed map of expressed genes in various tissues and diseases, but also be powerful to study the genetic variation of gene expression among individuals. With the increasing availability of time series gene expression data derived from faster and cheaper transcriptomic technologies, microarray technologies will also be helpful to address fundamental questions related to the genetic regulation of the developmental processes of tissues and diseases, such as vertebrate somitogenesis, cell cycle, hormonal signaling, and circadian rhythms (8–10). In this chapter, we have developed a statistical model for mapping specific genes or expression quantitative trait loci (eQTLs) that control temporal patterns of gene expression level in a time course.

The model integrates the Fourier series function to approximate periodic changes of gene expression for individual eQTL genotypes. By testing the differences in Fourier series parameters and their combination among eQTL genotypes, the model can detect the existence of a significant eQTL responsible for dynamic gene expression profiles and further study the developmental interplay between genetic action and gene expression. The model capitalizes on the parsimonious modeling of the covariance structure, increasing the power of the model to detect significant eQTLs. Simulation studies have been used to test and validate the usefulness and utilization of the model. It anticipates that the model has potential to handle of real data collected from a practical dynamic genomic study.

In this study, we founded the model on individual genes that are expressed periodically. For a given gene, the model is able to discern the eQTL that controls the dynamic expression of the gene. In practice, hundreds of thousands of genes are often measured simultaneously on timescale. Because of their distinct biological functions, these genes follow different dynamic patterns of gene expression. We can first implement the functional clustering of Kim et al. (22) to categorize genes into different groups and then use our functional mapping model to map eQTLs for a single gene from individual groups. As compared withy this two-stage mapping model, we can develop an alternative that combines functional clustering and functional mapping into a two-stage hierarchic mixture framework. We expect that the combining approach will have better power to test whether and how an eQTL determines the temporal expression of genes in a complex genetic network.

5. Supplementary Materials

The source code implementation of the methods of this chapter is accesible on the web at http://statgen.psu.edu/software.html.

Acknowledgments

This work is supported by NSF/NIH joint grant DMS/NIGMS-0540745 and NIH ARRA grant 09095.

Appendix

Here, we give the procedure for estimating parameters in $\Lambda = (\Theta, \Psi)$ within the EM algorithm framework. In the E-step, the posterior expectation of z_{ij} is evaluated as:

$$P_{j|i} = E[z_{ij}|\Lambda, y_i] = \Pr[z_{ij} = 1|\Lambda, y_i] = \frac{\omega_{j|i} f_j(y_i; \mu_j, \Sigma)}{\sum_{j'=0}^{2} \omega_{j'|i} f_{j'}(y_i; \mu_j', \Sigma_i)}, \quad (14)$$

where we assume that the covariance matrix is subject specific.

In the M-step, closed form solutions exist for ω (see Eqs. 8–11) and the parameters in Λ except for τ and σ^2.

Suppose the gene expression trajectory is approximated by the first K orders of the Fourier series, then $\Lambda_j = (c_j, \tau_j)$, where $c_j = (\alpha_{0j}, \alpha_{1j}, \beta_{1j}, \ldots, \alpha_{Kj}, \beta_{Kj})$. We have

$$\frac{\partial \log L_c(\Lambda|y)}{\partial c_j} = \left[\frac{\partial \log L_c(\Lambda|y)}{\partial \mu_{ij}}\right]\left[\frac{\partial \mu_j}{\partial c_j}\right]. \quad (15)$$

The parameter c_j can be updated by setting (Eq. 15) to zero. Since

$$\frac{\partial \log L_c(\Lambda|y)}{\partial \mu_j} = \sum_{i=1}^{n} P_{j|i}(y_i - \mu_j)^T \Sigma_i^{-1}$$

and $\partial \mu_j / \partial c_j = D_i(\tau_j)$, where

$$D_i(\tau_j) = \begin{pmatrix} 1 & \cos\left(\frac{2\pi t_{i1}}{\tau_j}\right) & \sin\left(\frac{2\pi t_{i1}}{\tau_j}\right) & \cdots & \cos\left(\frac{2\pi K t_{i1}}{\tau_j}\right) & \sin\left(\frac{2\pi K t_{i1}}{\tau_j}\right) \\ 1 & \cos\left(\frac{2\pi t_{i2}}{\tau_j}\right) & \sin\left(\frac{2\pi t_{i2}}{\tau_j}\right) & \cdots & \cos\left(\frac{2\pi K t_{i2}}{\tau_j}\right) & \sin\left(\frac{2\pi K t_{i2}}{\tau_j}\right) \\ \vdots & \vdots & \vdots & \vdots & \vdots & \vdots \\ 1 & \cos\left(\frac{2\pi t_{im_i}}{\tau_j}\right) & \sin\left(\frac{2\pi t_{im_i}}{\tau_j}\right) & \cdots & \cos\left(\frac{2\pi K t_{im_j}}{\tau_j}\right) & \sin\left(\frac{2\pi K t_{im_j}}{\tau_j}\right) \end{pmatrix},$$

we have

$$\hat{c}_j = \left[\sum_{i=1}^{n} P_{j|i} D_i(\tau_j)^T \Sigma_i^{-1} D_i(\tau_j)\right]^{-1} \left[\sum_{i=1}^{n} P_{j|i} y_i^T \Sigma_i^{-1} D_i(\tau_j)\right].$$

Since the analytical form of the inverse of Σ_i is not available, we use the recursive method to calculate the inverse matrix of ARMA(p,q) through its association with ARMA(p,$q-1$).

We can write $\Sigma_i = \sigma^2 R_i$, where R_i is the correlation matrix that is entirely determined by the ARMA parameters $\varphi_1, \ldots, \varphi_p, \theta_1, \ldots, \theta_q$. The variance σ^2 can be updated by:

$$\hat{\sigma}^2 = \frac{\sum_{i=1}^{n} \sum_{j=1}^{J} P_{j|i}(y_i - \mu_j)^T R_i^{-1}(y_i - \mu_j)}{\sum_{i=1}^{n} m_i}. \quad (16)$$

Again R_i^{-1} can be calculated by the method of Haddad (28).

Because there are no closed form solutions for τ_j and ARMA parameters $\varphi_1, \ldots, \varphi_p$ and $\theta_1, \ldots, \theta_q$, their estimates are updated using one-step Newton–Raphson method within each iteration. In particular, in the ($v+$ 1)th iteration, τ_j can be updated by:

$$\tau_j^{v+1} = \tau_j^v - \frac{\partial/\partial\tau_j \log L_c(\Lambda|y)|_{\Lambda=\Lambda^v}}{\partial^2/\partial\tau_j^2 \log L_c(\Lambda|y)|_{\Lambda=\Lambda^v}}, \quad (17)$$

where

$$\frac{\partial}{\partial\tau_j}\log L_c(\Lambda|y) = \sum_{i=1}^{n} P_{j|i}(y_i - \mu_j)^T \Sigma_i^{-1}\delta_{ij}$$

with δ_{ij} being a $m_i \times 1$ vector whose components

$$\delta_{ijl} = \sum_{k=1}^{K}\left[\alpha_{kj}\sin\left(\frac{2\pi kt_l}{\tau_j}\right)\frac{2\pi kt_l}{\tau_j^2} - \beta_{kj}\cos\left(\frac{2\pi kt_l}{\tau_j}\right)\frac{2\pi kt_l}{\tau_j^2}\right],$$

and

$$\frac{\partial^2}{\partial\tau_j^2}\log L_c(\Lambda|y) = \sum_{i=1}^{n}\left[-P_{j|i}\delta_{ij}^T\Sigma_i^{-1}\delta_{ij} + P_{j|i}(y_i - \mu_j)^T\Sigma_i^{-1}\frac{\partial^2}{\partial\tau_j^2}\mu_j\right].$$

Similarly, the parameters $\varphi_1, \ldots, \varphi_p$ and $\theta_1, \ldots, \theta_q$ can be updated by the one-step Newton–Raphson method outlined above. However, there are no analytical forms of the first and the second derivatives of the expected complete data log-likelihood with respect to the φ's and θ's, we use the numerical differentiation method to calculate these quantities. To ease the presentation of the method, denote the $(p+q)$ dimensional vector $\psi = (\varphi_1, \ldots, \varphi_p, \theta_1, \ldots, \theta_q)$. The first and the second derivatives with respect to the Kth component in Ψ are approximated, respectively, by:

$$\frac{E[\log L_c(\Lambda_{-\psi}, \psi + h_n e_K|y)] - E[\log L_c(\Lambda|y)]}{h_n}, \quad (18)$$

and

$$\frac{E[\log L_c(\Lambda_{-\psi},\psi + h_n e_K|y)] - 2E[\log L_c(\Lambda|y)] + E[\log L_c(\Lambda_{-\psi},\psi - h_n e_K|y)]}{h_n^2}, \quad (19)$$

where we use E to represent the posterior expectation of the complete data log-likelihood with respect to, $\Lambda_{-\psi}$ denotes the parameters in Λ other than ψ, the $(p+q)$ vector e has unity length with the Kth component set to 1, and h_n is the bandwidth chosen by the investigator. When h_n is small enough, the numerical differentiation approximates the true derivatives adequately. On the other hand, if h_n is too small, the random errors from the numerical computation may deteriorate the results.

References

1. Brem, R., Yvert, G., Clinton, R., and Kruglyak, L. (2002) Genetic dissection of transcriptional regulation in budding yeast. *Science* **296**, 752–755.
2. Cheung, V. and Spielman, R. (2002) The genetics of variation in gene expression. *Nature Genetics* **32**, 522–525.
3. Cheung, V. and Spielman, R. (2009) Genetics of human gene expression: mapping DNA variants that influence gene expression. *Nature Reviews Genetics* **10**, 595–604.
4. Cheung, V., Conlin, L., Weber, T., Arcaro, M., Jen, K., Morley, M., and Spielman, R. (2003) Natural variation in human gene expression assessed in lymphoblastoid cells. *Nature Genetics* **33**, 422–425.
5. Cookson, W., Liang, L., Abecasis, G., Moffatt, M., and Lathrop, M. (2009) Mapping complex disease traits with global gene expression. *Nature Reviews Genetics* **10**, 184–194.
6. Schadt, E., Monks, S., Drake, T., Lusis, A., Che, N., Colinayo, V., Ruff, T., Milligan, S., Lamb, J., Cavet, G., et al. (2003) Genetics of gene expression surveyed in maize, mouse and man. *Nature* **422**, 297–302.
7. Jansen, R. and Nap, J. (2001) Genetical genomics: the added value from segregation. *Trends in Genetics* **17**, 388–391.
8. Goldbeter, A. (2002) Computational approaches to cellular rhythms. *Nature* **420**, 238–245.
9. Rustici, G., Mata, J., Kivinen, K., Lió, P., Penkett, C., Burns, G., Hayles, J., Brazma, A., Nurse, P., and Bahler, J. (2004) Periodic gene expression program of the fission yeast cell cycle. *Nature Genetics* **36**, 809–817.
10. Swinburne, I., Miguez, D., Landgraf, D., and Silver, P. (2008) Intron length increases oscillatory periods of gene expression in animal cells. *Genes & Development* **22**, 2342–2346.
11. Holter, N., Maritan, A., Cieplak, M., Fedoroff, N., and Banavar, J. (2001) Dynamic modeling of gene expression data. *Proceedings of the National Academy of Sciences of the United States of America* **98**, 1693–1698.
12. Qian, J., Dolled-Filhart, M., Lin, J., Yu, H., and Gerstein, M. (2001) Beyond synexpression relationships: local clustering of time-shifted and inverted gene expression profiles identifies new, biologically relevant interactions. *Journal of Molecular Biology* **314**, 1053–1066.
13. Bar-Joseph, Z., Gerber, G., Gifford, D., Jaakkola, T., and Simon, I. (2003) Continuous representations of time-series gene expression data. *Journal of Computational Biology* **10**, 341–356.
14. Luan, Y. and Li, H. (2003) Clustering of time-course gene expression data using a mixed-effects model with B-splines. *Bioinformatics* **19**, 474–482.
15. Park, T., Yi, S., Lee, S., Lee, S., Yoo, D., Ahn, J., and Lee, Y. (2003) Statistical tests for identifying differentially expressed genes in time-course microarray experiments. *Bioinformatics* **19**, 694–703.
16. Wakefield, J., Zhou, C., and Self, S. (2003) Modelling gene expression over time: curve clustering with informative prior distributions. *Bayesian Statistics* 7, 721–732.
17. Ernst, J., Nau, G., and Bar-Joseph, Z. (2005) Clustering short time series gene expression data. *Bioinformatics* **21**, 159–168.
18. Storey, J., Xiao, W., Leek, J., Tompkins, R., and Davis, R. (2005) Significance analysis of time course microarray experiments. *Proceedings of the National Academy of Sciences of the United States of America* **102**, 12837–12842.
19. Ma, P., Castillo-Davis, C., Zhong, W., and Liu, J. (2006) A data-driven clustering method for time course gene expression data. *Nucleic Acids Research* **34**, 1261–1269.
20. Ng, S., McLachlan, G., Wang, K., Ben-Tovim Jones, L., and Ng, S. (2006) A mixture model with random-effects components for clustering correlated gene-expression profiles. *Bioinformatics* **22**, 1745–1752.
21. Inoue, L., Neira, M., Nelson, C., Gleave, M., and Etzioni, R. (2007) Cluster-based network model for time-course gene expression data. *Biostatistics* **8**, 507–525.
22. Kim, B., Zhang, L., Berg, A., Fan, J., and Wu, R. (2008) A Computational Approach to the Functional Clustering of Periodic Gene Expression Profiles. *Genetics* **180**, 821–834.
23. Wang, Z. and Wu, R. (2004) A statistical model for high-resolution mapping of quantitative trait loci determining human HIV-1 dynamics. *Statistics in Medicine* **23**, 3033–3051.
24. Ma, C., Casella, G., and Wu, R. (2002) Functional mapping of quantitative trait loci underlying the character process: a theoretical framework. *Genetics* **161**, 1751–1762.
25. Li, N., McMurry, T., Berg, A., Zhong, W., Berceli, S., and Wu. (2010 (accepted)) Functional clustering of periodic transcriptional profiles through arma(p,q). *PloS ONE* **5**, e9894.
26. Spellman, P., Sherlock, G., Zhang, M., Iyer, V., Anders, K., Eisen, M., Brown, P., Botstein, D., and Futcher, B. (1998) Comprehensive identification of cell cycle-regulated genes of the yeast Saccharomyces cerevisiae by microarray hybridization. *Molecular biology of the cell* **9**, 3273–3297.

27. Brockwell, P. and Davis, R. (1991) *Time series: theory and methods.* (Springer).
28. Haddad, J. (2004) On the closed form of the covariance matrix and its inverse of the causal ARMA process. *Journal of Time Series Analysis* **25**, 443–448.
29. Akaike, H. (1974) A new look at the statistical model identification. *IEEE Transactions on Automatic Control* **19**, 716–723.
30. Schwarz, G. (1978) Estimating the dimension of a model. *The Annals of Statistics* **6**, 461–464.

Chapter 13

Evolutionary Aspects of a Genetic Network: Studying the Lactose/Galactose Regulon of *Kluyveromyces lactis*

Alexander Anders and Karin D. Breunig

Abstract

The budding yeast *Kluyveromyces lactis* has diverged from the *Saccharomyces* lineage before the whole-genome duplication and its genome sequence reveals lower redundancy of many genes. Moreover, it shows lower preference for fermentative carbon metabolism and a broader substrate spectrum making it a particularly rewarding system for comparative and evolutionary studies of carbon-regulated genetic networks. The lactose/galactose regulon of *K. lactis,* which is regulated by the prototypic transcription activator Gal4 exemplifies important aspects of network evolution when compared with the model *GAL* regulon of *Saccharomyces cerevisiae.* Differences in physiology relate to different subcellular compartmentation of regulatory components and, importantly, to quantitative differences in protein–protein interactions rather than major differences in network architecture. Here, we introduce genetic and biochemical tools to study *K. lactis* in general and the lactose/galactose regulon in particular. We present methods to quantify relevant protein–protein interactions in that network and to visualize such differences in simple plate assays allowing for genetic approaches in further studies.

Key words: *Kluyveromyces lactis*, Lactose metabolism, Galactose, GAL4, GAL80, GAL1, Galactokinase, Transcription regulation, Yeast genetics

1. Introduction

Genetic networks reflect the adaptation of an organism to a particular environment. The architecture and plasticity of the network have evolved in response to the likelihood and amplitude of changes in the environment. One approach to get insight into the adaptive strategies and mechanisms of network evolution is based on qualitative and quantitative comparison of genetic interactions in homologous networks between organisms that have adapted to different environments. Comparative analysis in different ascomycetes has been particularly rewarding in this respect (1–4). This chapter focuses on the lactose/galactose regulon of the yeast *Kluyveromyces*

Attila Becskei (ed.), *Yeast Genetic Networks: Methods and Protocols*, Methods in Molecular Biology, Vol. 734, DOI 10.1007/978-1-61779-086-7_13,

lactis and its relationship to the intensely studied *GAL* regulon of *Saccharomyces cerevisiae.* It introduces readers acquainted with *S. cerevisiae* to working with *K. lactis.* Specifically, it presents methods to characterize genetically and biochemically the components of the Gal4-regulated transcriptional network.

The *S. cerevisiae GAL* regulon has been subject to several mathematical modeling approaches based on the wealth of experimental data that describe the regulation of the galactose metabolic genes (5–7). The regulatory module senses galactose and consists of the transcription activator Gal4, its inhibitor Gal80, and a galactose sensor molecule, Gal3 (see (8) for a recent review). The three components control each other's respective activity through protein–protein interactions and positive and negative feedback regulation of transcription. A related network is found in *K. lactis* where it not only regulates galactose but also lactose metabolic genes (Fig. 1).

K. lactis diverged from the *Saccharomyces* lineage before the whole-genome duplication (WGD) occurred, and several genetic redundancies found in *S. cerevisiae* are lacking in this organism (10, 11). A prominent example is the *GAL1–GAL3* gene duplication found in *S. cerevisiae* but not in *K. lactis.* Subfunctionalization of these highly conserved genes was revealed by comparison with *K. lactis* (12, 13). Whereas the *K. lactis GAL1* gene (*KlGAL1)* encodes a bifunctional protein, which on the one hand binds Gal80 and thereby activates Gal4 (Fig. 2), and on the other hand catalyzes galactose phosphorylation, the first step in galactose utilization, in *S. cerevisiae* these two functions are supplied by the *GAL3* and *GAL1* genes, respectively (see Fig. 1). Gal3 has lost enzymatic activity during evolution but could be partially reactivated by the insertion of two amino acids (15). Gal1

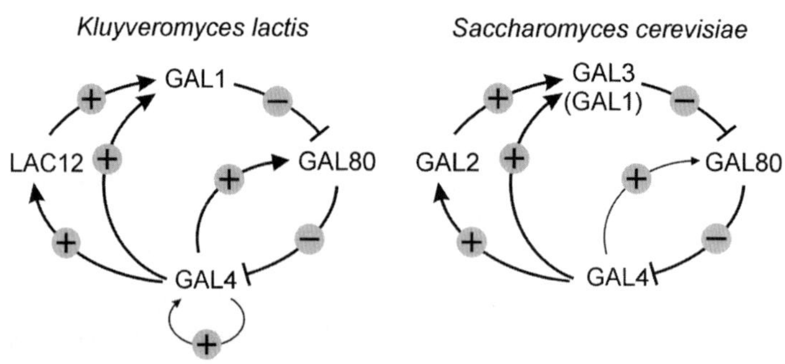

Fig. 1. The regulatory circuits that control Gal4 activity in *Kluyveromyces lactis* and *Saccharomyces cerevisiae.* Interconnected positive and negative feedback loops are important to adjust Gal4 activity to the availability of intracellular galactose in both yeasts. *K. lactis* differs from *S. cerevisiae* by a stronger negative feedback on Gal4 caused by *GAL80* gene induction indicated by the line width, by the active uptake of galactose and lactose via the *LAC12*-encoded lactose permease, by the nonredundant bifunctional *GAL1* gene product and autoinduction of the *KlGAL4* gene (9).

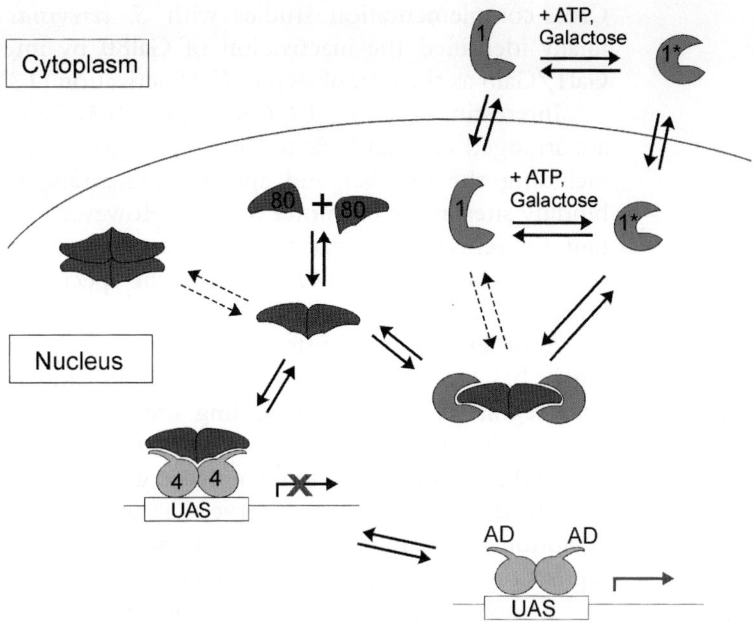

Fig. 2. Mechanistic model for the galactose switch in *K. lactis*. Monomeric KlGal1 (marked 1) enters the nucleus KlGal80 independently, where it can interact with dimeric KlGal80 (80). Binding of its substrates galactose and ATP increases the affinity for KlGal80. KlGal1–KlGal80 complex formation in the nucleus (14) prevents interaction of KlGal80 with the activation domain (AD) of KlGal4 (4) thus stimulating KlGal4-activated transcription. Binding of dimeric KlGal80 to KlGal4 is highly cooperative. Higher KlGal80 oligomerization occurs and is incompatible with KlGal1 interaction (but may be compatible with KlGal4 interaction). *Broken arrows* denote weak interactions. *UAS* upstream activating sequence (reproduced from ref. 14 with permission from Journal of Biological Chemistry).

has retained some regulatory function and can partially substitute for Gal3 at elevated protein concentrations.

As the name indicates, *K. lactis* has become specialized for growth in lactose-rich environments and can be isolated from dairy products. In contrast to bacteria, lactose utilization is a rare trait among eukaryotic microorganisms in general and hemiascomycetes in particular. Not only *S. cerevisiae* but most yeasts are unable to assimilate lactose, probably due to the loss of the metabolic genes. The genetic basis of lactose metabolism has been most extensively studied in *K. lactis* var. *lactis*. *K. lactis* var. *drosophilarum* is considered a separate variety based on its lactose-negative phenotype (16, 17). Two coupled genes *LAC12* (KLLA0B14861g) and *LAC4* (KLLA0B14883fg) encoding a lactose permease and a eukaryotic β-galactosidase, respectively, are key to lactose uptake and hydrolysis. These genes are located at a subtelomeric position on chromosome II of the *K. lactis* genome. This location may explain loss of the Lac^{+} phenotype in individual strains.

The *LAC* genes are coregulated with the *GAL* genes by the *K. lactis* homologue of Gal4, Lac9 (or KlGal4). KlGal4 regulation has been studied extensively due to the relative ease to quantify β-galactosidase in cell extracts or even in living cells (18–21).

Cross-complementation studies with *S. cerevisiae* genes have finally identified the inactivation of Gal80 by interaction with Gal1/Gal3 as the crucial step in Gal4 activation (13).

Interestingly, the *GAL1*, *GAL10*, and *GAL7* genes in *K. lactis* are arranged very similarly to the homologous *S. cerevisiae* genes including the number and approximate arrangement of Gal4 binding sites in the promoter regions. However, there are important differences in the spacing that apparently reflect the evolution of the bifunctional *KlGAL1* gene to the specialized and tightly repressible *ScGAL1* gene (22).

Another important difference in the transcriptional circuit is a higher basal level of *KlGAL4* gene expression and a positive feedback regulation via a Gal4-binding site in its promoter (comp. Fig. 1). This autoinduction contributes to efficient lactose utilization and facilitates induction in glucose-grown cells (23, 24).

These published data can serve as a starting point to study the evolution of dynamic aspects of the network. Furthermore, recent reports on co-crystals between the Gal80 proteins of *S. cerevisiae* and *K. lactis* with short peptides comprising the activation domain of Gal4 (25, 26) form the basis for detailed structure–function analyses of transcriptional switches. Influences of protein and RNA stability regulation, subcellular localization, and protein kinase and metabolite signaling await characterization.

In the following chapter, we would like to highlight *K. lactis* as a model system for comparative studies with *S. cerevisiae*. *K. lactis* has similar options for genetic interventions and can serve as a reference genome to study the consequences of the whole-genome duplication. We also provide methods for *in vitro* analysis of components of the *GAL* regulatory module to determine quantitative parameters essential for mathematical modeling.

1.1. Working with Kluyveromyces lactis

1.1.1. Physiological and Genetic Aspects

Despite the morphological and genetic similarities, there are considerable physiological differences between *S. cerevisiae* and *K. lactis*, which have been covered by several reviews (9, 10, 16, 17, 27–29). The following paragraph summarizes some general and some specific aspects related to the *LAC/GAL* regulon for those readers starting to work with *K. lactis*.

The name of the genus *Kluyveromyces* is derived from the Kluyver effect, a physiological trait that describes the inability to grow on certain fermentable sugars in the absence of oxygen (30, 31). As *S. cerevisiae*, *K. lactis* shows a positive Kluyver effect for galactose, which apparently results from sugar uptake rates being too low to support fermentative growth (31, 32). Note that the Kluyver effect for galactose in *K. lactis* is strain-dependent, probably reflecting variations in sugar transporter genes found among strains (33–35).

K. lactis is unable to grow in the complete absence of oxygen (36, 37) and prefers respiratory metabolism over fermentation when oxygen is abundant (29, 38, 39). This preference is related to the so-called Crabtree effect, which describes aerobic ethanol production. *K. lactis* is Crabtree negative, whereas *S. cerevisiae* is Crabtree positive. The Crabtree effect drastically reduces biomass yield due to ethanol production and disfavors many industrial processes. The use of Crabtree-negative yeasts is, therefore, of considerable biotechnological interest (40).

K. lactis is unstable as a diploid and sporulates even on rich media. To maintain diploids, they have to be kept under selective pressure. Although *K. lactis* can switch mating-type, there is no HO endonuclease to induce this process (1). Rather, switching occurs spontaneously and the frequency differs between strains. The reference strain CBS 2359 has a stable mating-type, which originally was arbitrarily assigned "*a*" (41) and later on confirmed to be MAT*a* by sequence comparison with *S. cerevisiae* (28).

1.1.2. Genetic Tools

Phenotypic characterizations and selections described for *S. cerevisiae* can be applied to *K. lactis.* Most *S. cerevisiae* selection markers can be used and most of its promoters work in the heterologous host. Episomal vectors have been established on the basis of the 2 μm-like plasmid pKD1, which was discovered in *K. lactis var. drosophilarum* (42, 43). In addition, centromeric vectors are avail able based on the centromere of chromosome 2 and *K. lactis* autonomously replicating sequences (named KARS) (44, 45).

Several transformation protocols have been reported specifically for *K. lactis* (46, 47), but most are only slight adaptations of *S. cerevisiae* protocols. An efficient method, based on (48), is described below.

In contrast to *S. cerevisiae*, *K. lactis*, in general, shows low gene targeting efficiency but high competence for nonhomologous recombination. Therefore, oligonucleotide-based gene targeting strategies creating PCR-generated selectable markers flanked by short (about 50 bp) homologous sequences to the target region, as extensively used in *S. cerevisiae*, are rarely successful in *K. lactis.* Efficiencies can be increased by adding a large excess of small DNA fragments to the transformation mixture, which probably titrates the Ku70/Ku80 protein complex involved in nonhomologous end-joining (49). However, since ectopic integration of DNA fragments is a frequent event elevated levels of secondary insertion mutations can be expected. Alternatively, a genetic intervention to inactivate Klku80 can be employed. A strain deleted for ***Klku80*** showed dramatically increased efficiencies of homologous recombination even with short homologous flanks (49). Again, elevated mutation rates can be expected in ***ku80*** mutants.

Transformation cassettes with long (more than 500 bp)-flanking homologous regions improve gene targeting and a two-step procedure *(knock-in–knock-out)* may be a somewhat lengthy but successful alternative in difficult cases (50).

A *K. lactis* vector allowing for regulated expression of the Cre-recombinase is available such that the Cre–lox system for site-specific recombination can be applied (51).

1.2. Analysis of the GAL/LAC Regulon of K. lactis

1.2.1. Selection of lac Regulatory Mutants

Most of the regulatory and structural genes involved in galactose/lactose utilization in *K. lactis* have been identified by the isolation of mutants not able to grow on lactose (see for example refs. 21, 52, 53). However, when streaked together with Lac$^+$ colonies efficient cross-feeding is observed. Sometimes Lac$^+$ and Lac$^-$ clones are even indistinguishable by colony size or spot assay. Therefore, when analyzing Lac$^-$ mutants, Lac$^+$ controls should be streaked on separate plates. Apparently, lactose-utilizing cells supply Lac$^-$ cells with one or more non-lactose carbon source(s). Since diffusion barriers in the agar do not compromise cross-feeding and KlSnf1 is required for the utilization of the "feed" (A. Anders and K.D. Breunig, unpublished observations), it has to be volatile; ethanol or ethyl acetate are likely candidates. It is possible that the ability of *K. lactis* to utilize "poor" carbon sources (e.g., ethanol) nearly as efficiently as "rich" carbon sources (e.g., glucose and lactose) contributes to the cross-feeding phenomenon. Supplementing lactose plates with 5-bromo-4-chloro-3-indolyl-galactopyranoside (X-Gal) allows a clear distinction between Lac$^-$ and Lac$^+$ cells by color assay to be made.

Klgal80 mutants can be identified most easily on X-Gal plates containing glucose. *Klgal4* mutants are defective for growth on lactose and on galactose.

As the lactose permease of *Escherichia coli*, the *LAC12* gene product of *K. lactis* is able to take up the chromogenic galactoside X-Gal. Upon cleavage by the endogenous β-galactosidase encoded by the *LAC4* gene, the products formed are galactose and 5,5′-dibromo-4,4′-dichloro-indigo, an insoluble blue product. Thus, as in *E. coli*, the expression of *LAC12* and *LAC4* can be assayed directly on X-Gal-containing agar plates.

1.2.2. In Vitro Analysis of GAL Proteins from K. lactis

A specific feature of *K. lactis* is the bifunctional protein KlGal1, which combines the functions of the *S. cerevisiae* galactokinase Gal1 and the galactose sensor protein Gal3. KlGal1 is enzymatically inactive when bound to KlGal80. This observation forms the basis of a convenient assay to measure the affinity between both proteins by monitoring galactokinase inhibition in the presence of KlGal80. Moreover, the influence of additional factors or proteins (e.g., Gal4) on KlGal1–KlGal80 interaction can be analyzed by including them in the galactokinase inhibition assay (14). The basic assay is described below.

2. Materials

2.1. Strains, Media and Supplements

1. *Strains*: Common *K. lactis* laboratory strains can be highly variable since many primary wild-type isolates have been entered into the research community. The European *Kluyveromyces* community has agreed on the strain CBS 2359 (NRRL Y-1140) as a reference strain. Its methionine auxotrophic derivative, CBS 2359/152 obtained by random mutagenesis, is the strain that has been sequenced in the genome project (10). JA6 is a strain that differs from CBS 2359 by being more sensitive to glucose repression and has been used by many labs studying glucose regulation (see Note 1). Some strains including JA6 possess double-stranded linear plasmids that encode zymocin, a heterotrimeric protein toxin, which is secreted and kills sensitive yeasts including *S. cerevisiae* (reviewed in ref. 9).
2. *Vitamin mix*: 1,000× stock solution with 12 g/L niacin, 4 g/L pantothenic acid (hemi-calcium salt), 1 g/L pyridoxine, 1 g/L thiamine–HCl, 1 g/L folic acid, 1 g/L *p*-aminobenzoic acid (54). Add the mix to growth media to obtain high cell densities.
3. *X-Gal stock*: 5 mg/mL 5-bromo-4-chloro-3-indolyl-galactopyranoside (X-Gal) in DMF. Solution can be stored about 2 weeks at −20°C.
4. *X-Gal-containing plates*: The standard concentration of X-Gal in solid media is 50 μg/mL (see Note 2). To prepare solid X-Gal media, dilute the appropriate volume of X-Gal stock solution into autoclaved medium at a temperature of about 50°C. Pour plates and store at 4°C no longer than 1 week before use.
5. *FOA-containing plates*: Use a concentration of 0.2 mg/mL fluoroorotic acid (FOA) as a starting point, but sensitivity may vary among strains.

2.2. Transformation

1. *PLAG solution*: 40% PEG4000, 0.1 M lithium acetate, 10 mM Tris–HCl, pH 7.5, 1 mM EDTA, 15% (v/v) glycerol. Autoclave to sterilize.
2. *Total RNA from E. coli*: Grow *E. coli* cells overnight in 100 mL LB medium. Prepare total RNA from the cells according to the procedure for manual plasmid preparation with the exception that RNase A is not added to any of the buffers. Dissolve the RNA pellet in 5 mL of nuclease-free water and determine the concentration by measuring the optical density of the solution at 260 nm. Adjust the

concentration to 10 mg/mL by diluting with nuclease-free water. Store aliquots at −20°C.

3. Sterile 50- and 1.5-mL tubes.
4. Thermomixer.

2.3. Expression and Purification of Recombinant KlGal80 Protein

1. *E. coli* strain Rosetta(DE3)-pLys (Novagen) bearing a pET vector (Novagen) coding for an N terminally or internally His_6-tagged KlGal80 protein (see Note 3).
2. *Antibiotics*: Prepare 1,000× stock solutions by dissolving 100 mg/mL ampicillin in 50% ethanol and 35 mg/mL chloramphenicol in ethanol. Store solutions at −20°C.
3. *Lactose stock solution* 30% (w/v): Heat to dissolve lactose in water, then filter-sterilize the solution using a filter with 0.2 μm pore size.
4. *LB-cam-amp medium*: 1% (w/v) tryptone, 0.5% (w/v) yeast extract, and 0.5% (w/v) NaCl in water. Autoclave to sterilize. After cooling to room temperature add chloramphenicol and ampicillin to final concentrations of 35 and 100 mg/L, respectively.
5. *Induction medium*: LB-cam-amp supplemented with 0.2 M KH_2PO_4 and 1.5% (w/v) lactose. To make 1 L of induction medium dissolve 10 g tryptone, 5 g yeast extract, 5 g NaCl, and 27 g KH_2PO_4 in 950 mL water. Autoclave and after cooling to room temperature, add 50 mL 30% (w/v) lactose solution and 1 mL of both chloramphenicol and ampicillin stock solutions.
6. *Protamine sulfate stock solution* (100×): Stock contains 5% (w/v) protamine sulfate in water, which is not completely soluble at room temperature. Thus, prior to use dissolve the precipitate by warming to 30°C.
7. *Lysis buffer*: 50 mM Tris, 20 mM Na citrate, 10 mM NaCl, 10 mM imidazole, 0.1% (w/v) Tween 20, pH 8.0 at 4°C. Store at 4°C (see Note 4).
8. *Wash buffer*: Lysis buffer supplemented with 50 mM imidazole. Store at 4°C.
9. *Elution buffer*: Lysis buffer with 250 mM imidazole. Store at 4°C.
10. *Storage buffer*: 20 mM Tris–HCl, 20 mM Na citrate, 10% (w/v) glycerol, pH 8.0.
11. Sonicator (Branson Sonifier 250).
12. Empty columns.
13. Ni-NTA sepharose (Qiagen Ni-NTA sepharose 6 fast flow).

2.4. Quantifying KlGal1–KlGal80 Interaction by Measuring Galactokinase Inhibition

1. Galactokinase assay components: Prepare individual solutions in water of 30 mg/mL bovine albumin fraction V (BSA; Roth), 100 mM fuctose-1,6-bisphosphate (FBP; Sigma), 25 mM NADH (Sigma), 200 mM phosphoenolpyruvate (PEP, Sigma), and 100 mM ATP (Sigma). Solutions are 100× stocks; store aliquots at −20°C.
2. 1 M galactose stock: Vortex vigorously to dissolve in water, filter sterilize the solution, and store at 4°C.
3. Solution of approximately 700 U/mL pyruvate kinase and 1,000 U/mL lactic dehydrogenase (PK/LDH enzymes; Sigma).
4. 2× Galactokinase (GK) raw buffer: 200 mM Tris–HCl, pH 7.9, 20 mM KCl, 10 mM $MgCl_2$, 200 mM potassium acetate (Kac) (see Note 5). Filter sterilize and store at room temperature.
5. Purified KlGal1 and KlGal80 proteins stored at −70°C.
6. A spectrophotometer capable of simultaneously measuring multiple samples and performing time course experiments (e.g. Beckman DU 640).
7. Cuvettes allowing passage of light with 340 nm wavelength.
8. Water bath.

3. Methods

3.1. Specific Growth Requirements and Genetic Selection

Researchers accustomed to work with *S. cerevisiae* will not have problems getting started with *K. lactis*. Cultivation at temperature optimum of 28–30°C in rich medium (YPD) or yeast nitrogen base (YNB) media (synthetic complete, SC or drop out media, SD) (55) is equivalent. However, *K. lactis* is slightly more sensitive to elevated temperature and normally does not grow above 37°C. The organism lacks the kynurenine pathway for de novo synthesis of NAD from tryptophan and the NAD salvage pathways provide the only endogenous source of the dinucleotide (56). Therefore, supplementation with niacin (a generic name for nicotinamide and nicotinic acid) is recommended, or may become essential, in minimal media. To obtain high cell densities, addition of a vitamin mix to the medium is recommended (for a recipe see Subheading 2) (54).

In general, *K. lactis* appears to be more sensitive to many drugs including fluoroorotic acid and gentamicin. Formation of spheroblasts requires less zymolyase or other cell wall-degrading enzymes and transformation is usually more efficient at lower cell densities compared with *S. cerevisiae*. There are, however, large differences between strains and protocols have to be adapted accordingly.

3.1.1. X-Gal Plate Assay to Quantify LAC Gene Expression

Spot, streak, or plate *K. lactis* cells or cell suspensions onto plates with the desired medium containing X-Gal. Incubate the plates at 30°C for about 2 days. Visual inspection of the blue coloring of the cells allows an estimation of *LAC4* expression which is a good measure for Gal4 activity and intracellular galactose levels. Since the blue dye precipitates inside the cells and thus does not diffuse through the agar minor subpopulation of blue cells can be detected in dotted suspension and even inside single colonies. Figure 3 shows a bistable switching behavior resulting from single amino acid exchanges in the Gal80 protein (a) and colony sectoring (b). It should be noted that the permease Lac12 is essential for X-Gal uptake and limiting Lac12 activity might require direct measurement of β-galactosidase activity (see Note 6).

Note that the use of the *E. coli* β-galactosidase (*lacZ*) reporter in gene expression studies requires a *lac4*-deficient strain.

3.2. Transformation

1. Inoculate 50 mL glucose complete or the appropriate selective medium in a 500-mL flask with 2 mL of an overnight culture.
2. Grow to an OD_{600} of 0.5 (about 2×10^7 cells/mL). Spin down the cells in a sterile 50-mL tube at 6,000 × *g* for 5 min at room temperature and decant the supernatant.
3. Resuspend the cell pellet in 2 mL PLAG solution by moderate vortexing.
4. Add 250 μL *E. coli* RNA and mix.
5. Distribute 200-μL portions of the suspension in sterile Eppendorf tubes. Store the aliquots at –70°C until use. The cells are transformation competent for several months with gradually decreasing transformation efficiency.

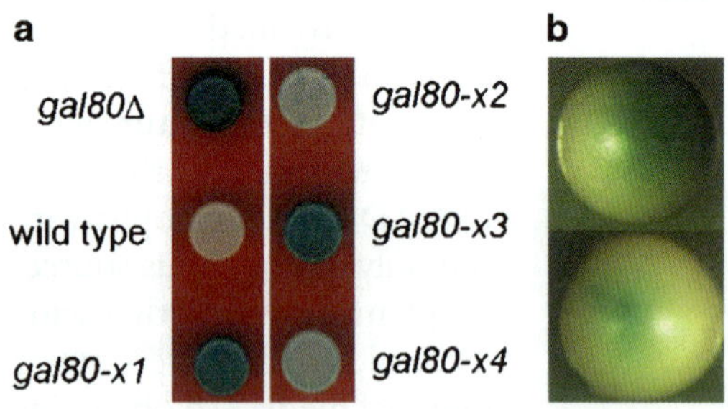

Fig. 3. Staining of *K. lactis* cells on X-Gal containing plates. (**a**) Suspensions of single colonies of wild-type and various *gal80* mutant strains were dotted on plates containing 2% glucose and X-Gal and incubated at 30°C for 2 days. Depending on the *gal80* allele, staining varies between *dark blue* (*gal80Δ*) and *white* (wild type). Note the heterogeneous staining of the *gal80-x2* mutant. *Blue dots* in the *white spot* indicate bistable *LAC* gene expression in this strain. The plates were scanned against a *red background* to improve the contrast. Photograph was kindly provided by D. Schmidt. (**b**) Example of single colonies showing mosaic patterns of *LAC* gene expression. The cells shown here express a *S. cerevisiae GAL4* variant instead of the endogenous *KlGAL4* gene and were restreaked from lactose onto glucose-containing plates.

6. Use about 100–500 ng of plasmid DNA and 1 μg of linear DNA. Add the DNA to an aliquot of frozen competent cells. The volume of the DNA must not exceed 20 μL.
7. Allow the cells to thaw with vigorous agitation in a thermomixer at 37°C for 3–5 min.
8. Incubate the cells for 1–1.5 h at 37°C and 15–30 min at 42°C.
9. Plate the cells on the appropriate selective medium.

3.3. Expression and Purification of KlGal80 Protein

The following procedures are adapted for a volume of 1 L *E. coli* culture for protein overproduction.

1. Expression: In the morning, inoculate 50–100 mL LB-cam-amp with *E. coli* cells carrying the KlGal80 expression plasmid, either using 0.5 mL cell suspension of an overnight culture (preferred) or cells directly from an agar plate. When the cells approach an OD_{600} of about 1–2 (usually in the evening), induce protein synthesis by diluting the cell suspension to OD_{600} of about 0.1 into 1 L induction medium prewarmed to 26°C. Let the cells grow overnight for 12–15 h at 26°C with constant shaking to a final OD_{600} 4–8. Harvest the cells by centrifugation, freeze the cell pellet in liquid nitrogen and store at –20°C. This procedure typically results in 20–40 mg purified KlGal80 protein per liter cell culture (see Note 7).
2. Cell lysis: Take care that the temperature of the protein solution does not rise significantly above 4°C during the following steps. Dissolve the cell pellet in 30–40 mL lysis buffer. Use the following settings for the sonicator: timer – hold, duty cycle – 50%, and output control – 5. Sonicate the cells three times for 30 s with 1 min intervals to prevent overheating of the sample. Keep the cells on ice during the whole procedure.
3. Extract preparation: Add protamine sulfate stock solution to a final concentration of 0.05% (w/v) for precipitation of nucleic acids and mix gently. Centrifuge at 35,000 × *g* for 20 min and transfer the supernatant to a new tube; be careful to avoid takeover of the cell debris. The extract solution should be clear at this point. If necessary, include a second centrifugation step.
4. Protein purification by batch procedure: Transfer 6 mL of a 50% (w/v) Ni-NTA slurry into a fresh 50-mL tube. Wash the sepharose twice by suspending in lysis buffer, pelleting at 2,000 × *g* for 3 min and decanting the supernatant. Add the cleared lysate to the sepharose and mix gently by shaking (200 rpm on a rotary shaker) at 4°C for 30 min. Load the suspension into an empty column, if desired collect the flow

through for later analysis. Wash twice with 5 mL wash buffer and eventually collect the wash fractions. Elute the protein with four times 1 mL elution buffer. Collect the eluates in four tubes and analyze by SDS–PAGE.

5. Dialyze the protein solution overnight against storage buffer. Freeze the samples in liquid nitrogen and store at –70°C until use.

3.4. Quantifying KlGal1–KlGal80 Interaction by Measuring Galactokinase Inhibition

The following protocol describes the use of a spectrophotometric assay for measuring the enzymatic galactokinase activity of KlGal1 (see Note 8 for explanation of the principle). Addition of increasing amounts of KlGal80 leads to gradual inhibition of the galactokinase activity which allows to quantify the interaction between KlGal1 and KlGal80. The basic assay described here can be easily extended to determine the influence of additional factors (e.g., KlGal4) on KlGal1–KlGal80 interaction (see, for example, ref. 14), to analyze the homologous *S. cerevisiae* proteins, and to perform cross-complementation studies with proteins originating from different yeasts (see Note 9).

1. Prepare buffer B used for the measurements. For a final volume of 10 mL, which is sufficient for about 20 single measurements, combine the following components in a falcon tube: 5 mL 2× GK raw buffer, 100 μL PK/LDH enzyme solution, 100 μL of each BSA, FBP, NADH, PEP, and ATP stock solution. Add water to 9.5 mL. The missing volume of 0.5 mL is reserved for the galactose solution which must not be added at this point. Mix the buffer by inverting the tube or by pipetting up and down. Place the buffer on ice where it is stable for several hours.
2. Prewarm the water bath to 30°C. Prewarm the cuvette holder/carriage to 30°C.
3. Thaw the appropriate amounts of purified KlGal1 and other purified proteins (e.g., KlGal80) and place them on ice. We recommend to incubate KlGal80 at room temperature some time before the experiment starts (see Note 10).
4. Blank the spectrophotometer against a cuvette containing water or buffer without NADH at a wavelength of 340 nm. Before the actual galactokinase inhibition measurements are performed, we recommend to titrate KlGal1 amounts to find an appropriate concentration (see Note 11). Also, test for background ATPase activities in preparations of KlGal1 and additional proteins to be used in the assay (see Note 12).
5. The following instructions assume use of a spectrophotometer which allows simultaneous measurement of six samples with minimal reaction volumes of 500 μL per cuvette. Pipette 3 mL of buffer B into a new Falcon tube and place in a preheated water bath. Wait until the buffer attains a temperature of 30°C.

6. Add KlGal1 protein to preheated buffer to achieve an appropriate final concentration (see Note 11). Mix gently.
7. Pipette 475 μL of the buffer mixture into each of six cuvettes.
8. The reaction in one of the cuvettes serves as a control for the determination of galactokinase activity in the absence of any further protein. For determination of a KlGal80 dose dependency of galactokinase inhibition, add increasing amounts of purified KlGal80 protein to the remaining cuvettes. Do not add more than 50 μL additional protein solution to each cuvette.
9. Start the galactokinase reaction by adding 25 μL of the galactose stock solution to each cuvette to reach a final concentration of 50 mM (see Note 13). Mix gently but thoroughly by either pipetting up and down or by using disposable stirring rods.
10. Place the cuvettes into the cuvette holder of the spectrophotometer.
11. Start data acquisition: Measure the absorbance at 340 nm over a period of 3 min at intervals of 30 s. Ensure measurement in a time window in which the decrease in absorbance is linear in each cuvette. In some cases, establishment of a steady state with linear progression of the optical density might take several minutes (see Note 14). Record measured reaction rate (decrease in optical units per minute) for each sample.
12. Data analysis: For each concentration of KlGal80, calculate the relative galactokinase activity. For this purpose, divide the measured rate by the rate determined in the control reaction without KlGal80 protein. Plot the calculated relative galactokinase activity against KlGal80 concentration which should result in a hyperbolic graph. Fit the equation "$y = K_d/(K_d + x)$" to the data points to find a value for the parameter K_d which best explains the experimental data. In this equation, y is the relative galactokinase activity, x the KlGal80 concentration in the cuvette, and K_d reflects the apparent steady-state binding constant (K_d value) for KlGal1–KlGal80 interaction in units as chosen for KlGal80 concentration. Check whether the resulting graph properly reflects the measured data.

4. Notes

1. Differences in glucose sensitivity of different strains are due to differences in glucose transporter gene content (33, 57, 58). Strain JA6 also has a weaker *KlGAL4* (*LAC9-2* allele) promoter than CBS2359, which affects glucose inhibition of the LAC/GAL regulon (59).

2. The X-Gal plate assay allows the detection of even very low Gal4 activity and the intensities of blue staining of colonies vary in the proportion to Gal4 activation over some two orders of magnitude. Depending on the concentration of X-Gal in the plates, the dynamic range can be adapted to lower or higher activities. At the standard concentration of 50 μg/mL above 1,000 mU/mg of β-galactosidase activity colonies are dark blue and differences can no longer be distinguished.

3. We have used either N terminally or internally His_6-tagged variants. The N terminally tagged variant has a somewhat lower affinity for KlGal1 when compared with the internally tagged protein. A *K. lactis* strain expressing the latter variant shows a galactose induction kinetics indistinguishable from wild type, whereas the expression of the N terminally tagged variant abolishes galactose induction almost completely. We interprete these observations to reflect the strong sensitivity of the galactose switch performance to relatively small quantitative parameter changes.

4. A specific feature of the protocol for KlGal80 purification is the inclusion of citrate in the buffers, which prevents the precipitation of KlGal80. EDTA has a similar and even somewhat stronger influence on KlGal80. The stabilizing effect is probably due to an unidentified mechanism that results in a monodisperse population of homodimers from a heterogeneous protein mixture rather than the metal chelating capabilities of both agents. Note that higher amounts of citrate or EDTA might interfere with the Ni-NTA metal-affinity chromatography.

5. In this protocol, a final concentration of 100 mM potassium acetate (Kac) is used to increase the stringency of the binding between the proteins to be tested (e.g., KlGal1 and KlGal80). We have tested different salts, including NaCl and KCl, at a concentration of 100 mM. Out of those, Kac was the only one which had no negative influence on the galactokinase activity itself.

6. Lac12 deficient cells stain white and discrepancies between β-galactosidase activities quantified in cell lysates and in X-Gal plate assays might reflect limiting Lac12 activity. Direct measurement of β-galactosidase activity requires either the preparation of cell extracts or permeabilization of the cells. The latter can be achieved by treatment with toluene or chloroform + SDS (27) or repeated freeze–thaw cycles (Yeast Protocols Handbook, Clontech Laboratories). Preparation of cell extracts is usually based on cell disruption by vortexing with glass beads and does not differ from protocols established for *S. cerevisiae*.

7. We have tested different conditions for the expression of KlGal80 by varying several growth and induction parameters as suggested in (60). The protocol given here results in both high cell density and good overproduction of KlGal80. However, production of mutant KlGal80 proteins, even with single amino acid exchanges, often leads to much lower protein yields. A lower protein yield usually correlates with a lower stability of the purified protein variant as indicated, for example, by precipitation at relatively low concentration.
8. KlGal1 catalyzes the phosphorylation of D-galactose to D-galactose-1-phosphate by hydrolyzing ATP and releasing ADP. ADP production (or ATP hydrolysis, respectively) by KlGal1 can be coupled to the oxidation of NADH (61). The "coupling" enzyme pyruvate kinase (PK) catalyzes the reaction of ADP and phosphoenolpyruvate (PEP) to ATP and pyruvate. The latter serves as a substate for the second "coupling" enzyme lactic dehydrogenase (LDH) and is reduced to lactate, whereas NADH is oxidized to NAD^+. As a net consequence of this reaction, hydrolysis of each molecule of ATP by the galactokinase and its recycling by PK results in oxidation of a molecule NADH to NAD^+. Thus, the decrease in optical density at 340 nm resulting from NADH consumption reflects the rate of ATP hydrolysis.
9. *S. cerevisiae* Gal80 is able to inhibit the enzymatic activities of both KlGal1 (our unpublished results) and ScGal1 (62). KlGal80, however, cannot inhibit ScGal1 (62), probably due to the low affinity between these proteins (13).
10. Adding KlGal80 to the KlGal1-containing reaction mixture directly from ice, we sometimes observed a delay of up to several minutes in the establishment of a constant galactokinase reaction rate. This is probably due to a delay in attaining a steady state in KlGal1–KlGal80 interaction. Pre-equilibration of KlGal80 to room temperature circumvented this problem.
11. NADH has an extinction coefficient of 6,300 M^{-1}/cm at 340 nm. Thus, the concentration of 0.25 mM given in this protocol results in an initial absorbance of about 1.6. We recommend choosing a concentration of KlGal1 which results in a decrease in absorbance of about 0.1–0.2/min. The resulting relatively long-lasting linear decrease in absorbance can then be easily "trapped" before reaching NADH limitation. According to our measurements, KlGal1, either purified from yeast or *E. coli*, has a turnover rate k_{cat} of about 60–70 s^{-1} at the conditions given here. Thus, a concentration of about 4 nM KlGal1 in the final mixture should result in a decrease in the absorbance of 0.1 per min.

12. Since the assay measures hydrolysis of ATP, control experiments have to be performed to exclude ATPase activities in the KlGal1 preparation besides galactokinase itself. This is best achieved by measuring the ATPase activity of the protein preparation in the absence of galactose (use water instead of galactose). Likewise, one has to exclude galactokinase and background ATPase activities in the preparations of proteins (e.g., KlGal80) to be analyzed in the assay. In the case of significant background ATPase activity in the KlGal1 preparation, rates have to be corrected by subtracting values determined in the absence of galactose from the rates measured in the presence of galactose for each concentration of KlGal1 used.
13. KlGal1 has a K_M value of about 3.5 mM for galactose (63). Thus, the protocol given here results in complete saturation of KlGal1 with galactose.
14. Two main reasons can delay the establishment of a steady state and a linear decrease in absorbance. Either the fluxes through the enzymatic reactions including the coupling reactions are limited by the substrate concentrations or suboptimal buffer conditions (e.g., omitting KCl from the buffer results in a delayed establishment of the maximal reaction rate) or formation of the KlGal1–KlGal80 (or other relevant) complex(es) takes time resulting in a continuous decrease in reaction rate during the measurements. When performing long-term measurements (eventually with lower -KlGal1 concentrations) in initial experiments, we suggest testing when the steady state in enzymatic fluxes and protein interactions is reached. Data acquisition should only start after steady-state conditions have been reached (see also Note 10).

References

1. Tsong, A. E., Tuch, B. B., Li, H., and Johnson, A. D. (2006) Evolution of alternative transcriptional circuits with identical logic, *Nature* **443**, 415–420.
2. Gasch, A. P., Moses, A. M., Chiang, D. Y., Fraser, H. B., Berardini, M., and Eisen, M. B. (2004) Conservation and evolution of cis-regulatory systems in ascomycete fungi, *PLoS. Biol.* **2**, e398.
3. Usaite, R., Jewett, M. C., Oliveira, A. P., Yates, J. R., III, Olsson, L., and Nielsen, J. (2009) Reconstruction of the yeast Snf1 kinase regulatory network reveals its role as a global energy regulator, *Mol Syst. Biol* **5**, 319.
4. Butler, G., Kenny, C., Fagan, A., Kurischko, C., Gaillardin, C., and Wolfe, K. H. (2004) Evolution of the MAT locus and its Ho endonuclease in yeast species, *Proc Natl Acad Sci U. S. A* **101**, 1632–1637.
5. Verma, M., Bhat, P. J., Bhartiya, S., and Venkatesh, K. V. (2004) A steady-state modeling approach to validate an in vivo mechanism of the *GAL* regulatory network in *Saccharomyces cerevisiae*, *Eur. J. Biochem.* **271**, 4064–4074.
6. Verma, M., Bhat, P. J., and Venkatesh, K. V. (2003) Quantitative analysis of GAL genetic switch of Saccharomyces cerevisiae reveals that nucleocytoplasmic shuttling of Gal80p results in a highly sensitive response to galactose, *J. Biol. Chem.* **278**, 48764–48769.
7. Acar, M., Mettetal, J. T., and van Oudenaarden, A. (2008) Stochastic switching as a survival strategy in fluctuating environments, *Nat Genet.* **40**, 471–475.

8. Campbell, R. N., Leverentz, M. K., Ryan, L. A., and Reece, R. J. (2008) Metabolic control of transcription: paradigms and lessons from *Saccharomyces cerevisiae*, *Biochem. J.* **414**, 177–187.
9. Schaffrath, R. and Breunig, K. D. (2000) Genetics and molecular physiology of the yeast *Kluyveromyces lactis*, *Fungal. Genet. Biol.* **30**, 173–190.
10. Dujon, B., Sherman, D., Fischer, G., Durrens, P., Casaregola, S., Lafontaine, I., De, M. J., Marck, C., Neuveglise, C., Talla, E., Goffard, N., Frangeul, L., Aigle, M., Anthouard, V., Babour, A., Barbe, V., Barnay, S., Blanchin, S., Beckerich, J. M., Beyne, E., Bleykasten, C., Boisrame, A., Boyer, J., Cattolico, L., Confanioleri, F., De, D. A., Despons, L., Fabre, E., Fairhead, C., Ferry-Dumazet, H., Groppi, A., Hantraye, F., Hennequin, C., Jauniaux, N., Joyet, P., Kachouri, R., Kerrest, A., Koszul, R., Lemaire, M., Lesur, I., Ma, L., Muller, H., Nicaud, J. M., Nikolski, M., Oztas, S., Ozier-Kalogeropoulos, O., Pellenz, S., Potier, S., Richard, G. F., Straub, M. L., Suleau, A., Swennen, D., Tekaia, F., Wesolowski-Louvel, M., Westhof, E., Wirth, B., Zeniou-Meyer, M., Zivanovic, I., Bolotin-Fukuhara, M., Thierry, A., Bouchier, C., Caudron, B., Scarpelli, C., Gaillardin, C., Weissenbach, J., Wincker, P., and Souciet, J. L. (2004) Genome evolution in yeasts, *Nature* **430**, 35–44.
11. Wolfe, K. H. and Shields, D. C. (1997) Molecular evidence for an ancient duplication of the entire yeast genome, *Nature* **387**, 708–713.
12. Meyer, J., Walker-Jonah, A., and Hollenberg, C. P. (1991) Galactokinase encoded by *GAL1* is a bifunctional protein required for induction of the *GAL* genes in *Kluyveromyces lactis* and is able to suppress the *gal3* phenotype in *Saccharomyces cerevisiae*, *Mol. Cell. Biol.* **11**, 5454–5461.
13. Zenke, F. T., Engels, R., Vollenbroich, V., Meyer, J., Hollenberg, C. P., and Breunig, K. D. (1996) Activation of Gal4p by galactose-dependent interaction of galactokinase and Gal80p, *Science* **272**, 1662–1665.
14. Anders, A., Lilie, H., Franke, K., Kapp, L., Stelling, J., Gilles, E. D., and Breunig, K. D. (2006) The galactose switch in *Kluyveromyces lactis* depends on nuclear competition between Gal4 and Gal1 for Gal80 binding, *J. Biol. Chem.* **281**, 29337–29348.
15. Platt, A., Ross, H. C., Hankin, S., and Reece, R. J. (2000) The insertion of two amino acids into a transcriptional inducer converts it into a galactokinase *Proc. Natl. Acad. Sci. USA.* **97**, 3154–3159.
16. Kurtzman, C. P. (2003) Phylogenetic circumscription of *Saccharomyces*, *Kluyveromyces* and other members of the *Saccharomycetaceae*, and the proposal of the new genera *Lachancea*, *Nakaseomyces*, *Naumovia*, *Vanderwaltozyma* and *Zygotorulaspora*, *FEMS Yeast Res* **4**, 233–245.
17. Lachance, M. A. (2007) Current status of *Kluyveromyces* systematics, *FEMS Yeast Res* **7**, 642–645.
18. Dickson, R. C. and Markin, J. S. (1980) Physiological Studies of β-galactosidase Induction in *Kluyveromyces lactis*, *J. Bacteriol.* **142**, 777–785.
19. Das, S., Breunig, K. D., and Hollenberg, C. P. (1985) A positive regulatory element is involved in the induction of the β-galactosidase gene from *Kluyveromyces lactis*, *EMBO J.* **4**, 793–798.
20. Gödecke, A., Zachariae, W., Arvanitidis, A., and Breunig, K. D. (1991) Coregulation of the *Kluyveromyces lactis* lactose permease and β-galactosidase genes is achieved by interaction of multiple LAC9 binding sites in a 2.6 kbp divergent promoter, *Nucleic Acids Res.* **19**, 5351–5358.
21. Zenke, F., Zachariae, W., Lunkes, A., and Breunig, K. D. (1993) Gal80 proteins of *Kluyveromyces lactis* and *Saccharomyces cervisiae* are highly conserved but contribute differently to glucose repression of the galactose regulon, *Mol. Cell. Biol.* **13**, 7566–7576.
22. Hittinger, C. T. and Carroll, S. B. (2007) Gene duplication and the adaptive evolution of a classic genetic switch, *Nature* **449**, 677–681.
23. Zachariae, W. and Breunig, K. D. (1993) Expression of the transcriptional activator LAC9 (KlGAL4) in *Kluyveromyces lactis* is controlled by autoregulation, *Mol. Cell. Biol.* **13**, 3058–3066.
24. Czyz, M., Nagiec, M. M., and Dickson, R. C. (1993) Autoregulation of *GAL4* transcription is essential for rapid growth of *Kluyveromyces lactis* on lactose and galactose, *Nucleic Acids Res.* **21**, 4378–4382.
25. Kumar, P. R., Yu, Y., Sternglanz, R., Johnston, S. A., and Joshua-Tor, L. (2008) NADP regulates the yeast GAL induction system, *Science* **319**, 1090–1092.
26. Thoden, J. B., Ryan, L. A., Reece, R. J., and Holden, H. M. (2008) The interaction between an acidic transcriptional activator and its inhibitor: The molecular basis of Gal4p recognition by Gal80p, *J. Biol.* Chem. **283**, 30266–72.
27. Wésolowski-Louvel, M., Breunig, K. D., and Fukuhara, H. (1996) *Kluyveromyces lactis*, in *Non-Conventional Yeasts in Biotechnology* (Wolf, K., Ed.) Springer-Verlag, Heidelberg.
28. Bolotin-Fukuhara, M., Toffano-Nioche, C., Artiguenave, F., Duchateau-Nguyen, G.,

Lemaire, M., Marmeisse, R., Montrocher, R., Robert, C., Termier, M., Wincker, P., and Wesolowski-Louvel, M. (2000) Genomic exploration of the hemiascomycetous yeasts: 11. Kluyveromyces lactis, *FEBS Lett.* **487**, 66–70.
29. Breunig, K. D., Bolotin-Fukuhara, M., Bianchi, M. M., Bourgarel, D., Falcone, C., Ferrero, I., Frontali, L., Goffrini, P., Krijger, J. J., Mazzoni, C., Milkowski, C., Steensma, H. Y., Wesolowski-Louvel, M., and Zeeman, A. M. (2000) Regulation of primary carbon metabolism in *Kluyveromyces lactis*, *Enzyme Microb. Technol.* **26**, 771–780.
30. Kluyver, A. J. and Clusters, M. T. J. (1940) The suitability of disaccharides as respiratory and assimilation substrates for yeasts which do not ferment these sugars, 6 ed., pp 121–162.
31. Fukuhara, H. (2003) The Kluyver effect revisited, *FEMS Yeast Res.* **3**, 327–331.
32. Goffrini, P., Ferrero, I., and Donnini, C. (2002) Respiration-dependent utilization of sugars in yeasts: a determinant role for sugar transporters, *J. Bacteriol.* **184**, 427–432.
33. Weirich, J., Goffrini, P., Kuger, P., Ferrero, I., and Breunig, K. D. (1997) Influence of mutations in hexose-transporter genes on glucose repression in *Kluyveromyces lactis*, *Eur. J. Biochem.* **249**, 248–257.
34. Baruffini, E., Goffrini, P., Donnini, C., and Lodi, T. (2006) Galactose transport in *Kluyveromyces lactis*: major role of the glucose permease Hgt1, *FEMS Yeast Res.* **6**, 1235–1242.
35. Billard, P., Ménart, S., Blaisonneau, J., Bolotin-Fukuhara, M., Fukuhara, H., and Wésolowski-Louvel, M. (1996) Glucose uptake in *Kluyveromyces lactis*: Role of the *HGT1* gene in glucose transport., *J. Bacteriol.* **178**, 5860–5866.
36. Visser, W., Scheffers, W. A., Batenburg-van der Vegte, W. H., and Van Dijken, J. P. (1990) Oxygen requirements of yeasts, *Appl. Environ. Microbiol.* **56**, 3785–3792.
37. Snoek, I. S. and Steensma, H. Y. (2006) Why does *Kluyveromyces lactis* not grow under anaerobic conditions? Comparison of essential anaerobic genes of *Saccharomyces cerevisiae* with the *Kluyveromyces lactis* genome, *FEMS Yeast Res.* **6**, 393–403.
38. Kiers, J., Zeeman, A.-M., Luttik, M., Thiele, C., Castrillo, J. I., Steensma, H. Y., Van Dijken, J. P., and Pronk, J. T. (1998) Regulation of alcoholic fermantation in batch and chemostat cultures of *Kluyveromyces lactis* CBS 2359, *Yeast* **14**, 459–469.
39. Zeeman, A. M., Kuyper, M., Pronk, J. T., Van Dijken, J. P., and Steensma, H. Y. (2000) Regulation of pyruvate metabolism in chemostat cultures of Kluyveromyces lactis CBS 2359, *Yeast* **16**, 611–620.
40. Gellissen, G. and Hollenberg, C. P. (1997) Application of yeasts in gene expression studies: a comparison of Saccharomyces cerevisiae, Hansenula polymorpha and Kluyveromyces lactis – a review, *Gene* **190**, 87–97.
41. Herman, A. and Roman, H. (1966) Allele specific determinants of homothallism in *Saccharomyces lactis*, *Genetics* **53**, 727–740.
42. Chen, X. J., Saliola, M., Falcone, C., Bianchi, M. M., and Fukuhara, H. (1986) Sequence organization of the circular plasmid pKD1 from the yeast *Kluyveromyces drosophilarum*, *Nucleic Acids Res.* **14**, 4471–4481.
43. Bianchi, M. M., Falcone, C., Jie, C. X., Wésolowski-Louvel, M., Frontali, L., and Fukuhara, H. (1987) Transformation of the yeast *Kluyveromyces lactis* by new vectors derived from the 1.6 μm circular plasmid pKD1, *Curr Genet* **12**, 185–192.
44. Heus, J. J., Zonneveld, B. J. M., Steensma, H. Y., and Van den Berg, J. A. (1993) The consensus sequence of *Kluyveromyces lactis* centromeres shows homology to functional centromeric DNA from *Saccharomyces cerevisiae*, *Mol. Gen. Genet.* **236**, 355–362.
45. Heus, J. J., Zonneveld, B. J., Steensma, H. Y., and Van den Berg, J. A. (1994) Mutational analysis of centromeric DNA elements of *Kluyveromyces lactis* and their role in determining the species specificity of the highly homologous centromeres from K. lactis and *Saccharomyces cerevisiae*, *Mol. Gen. Genet.* **243**, 325–333.
46. Das, S. and Hollenberg, C. P. (1982) A High-Frequency Transformation System for the Yeast *Kluyveromyces lactis*, *Curr Genet* **6**, 123–128.
47. Dohmen, R. J., Strasser, A. W. M., Höner, C. B., and Hollenberg, C. P. (1991) An efficient transformation procedure enabling long-term storage of competent cells of various yeast genera, *Yeast* 7, 691–692.
48. Akada, R., Kawahata, M., and Nishizawa, Y. (2000) Elevated temperature greatly improves transformation of fresh and frozen competent cells in yeast, *Biotechniques* **28**, 854–856.
49. Kooistra, R., Hooykaas, P. J., and Steensma, H. Y. (2004) Efficient gene targeting in *Kluyveromyces lactis*, *Yeast* **21**, 781–792.
50. Rothstein, R. J. (1983) One-Step Gene Disruption in Yeast, *Methods in Enzymology* **101**, 202–211.
51. Steensma, H. Y. and Linde, J. J. (2001) Plasmids with the Cre-recombinase and the dominant *nat* marker, suitable for use in prototrophic strains of *Saccharomyces cerevisiae* and *Kluyveromyces lactis*, *Yeast* **18**, 469–472.

52. Sheetz, R. M. and Dickson, R. C. (1980) Mutations affecting synthesis of β-galactosidase activity in the yeast *Kluyveromyces lactis*, *Genetics* **95**, 877–890.
53. Sheetz, R. M. and Dickson, R. C. (1981) *LAC4* Is the structural gene for β-galactosidase in *Kluyveromyces lactis*, *Genetics* **98**, 729–745.
54. Inchaurrondo, V. A., Flores, M. V., and Voget, C. E. (1998) Growth and β-galactosidase synthesis in aerobic chemostat cultures of *Kluyveromyces lactis*, 20 ed., pp 291–298.
55. Sherman, F., Fink, G. R., and Hicks, J. B. (1986) *Laboratory Course Manual for Methods in Yeast Genetics* Cold Spring Harbor Laboratory, Cold Spring Harbor, N.Y.
56. Li, Y. F. and Bao, W. G. (2007) Why do some yeast species require niacin for growth? Different modes of NAD synthesis, *FEMS Yeast Res* 7, 657–664.
57. Breunig, K. D. (1989) Glucose repression of LAC gene expression in yeast is mediated by the transcriptional activator LAC9, *Mol. Gen. Genet.* **216**, 422–427.
58. Milkowski, C., Krampe, S., Weirich, J., Hasse, V., Boles, E., and Breunig, K. D. (2001) Feedback regulation of glucose transporter gene expression in *Kluyveromyces lactis* by glucose uptake, *J. Bacteriol* **183**, 000–001.
59. Zachariae, W., Kuger, P., and Breunig, K. D. (1993) Glucose repression of lactose/galactose metabolism in *Kluyveromyces lactis* is determined by the concentration of the transcriptional activator LAC9 (KlGAL4), *Nucl. Acids Res.* **21**, 69–77.
60. Kopetzki, E., Schumacher, G., and Buckel, P. (1989) Control of formation of active soluble or inactive insoluble baker's yeast alpha-glucosidase PI in *Escherichia coli* by induction and growth conditions, *Mol. Gen. Genet.* **216**, 149–155.
61. Kornberg, A. and Priver, W. E., Jr. (1951) Enzymatic phosphorylation of adenosine and 2,6-diaminopurine riboside, *J. Biol. Chem.* **193**, 481–495.
62. Sellick, C. A., Jowitt, T. A., and Reece, R. J. (2009) The effect of ligand binding on the galactokinase activity of yeast Gal1p and its ability to activate transcription, *J Biol Chem* **284**, 229–236.
63. Engels, R. (1999) PhD thesis, Heinrich-Heine-Universitaet Duesseldorf.

Chapter 14

Analysis of Subtelomeric Silencing in *Candida glabrata*

Alejandro Juárez-Reyes, Alejandro De Las Peñas, and Irene Castaño

Abstract

Analysis of gene function often involves detailed studies of when a given gene is expressed or silenced. Transposon mutagenesis is a powerful tool to generate insertional mutations that provide with a selectable marker and a reporter gene that can be used to analyze the transcriptional activity of a specific locus in a variety of microorganisms to study gene regulation. Then the reporter gene expression can be easily measured under different conditions to gain insight into the regulation of the particular locus of interest. We have used transposon mutagenesis as a tool to generate insertional mutations with a modified Tn*7* transposon containing the reporter gene *URA3* (Tn*7*-*URA3*) to study subtelomeric silencing in the opportunistic fungal pathogen *Candida glabrata*. This method consists of two major steps: an in vitro Tn*7*-*URA3* mutagenesis of a plasmid containing the desired subtelomeric region to be analyzed, followed by homologous recombination into the target region of the *C. glabrata* genome. As an alternative, a fusion PCR protocol can also be used in which the *URA3* reporter gene can be "fused" together with the 5′ and 3′ regions of the desired insertion point by a two step PCR protocol. This fusion product can be introduced into the *C. glabrata* genome by homologous recombination after transformation in the same way as the Tn*7*-*URA3* mutagenesis products. Once the *URA3* reporter gene has been introduced in the desired locus in the *C. glabrata* genome, a simple plate growth assay is performed to assess the expression of the reporter gene.

Key words: Insertional mutagenesis, Modified Tn*7*-*URA3* transposon, Fusion PCR, Subtelomeric silencing, *Candida glabrata*, *URA3* reporter gene

1. Introduction

The genomes of eukaryotes are organized into two distinct domains, the transcriptionally active regions known as euchromatin and the transcriptionally inactive domains or heterochromatin. In heterochromatic regions, a specialized chromatin structure is assembled which maintains the silencing of the majority of genes present at this location. In *Saccharomyces cerevisiae* and many other microorganisms, there are several silenced regions throughout the genome. One example is the region adjacent to the telomeres (subtelomeric region) where the chromatin is assembled into a

Attila Becskei (ed.), *Yeast Genetic Networks: Methods and Protocols*, Methods in Molecular Biology, Vol. 734,
DOI 10.1007/978-1-61779-086-7_14, © Springer Science+Business Media, LLC 2011

compact structure that is repressive for transcription. By introducing a reporter gene in a region close to a telomere in *S. cerevisiae*, it was shown that it is subject to transcriptional silencing, a phenomenon termed telomere position effect or TPE (1–3). In *S. cerevisiae* and the opportunistic fungal pathogen *Candida glabrata*, chromatin-based transcriptional silencing at telomeres and other silenced sites (the silent mating-type loci *HML* and *HMR* and the rDNA array of *S. cerevisiae*) depends on several proteins, some of which bind directly to *cis*-acting sequences adjacent to each region. These proteins then recruit the other silencing proteins to establish and propagate the repressive chromatin structure.

A convenient reporter gene to study subtelomeric silencing in *C. glabrata* is the *URA3* gene. Expression of this reporter gene can be assessed both positively and negatively since orotidine-5-phosphate decarboxylase, the product of *URA3*, required for uracil biosynthesis, converts the compound 5-FOA (5-fluoro-orotic acid) into a toxic metabolite (4). Therefore, cells that express the *URA3* gene cannot grow on plates containing 5-FOA, whereas they grow on plates lacking uracil. Conversely, cells that maintain the *URA3* gene silenced, grow on 5-FOA plates but not on plates without uracil. This plate growth assay provides a simple and convenient way to determine the expression of the reporter gene.

Studies in *S. cerevisiae* have generally placed the *URA3* gene directly adjacent to the telomere repeats, but deleting the subtelomeric regions, resulting in so-called truncated telomeres. At these telomeres, the silenced chromatin structure spreads from the telomeres toward the centromere in a linear fashion and decreases rapidly with increasing distance from the telomere (1–3). At native telomeres, however, introduction of the reporter *URA3* gene at increasing distances from different telomeres shows that silencing is discontinuous, and the level of silencing differs between telomeres and decreases precipitously as the distance between the reporter and the telomere is increased (5).

To study whether a family of subtelomeric genes encoding cell wall protein adhesins in *C. glabrata* is subject to chromatin-based silencing, we introduced the *URA3* reporter systematically at increasing distances from several telomeres of *C. glabrata* (6–8).

In these studies, we found that subtelomeric silencing varies from telomere to telomere and is differentially regulated by some of the silencing proteins (8). In addition, subtelomeric silencing in *C. glabrata* seems to propagate much longer distances from the telomere than in *S. cerevisiae* (over 20 kb in some cases compared with 4–8 kb). It seems clear then that systematic analysis using the reporter *URA3* gene integrated at native telomeres (rather than truncated ones) is a convenient and powerful tool to analyze subtelomeric silencing in *C. glabrata*.

We have successfully used a protocol that consists of four separate procedures:

1. The first step consists of an in vitro mutagenesis protocol using a modified mini-Tn*7* (Tn*7*-*URA3*) that contains the *S. cerevisiae URA3* gene with its own promoter to efficiently generate random reporter gene insertions in a previously cloned fragment of interest (a subtelomeric region of *C. glabrata* in this case) (9, 10). Another method for obtaining *URA3* insertions into desired chromosomal locations in *C. glabrata* is by "fusion PCR" (also described as an alternative strategy).
2. Transformation into *Escherichia coli* of the in vitro mutagenized plasmids (Tn*7*-*URA3* insertions) and screening of individual colonies by colony PCR to map and identify the plasmids containing insertions at positions of interest within the cloned subtelomeric region.
3. Digestion of the mutagenized plasmids and transformation into *C. glabrata* where homologous recombination occurs. In this step, the wild-type subtelomeric region is replaced by the same region containing the Tn*7*-*URA3* insertion.
4. Plate growth assay to assess *URA3* reporter expression using 5-FOA containing media to assess subtelomeric silencing.

S. cerevisiae is recognized as an ideal eukaryotic organism for studying molecular biology and genetic problems. Many molecular biology techniques have been developed to genetically manipulate *S. cerevisiae*, some of these include mutant generation by allele replacement, gene cloning, construction of plasmid libraries, transformation, etc. (11, 12). *C. glabrata* is closely related phylogenetically to *S. cerevisiae* and a high degree of synteny is found between both genomes (13). Also, close orthologues of the vast majority of *S. cerevisiae* genes can be found in *C. glabrata*. In addition, *C. glabrata* is easily manipulated genetically since most of the molecular biology techniques developed for *S. cerevisiae* can be used successfully in *C. glabrata* with only minor modifications. In Subheading 3.4, we point out some of the differences in growth rate and recombination frequencies between *C. glabrata* and *S. cerevisiae*.

2. Materials

2.1. In Vitro Tn7-URA3 Mutagenesis

1. Target plasmid: The integrative plasmid containing the subtelomeric region where *URA3* insertions will be isolated. This plasmid should be prepared using the Wizard miniprep kit from Promega (see Note 1) (Promega, Madison, WI).

2. Donor plasmid: pIC6 (10): vector where the modified Tn*7*-*URA3* is cloned (source of the *URA3* gene).
3. Agarose.
4. 10× Bromophenol blue loading buffer: 20% (w/v) ficoll 400, 0.1% (w/v) disodium EDTA, 1% SDS, and 0.25% (w/v) bromophenol blue.
5. DNA molecular weight markers: 1 kb Plus Ladder, (Invitrogen).
6. TransposaseABC* (New England Biolabs, Ipswich, MA, USA). The package comes with the Transposition Buffer and Start solution. 10× Buffer: 250 mM Tris–HCl pH 8.0, 20 mM DTT, and 20 mM ATP. Start solution: 300 mM magnesium acetate.
7. 3 M Sodium acetate.
8. Phenol:chloroform:isoamilic alcohol mix 25:24:1 (Sigma, St. Louis, MO, USA).
9. Glycogen (Roche, Manheim, Germany).
10. Ethanol (100 and 70% v/v).
11. Sterile TE solution: 10 mM Tris–HCl and 0.1 mM EDTA pH 8.0.

2.2. Fusion PCR

1. Three pairs of oligonucleotides at 5 μM [primers 1 and 2 to amplify fragment A, primers 3 and 4 to amplify fragment B (*URA3* reporter gene), and primers 5 and 6 to amplify fragment C].
2. Plasmid containing the *URA3* gene from *S. cerevisiae* with its own promoter and 3′ UTR to use as template (such as pRS306 or YIp*lac*211 (14, 15)).
3. Genomic DNA from wild-type *C. glabrata* for template.
4. A high-fidelity DNA polymerase should be used, such as *Pfu* DNA Polymerase (Promega, Madison, WI). Comes with 10× buffer: 200 mM Tris–HCl (pH 8.8), 100 mM KCl, 100 mM $(NH_4)_2SO_4$, 1% Triton X-100, and 1 mg/mL BSA or Phusion high-fidelity DNA polymerase (New England Biolabs), in case no PCR product is obtained with *Pfu*.
5. Expand Long Template PCR System (Roche).
6. 2.5 mM dNTPs.
7. Agarose.
8. 1× TAE Buffer: 4.84 g Tris base, 1.142 mL glacial acetic acid, 0.744 g disodium EDTA, bring to 1 L with distilled water.
9. 10× Bromophenol blue loading buffer: (same as Subheading 2.1, item 4), or 10× xylene cyanol loading buffer: 20% (w/v) ficoll 400, 0.1% (w/v) disodium EDTA, 1% SDS, and 0.25% (w/v) xylene cyanol.
10. Qiaquick gel extraction kit (Qiagen CA, USA).

2.3. Transformation into E. coli DH10B Electrocompetent Cells and Mapping of Insertions by Colony PCR

1. *E. coli* electrocompetent cells DH10B (Invitrogen San Diego, CA).
2. Electroporation cuvettes 0.1 cm gap (MicroPulser cuvettes) (Bio-Rad, Hercules, CA, USA).
3. Disposable sterile Petri dishes (95 × 15 mm) (Fisher Scientific).
4. SOC liquid media for recovering *E. coli* transformants after electroporation: for 100 mL: 2 g triptone, 0.55 g yeast extract, 1 mL stock 1 M NaCl, and 1 mL stock 1 M KCl. Bring to 97 mL with double-distilled water and autoclave. Add filter sterilized: 1 mL 1 M magnesium chloride, 1 mL 1 M magnesium sulfate, and 1 mL 2 M glucose.
5. Kanamycin stock solution: 30 mg/mL kanamycin sulfate (Sigma) in water, filter sterilized (keep frozen at −20°C can be stored for several months).
6. LB + kanamycin (30 μg/mL) agar plates (1 L): 10 g tryptone (Fisher Scientific), 5 g yeast extract (Fisher Scientific), 5 g NaCl, 15 g agar (Fisher Scientific), bring to 1 L with double-distilled water. Autoclave 20 min. Let cool to 45°C and add 1 mL of 30 mg/mL filter sterilized kanamycin stock solution. Pour in sterile disposable Petri dishes. Let cool and solidify. Keep at 4°C wrapped in plastic bags or in a plastic hermetic container. These plates last several months when stored in this way.
7. LB + Kanamycin liquid media for liquid cultures (same as plates but without agar).
8. Tn*7-URA3* outward primers: (1) 5′-ATAATCCTTAAAAA-CTCCATTTCCACCCCT-3′; (2) 5′-ACTTTATTGTCATA-GTTTAGATCTATTTTG-3′. 5 μM solution in 10 mM Tris–HCl pH 8.0.
9. Vector primers: polylinker primers (pUC universal primers or vector-specific primers) directed toward the insert sequences. 5 μM stock solution in 10 mM Tris–HCl pH 8.0.
10. Taq polymerase (Invitrogen).
11. 2.5 mM dNTP mix (Invitrogen).
12. Agarose.
13. 10× Xylene cyanol loading buffer (same as Subheading 2.2, item 9 above).
14. DNA molecular weight markers: 1 kb Plus Ladder, (Invitrogen).
15. Miniprep DNA extraction kit (any commercial kit works well).
16. Appropriate restriction enzymes to liberate the Tn*7-URA3* insertion for homologous recombination in *C. glabrata*.

2.4. Homologous Recombination of the Tn7 Insertion into the Subtelomeric Region of C. glabrata

1. YPD liquid media: 10 g yeast extract, 20 g peptone (Fisher), and 2% glucose (see Note 2).
2. SC w/o uracil agar plates: 1.7 g yeast nitrogen base (YNB) w/o aminoacids (Difco, Lawrence, KS), 5 g ammonium sulfate, 6 g casamino acids (Fisher Scientific), 2% glucose, and 20 g agar. Keep at 4°C wrapped in plastic bags or in a plastic hermetic container. These plates last several months when stored in this way.
3. Freshly diluted sterile 100 mM lithium acetate (see Note 3).
4. Filter sterilized 50% polyethylene glycol (PEG) (Fluka) (MW range 3,500–4,500).
5. 10 mg/mL Salmon sperm DNA stock solution (Invitrogen) (see Note 4).
6. Dimethyl sulfoxide (DMSO Hybrid max, Sigma).
7. *C. glabrata* subtelomeric-specific primers (5 μM solution in 10 mM Tris–HCl pH 8.0) (see Note 5).
8. Tn*7*-*URA3* outward primers (same as Subheading 2.3, item 8 above).
9. 20 mM Sodium hydroxide.
10. PCR reagents (same as Subheading 2.3, items 9–14 above).

2.5. Plate Growth Assay to Assess URA3 Gene Expression on 5-FOA-Containing Plates

1. YPD liquid media (same as Subheading 2.4, item 1).
2. 96-Well plates: 96-Well Cell Culture Cluster (Corning, NY, USA).
3. YPD agar plates. Same as above (Subheading 2.4, item 1) adding 20 g agar per L. Autoclave. Plates last several months when stored at 4°C wrapped in plastic bags or in a plastic hermetic container.
4. SC w/o uracil plates. Same as above (Subheading 2.4, item 2).
5. 5-FOA agar plates (1 L): 6 g casamino acids, 5 g ammonium sulfate, 50 mg uracil (Sigma), 1.7 g YNB w/o amino acids, 2% glucose, 0.9 g of 5-fluoro-orotic acid, monohydrate (5-FOA) (Toronto Research Chemicals, Inc.), and double-distilled water (see Note 6). These plates can be stored for many months wrapped in plastic bags or in a plastic hermetic container.
6. Multichannel pipette (12 channels).

3. Methods

Transcriptional silencing is a process by which a repressive chromatin structure is established at several locations throughout the genome of *S. cerevisiae* and *C. glabrata* as well as higher eukaryotes. This repressive structure is heritable but relatively unstable since in a given population of cells in which a reporter *URA3* gene is placed adjacent to a telomere, some cells may escape silencing. It is possible that upon replication the reassembly of silent chromatin could compete with the binding of transcriptional activators, this would result in some cells expressing the reporter and the rest of them maintaining it silenced (16). *C. glabrata* cultures of cells containing the *URA3* gene at subtelomeric positions are composed of two distinct populations, those that express the reporter gene and, therefore, can grow on media lacking uracil (and not on 5-FOA media), and those that maintain silenced the reporter and, therefore, grow on 5-FOA plates but require uracil for growth.

3.1. In Vitro Mutagenesis

The strategy of the method is as follows. The Tn*7*- *URA3*-containing plasmid (pIC6) (10) contains only one origin of replication located within the Tn*7*- *URA3* element. This is the conditional origin of replication R6Kγ that requires the *E. coli* protein π (the product of the gene *pir*), not present in the strain DH10B, so pIC6 cannot replicate in this strain. By selecting for kanamycin resistance after transformation in DH10B *E. coli* cells, only those events of transposition of the Tn*7*- *URA3* into the target plasmid will be selected since only these will be able to replicate in the host strain and give rise to Km^R colonies (see Fig. 1a).

1. Prepare plasmid DNA: target plasmid (containing subtelomeric sequences) and donor Tn*7*- *URA3* plasmid DNA (see Fig. 1a) using Promega Wizard miniprep kit following manufacturer's instructions.
2. Quantify the amount of DNA in each sample with a spectrophotometer and assess integrity by agarose gel electrophoresis: Prepare minigels with 0.8% agarose in 1× TAE Buffer.
3. Set up mutagenesis reactions using TnsABC* from NEB. The donor to target plasmid ratio should be ~0.5–2 donor:target molar ratio using ~250 ng total DNA in the reaction mix.
4. Set up two reactions:
   (a) Negative control: no transposase (instead, add 1 μL sterile water).
   (b) Donor (Tn*7* carrying plasmid) + target (subtelomeric sequences) plasmid.

Buffer GPS 10×	2.0 μL
Tn7 donor plasmid (pIC6)	*X* μL (~100 ng)
Target plasmid	*X* μL (~100–200 ng)
dH_2O	*X* μL
Total volume	18 μL

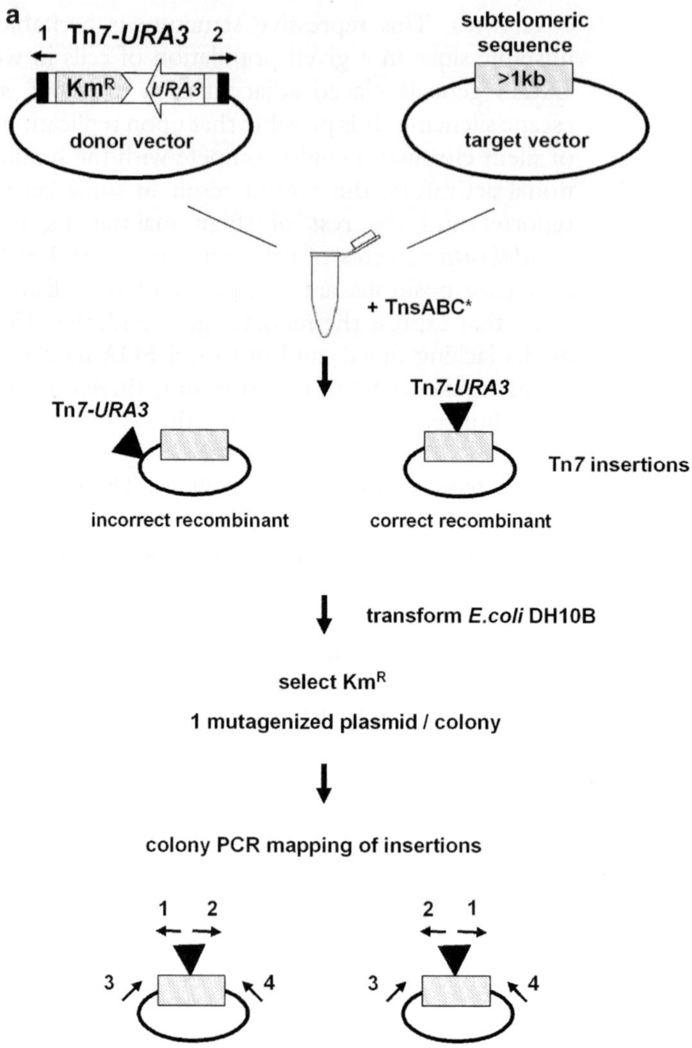

Fig. 1. Generation of *URA3* reporter insertions in vitro. (**a**) Schematic representation of in vitro mutagenesis with Tn*7-URA3* into cloned genomic fragments using purified transposaseABC*. Tn*7-URA3* element consists of a mini-Tn*7* derivative that contains *S. cerevisiae URA3* gene with its own promoter, a selectable marker for *E. coli* transformation, the Tn*7* end sequences, and a conditional origin of replication that requires the π protein. Target plasmid contains the genomic sequence (subtelomeric region) to be mutagenized with Tn*7-URA3* ($\geq$1 kb to promote homologous recombination). The transposition reaction proceeds in vitro and the Tn*7-URA3* element transposes efficiently into the target plasmid obtaining many individual insertions in both orientations. This mix is transformed into electrocompetent *E. coli* DH10B cells and transposition events are obtained by selecting Km^R colonies. In this strain, donor Tn*7-URA3* plasmid cannot replicate because it does not express the π protein. Insertions in the subtelomeric sequences of the target plasmid (correct insertions) are identified by colony PCR. (**b**) Generation of *URA3* insertions into genomic DNA fragments by fusion PCR. Three separate PCR products are amplified using a high-fidelity Taq DNA polymerase. A 5′ flanking fragment of the desired site of *URA3* insertion (fragment A), the *URA3* gene with its own promoter and its 3′ UTR (fragment B), and a 3′ flanking fragment (fragment C). Primers 2 and 5 contain a "tail" of 20–25 nucleotides annealing with the promoter region and 3′ UTR of the *URA3* gene, respectively. Fragments A and B share a region of homology and fragments B and C share another region of homology. These regions will allow these fragments to overlap such that in a second round of PCR with the three gel-purified fragments as templates and primers 1 and 6, a fusion product can be amplified. The amplified fusion product is gel purified and used directly to transform *C. glabrata* selecting for Ura^+ colonies. Transformants should be checked by PCR to verify that homologous recombination occurred and the *URA3* gene is inserted at the desired location in the genome.

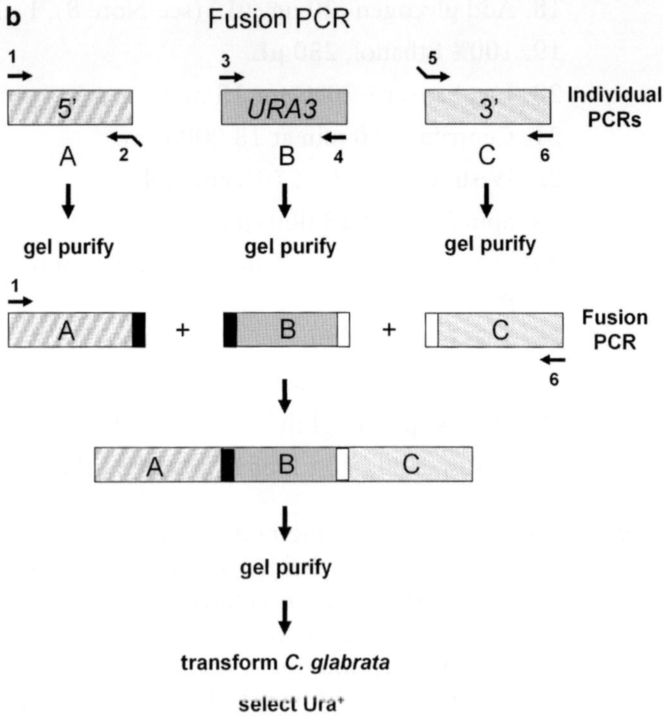

Fig. 1. (continued)

5. Mix thoroughly by pipetting up and down a few times.
6. Add: Transposase (TnsABC*), 1 μL (except no TnsABC* control).
7. Mix and incubate 10 min at 37°C to allow for site selection.
8. Add: Start solution, 1 μL.
9. Mix thoroughly pipetting up and down a few times.
10. Incubate 1 h at 37°C.
11. Stop Reaction by phenol extraction followed by precipitation of the DNA.
12. Add: sterile H_2O, 80.0 μL.
13. 3 M sodium acetate (see Note 7), 10.0 μL.
14. Phenol:chloroform:isoamilic alchohol, 110.0 μL.
15. Vortex thoroughly.
16. Centrifuge 3 min at g-force and in subsequent occurrences and also for 3,500 rpm.
17. Transfer the aqueous phase to a clean tube, ~100 μL.

18. Add glycogen (20 μg/μL) (see Note 8), 1 μL.
19. 100% Ethanol, 250 μL.
20. Incubate at −70°C for 15 min.
21. Centrifuge 10 min at 13,000 rpm.
22. Wash with 1 mL of 70% ethanol.
23. Spin 3 min at 13,000 rpm.
24. Decant ethanol carefully taking care not to through away the pellet.
25. Spin briefly 1 min.
26. Remove all the ethanol with a micropipette.
27. Dry the pellet lightly.
28. Resuspend in 20 μL of 10 mM Tris-HCI, 0.1 mM EDTA.

3.2. Alternate Method: Fusion PCR

This is an alternative method to generate a single *URA3* insertion within a previously defined subtelomeric sequence that will be later transformed and integrated by homologous recombination into the same region in the *C. glabrata* genome. This is an adaptation of previous methods used to generate mutants without cloning the gene of interest (17, 18). It consists of two steps of PCR and requires the designing of a total of six oligonucleotides. Two oligonucleotides to PCR amplify a 5′ flanking sequence to the insertion (oligonucleotides 1 and 2, Fig. 1b). Oligonucleotide 2 should contain a "tail" of approximately 20–25 nucleotides complementary to the 5′ of the *URA3* promoter region (see Fig. 1b). The second set of two oligonucleotides (oligonucleotides 5 and 6, Fig. 1b) is designed to amplify a 3′ flanking sequence, oligonucleotide 5 should contain a "tail" of about 20–25 nucleotides complementary to the 3′UTR of the *URA3* gene (see Fig. 1b). The third set (oligonucleotides 3 and 4, Fig. 1b) is designed to PCR amplify the *URA3* gene with its promoter and 3′ UTR. These oligonucleotides are the same for any *URA3* reporter gene insertion, regardless of the insertion site desired. Only the oligonucleotides to PCR amplify the 5′ and 3′ flanking sites of the insertion need to be designed specifically for each insertion site (see Note 9).

In the second PCR, the three fragments amplified in step 1 are gel purified, amplified, and fused together using the outermost primers 1 and 6. This is possible because primers 2 and 5 contain a "tail" homologous to the 5′ and 3′ ends of *URA3* gene, respectively, after the first PCR reaction, the 5′ end of fragment B is homologous to the 3′ end of fragment A and can be annealed in the following PCR step. The 3′ end of fragment B and the 5′ end of fragment C are homologous after the first PCR reaction (see Fig. 1b).

1. Prepare three PCR mixes to separately amplify fragments A, B, and C (see Fig. 1b):

   For fragment A use primers 1 and 2: 5′ flanking sequence

   For fragment B use primers 3 and 4: *URA3* reporter gene

   For fragment C use primers 5 and 6: 3′ flanking sequence (see Note 10).

Template	50–200 ng
10× PCR buffer (Pfu)	5 μL
5 μM primer A	100–500 nM
5 μM primer B	100–500 nM
2.5 mM dNTPs	100–250 μM
Pfu polymerase	2.5 U
H_2O	to 50 μL final volume

   After the PCR reaction has finished, keep the tubes at 4°C or frozen.

2. Run a 0.8% agarose gel to verify that fragments were efficiently amplified (see Note 11).
3. Gel purify the three fragments using Qiagen gel extraction kit following the manufacturer's instructions.
4. Set up a second PCR mix using Expand Long Template PCR system (special for longer templates) and follow manufacturer's instructions (see Note 12):

   Prepare three tubes for PCR mix (one for each of the three buffers provided by the manufacturer).

   Mix all three PCR fragments from step 1. The molar ratio of fragments A, B, and C should be 1:3:1.

   Set up the PCR reactions according to the manufacturer's instructions.

   Run a 0.8% agarose gel with an aliquot of each PCR mix.

   Gel purify the fusion product using Qiagen gel extraction kit.

5. This gel purified fusion product is used to directly transform *C. glabrata* (Subheading 3.4 below).

These two alternative methods of generating *URA3* insertions at desired positions throughout the *C. glabrata* genome offer different advantages. In vitro transposition of the Tn*7*-*URA3* transposon generates in one step many different insertions in both orientations at any position in the target plasmid. By fusion PCR, however, only one insertion at a time can be obtained, but the advantage is that no fragment needs to be cloned, whereas for Tn*7*-*URA3* insertions the target sequence needs to be cloned first. Both methods work well and the choice mainly depends on the availability of reagents or the number of insertions needed.

3.3. Transformation into E. coli DH10B Electrocompetent Cells and Mapping of Insertions by Colony PCR

The transposition reaction in vitro is very efficient and hundreds of transformants are routinely obtained, each one represents one insertion in the target plasmid (in the vector or the insert). After transformation of the mutagenesis mix into DH10B *E. coli* cells is performed, insertions should be mapped by PCR to identify the ones that are within the subtelomeric sequences of interest.

1. Thaw on ice 90 μL of electrocompetent *E. coli* DH10B cells (see Note 13).
2. Set up three microcentrifuge tubes on ice and label them. Add 30 μL of electrocompetent DH10B cells to each tube.
3. To first tube, add 2 μL of the mutagenesis reaction (Subheading 3.1, item 4). To the second tube, add 2 μL of the "no TnsABC*" control. To the third tube, add 2 μL of TE (10 mM/0.1 mM). Mix each tube well by pipetting up and down.
4. Label three electroporation cuvettes and place them on ice. Transfer the corresponding mix of electrocompetent DH10B cells and DNA (or TE) to the corresponding cuvette.
5. Electroporate at 1.8 V, 200 Ω, 25 μF and immediately add 1 mL of sterile SOC.
6. Transfer the entire cell suspension to microcentrifuge tubes.
7. Incubate at 37°C for 1 h.
8. Plate on LB + 30 μg/mL kanamycin plates.
9. Incubate at 37°C overnight.

   Colony PCR to map Tn*7* (*URA3*) insertions (see Note 14):

   After ~12 h incubation at 37°C, many Km^R transformants are routinely obtained, each colony contains one individual insertion per molecule of target DNA (see Note 15).
10. Prepare two PCR mixes differing in the particular combination of oligonucleotides to amplify the fragments between each of the Tn*7*-*URA3* ends and the polylinker of the vector. Select insertions that leave at least 500 bp on either side of subtelomeric sequences to allow for homologous recombination (see Note 16). Example of PCR reactions to map 22 insertions.
11. Prepare 48 PCR tubes to map 22 colonies with two different pairs of primers (see Fig. 1a). Add a positive control using a colony containing Tn*7*-*URA3* donor plasmid pIC6 and a no template negative control for each PCR mix (see Note 17).

12. Set up cocktail mix 1 for colony PCR (24 reactions: 22 colonies and 2 controls).

    E. coli cells from individual colony

10× PCR buffer	3 μL	25×	75 μL
20× $MgCl_2$	1.5 μL	37.5 μL	
5 μM Tn*7* outward primer 1	3 μL	75 μL	
5 μM Tn*7* polylinker primer 3	3 μL	75 μL	
2.5 mM dNTPs	3 μL	75 μL	
Taq polymerase	0.5 μL	12.5 μL	
Sterile H_2O	16 μL	400 μL	
	–	–	
	30 μL	750 μL	

    Prepare a second mix using Tn*7*-*URA3* outward primer 2 and polylinker primer 3 using the same 22 colonies.

13. Add 30 μL of the mix to each prelabeled PCR tube.
14. Using a sterile yellow pipette tip take a small amount of cells from individual Km^R colonies and resuspend the cells in each tube containing the PCR mix, taking care to inoculate each colony in a fresh Km plate before resuspending in the PCR mix, so the desired colonies can be recovered after the PCR.
15. Immediately run the PCR program using the appropriate annealing temperature for the pair of primers used.
16. Run a 0.8% agarose gel with the PCR products and select insertions of interest, if possible in both orientations, not too close to the ends of the fragment (see Note 16).
17. Extract plasmid DNA from several insertions along the subtelomeric region of interest using any commercial DNA extraction kit.
18. Digest the plasmid containing the insertion with the appropriate restriction enzyme so that the ends of the fragment that contains the Tn*7*-*URA3* insertion should be 100% homologous to the desired *C. glabrata* integration site (see Fig. 2 and Note 18). After checking that the digestion is complete, inactivate the restriction enzyme (see Note 19).

3.4. Homologous Recombination of the Tn7 Insertion into the Subtelomeric Region of C. glabrata

The protocol for transformation of *C. glabrata* is very similar to the one used for *S. cerevisiae* with only a few modifications. Here we specify some of the practical differences between both organisms. *C. glabrata* grows faster in vitro than *S. cerevisiae* (duplication time for wild-type strains approximately 60 vs. 90 min in YPD-rich media), and overnight cultures of *C. glabrata* generally

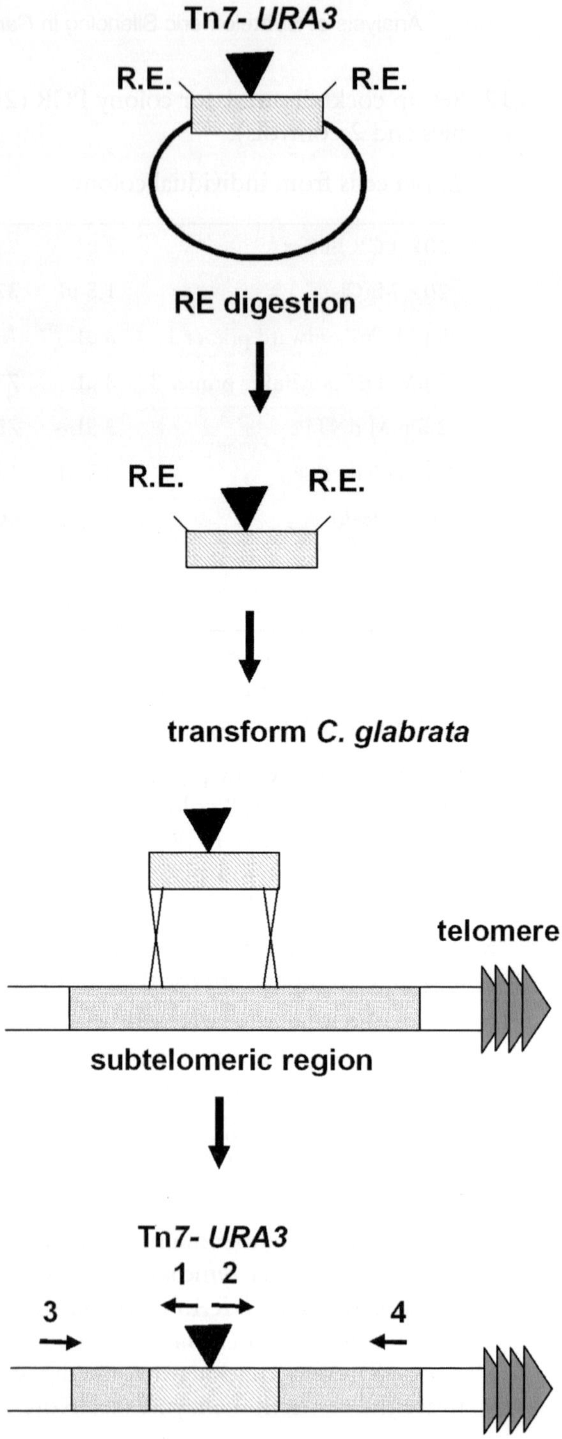

Fig. 2. Homologous recombination of Tn*7*-*URA3* insertions into the *C. glabrata* genome. Plasmid DNA is extracted from the colonies containing the desired insertions of Tn*7*-*URA3* in the target plasmids. The plasmid is digested with restriction enzymes that cut in DNA flanking the Tn*7*-*URA3* and within the *C. glabrata* sequences to direct the integration of the fragment at the correct locus by homologous recombination. The digested DNA mix is used to transform *C. glabrata* directly and selecting for Ura^+ transformants. Candidate transformants must be checked by PCR to confirm that integration occurred at the desired location in the *C. glabrata* genome, using Tn*7*-*URA3* primers and primers that anneal in the chromosome *outside* the fragment that was recombined.

reach higher OD (OD_{600} 30–40 for *C. glabrata* vs. OD_{600} 10–12 for *S. cerevisiae* in YPD; and OD_{600} 15–16 vs. OD_{600} 3–4, respectively, in SC media).

In *C. glabrata*, the rate of nonhomologous recombination is higher than in *S. cerevisiae* (we obtain about 10% of transformants that have undergone nonhomologous recombination even when long fragments of homology are used (10), compared with ~1% in *S. cerevisiae*). To achieve efficient homologous recombination after transformation in *C. glabrata*, the region of homology needs to be relatively large: at least 500 bp on either end compared with 50 bp used in *S. cerevisiae* (see Notes 16 and 18).

For transformation into *C. glabrata*, it is not required to gel purify the DNA fragment containing the Tn*7-URA3* insertion. We use a modified version of the lithium acetate method (19), it should be noted that *C. glabrata* is more sensitive to lithium acetate, therefore this solution should be carefully made, and cells should not be incubated for very long periods in solutions containing this compound.

1. Inoculate 5 mL YPD with the strain to be transformed and incubate overnight at 30°C in a roller or air shaker.
2. Inoculate a 250 mL sterile flask containing 30 mL YPD with 300 μL of the overnight culture and shake at 30°C until the culture reaches OD_{600} 1 (about 3.5–4 h).
3. Spin all the culture for 5 min at 3,500 rpm in a table-top centrifuge in 50 mL conical tubes at room temperature.
4. Wash the cell pellet once with 30 mL sterile water.
5. Resuspend in 1 mL 100 mM lithium acetate (freshly diluted from a 1 M solution) (see Note 3) and transfer to a microcentrifuge tube (do not add the 1 M stock lithium acetate solutions directly to *C. glabrata* cells).
6. Centrifuge at 13,000 rpm for 30″ and resuspend the cell pellet in 300 μL of 100 mM lithium acetate.
7. Label as many sterile centrifuge tubes as DNAs to be transformed plus one for a negative control without DNA. Add 50 μL of cell suspension to each clean sterile microcentrifuge tube, including the negative control.
8. In a separate sterile tube, prepare a transformation cocktail as follows:

50% PEG	240 μL
1 M lithium acetate	36 μL
2 mg/mL salmon sperm	DNA
(Single stranded, see Note 4)	25 μL

Mix thoroughly and add 301 μL of this mix to each transformation tube (see Note 20).

9. Prepare each DNA to be transformed: ~500 ng of digested DNA containing subtelomeric sequences with different Tn*7*-*URA3* insertions (Subheading 3.3, step 18) in 50 μL of TE. Add the DNA solution to each tube with the cell suspension. To the negative control add 50 μL of TE.
10. Incubate at 30°C for 45 min in a roller (or mix the tubes by inverting them every 10 min to prevent the cells from decanting).
11. Add 43 μL of DMSO (hybrid max) to each tube and mix gently.
12. Incubate at 42°C for 15 min.
13. Immediately centrifuge at 13,000 rpm for 30″. Discard the supernatant by decanting carefully (cells may be loose and could be thrown away inadvertently).
14. Resuspend in 600 μL of sterile water by pipetting up and down carefully (see Note 21). Plate 300 μL of the cell suspension per SC w/o ura plate (two in total per transforming DNA plus a negative control).
15. Incubate plates at 30°C for 48 h.
16. After this incubation period, anywhere from 50 to several hundreds of Ura^+ transformants are routinely obtained (the no DNA control should have no colonies). To streak purify transformants, pick six individual Ura^+ colonies from each transformation and streak them on fresh SC w/o ura plates to obtain single colonies and incubate at 30°C for 48 h.
17. Repeat the single colony purification.
18. To verify that homologous recombination has occurred, perform two PCR reactions on each purified colony. The first reaction using one of the Tn*7*-*URA3* outward primers and another primer in the *C. glabrata* genome outside of the cloned fragment that was recombined (primers 1 and 3, see Note 22 and Fig. 2 bottom). The second PCR reaction is performed using the other Tn*7*-*URA3* outward primer and the other primer flanking the recombined region (primers 2 and 4, see Fig. 2 bottom). A fast genomic DNA extraction on individual colonies is performed to obtain the template DNA for diagnostic PCR of the insertions.
19. Resuspend thoroughly a small amount of cells from a *C. glabrata* colony with a sterile pipette tip in 20 μL of 20 mM NaOH.
20. Freeze at −80°C for 5–10 min.
21. Heat at 100°C for 10 min (in a PCR machine).
22. Centrifuge at 13,000 rpm for 1 min.

23. Use 1.5 μL of supernatant for PCR reaction as template DNA (follow the steps for *E. coli* colony PCR, Subheading 3.3, steps 10–15 above).

Once the transformants have been checked, make glycerol stocks of at least two transformants from each insertion for further studies. To make glycerol stocks, make a thick patch of cells from each transformant on an SC w/o ura plate covering the entire plate and incubate at 30°C for 48 h. Prepare cryovials with 1.8 mL 10% sterile glycerol. Scrape with a long-sterile toothpick a large amount of cells and resuspend them in the glycerol. The suspension must be very thick with cells and turn white. Mix them thoroughly and freeze them at −80°C.

3.5. Plate Growth Assay to Assess URA3 Gene Expression on 5-FOA-Containing Plates

To assess silencing of the *URA3* reporter gene carried in the Tn*7*-*URA3* insertion, a simple and convenient plate growth assay is performed. Cells must be pregrown in an overnight culture in rich media (YPD) to avoid selecting for cells that express the *URA3* gene.

1. Start an overnight culture of each of the strains containing each insertion to be assayed in 5 mL-rich media (YPD) and incubate at 30°C in a roller for 12 h.
2. Measure OD_{600} and dilute in sterile water to obtain an OD_{600} 1.
3. Make tenfold serial dilutions of each cell culture at OD_{600} 1: Use a sterile 96-well plate, to dilute each strain into six wells of one row. For this, dispense 180 μL of sterile H_2O per well starting from the second well through the sixth well of each row. In the first well of each row, put 200 μL of the OD_{600} 1 culture (10^0 culture). Mix this cell suspension thoroughly by pipetting up and down and take 20 μL from this well and put it into the next well that already contains 180 μL water. Mix thoroughly pipetting up and down (10^{-1} dilution). Change the pipette tip and take 20 μL of this 10^{-1} dilution and put it in the next well (10^{-2} dilution). Repeat this procedure until the sixth well to obtain a 10^{-5} dilution. Repeat this for all cultures to be tested. When diluting several cultures, a multichannel pipette can be used to dilute all of them at the same time.
4. Arrange the three different types of agar plates: YPD (rich media), SC w/o ura and 5-FOA plates (see Note 23). To obtain well-separated spots of each culture, only spot five different strains per plate. On the back of the plate draw with a marker five lines and a small dot between the lines to align the first tip of the multichannel pipette. Using this pipette, mix the six wells corresponding to one strain (one row) by pipetting up and down and take 5 μL from each well and drop the cell suspension from the six wells of each row onto the first agar plate. Repeat the process for the other two agar plates, so

that 5 μL of each dilution from each strain are spotted on all the plates. When all the strains have been spotted, allow them to dry perfectly. Incubate the plates for 48 h taking photographs every 24 h.

An example of a similar experiment is shown in Fig. 3, where we introduced the Tn*7*-*URA3* insertion at three different positions in the subtelomeric region of the right end of chromosome E in *C. glabrata*. Growth on 5-FOA medium indicates silencing of the *URA3* reporter gene, while growth on SC w/o ura indicates expression of this gene. Growth on both types of plates is indicative of two different populations of cells in the culture: those that express the *URA3* gene (that grow on SC w/o ura plates) and those that keep it in a silenced conformation (5-FOA^{R} cells). Growth on YPD plates reflects the viable count of the culture since both populations can grow on this medium. In the experiment shown in Fig. 3, it can be seen that the two insertions closest to the telomere (1.3 and 14.8 kb from the telomeric repeats, respectively) are the most strongly silenced insertions, since there are many cells that grow on 5-FOA. As the distance between the reporter gene and the telomere increases (line 3, Fig. 3) the amount of cells growing on 5-FOA decreases dramatically, suggesting that at this position, silencing coming from the telomere decreases and, therefore only very few cells in this culture maintained the reporter gene silenced. At the furthest position from the telomere tested in

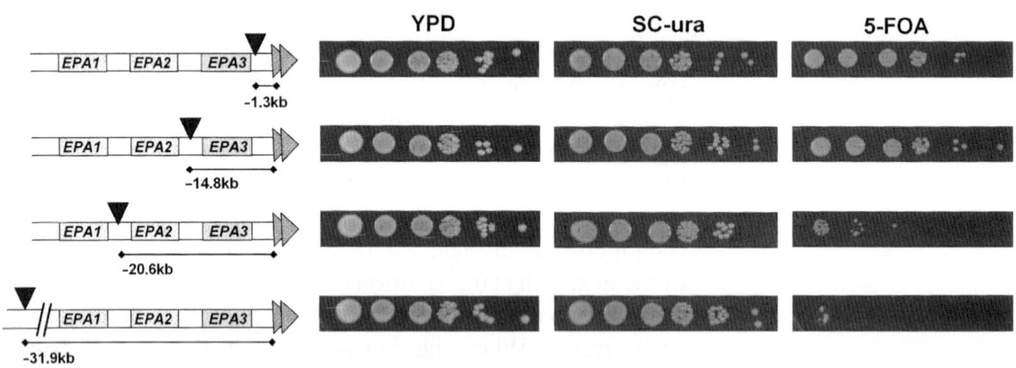

Fig. 3. Subtelomeric region where *EPA* genes are localized is subject to silencing. Plate growth assay to assess expression of the *URA3* reporter placed at four different positions in the subtelomeric region of the right telomere of chromosome E of *C. glabrata* (shown on the left side of the figure). Inverted *black triangles* represent the Tn*7-URA3* element, *gray boxes* are native subtelomeric genes encoding Epa1, 2, 3 proteins. The two overlapping *dark gray triangles* represent the telomeric DNA repeats. Distance of the Tn*7-URA3* element to the telomeric repeats is indicated. Strains carrying each one of the Tn*7-URA3* insertions were grown to stationary phase in YPD (non-selective rich media). The OD_{600} was adjusted to 1 with water (10^{0} cell suspension). Cell suspensions from all cultures were serially diluted tenfold up to 10^{-5}. Equal numbers of cells of each dilution were spotted onto YPD, SC w/o ura, and 5-FOA plates using a multichannel pipette. Growth on YPD reflects the total number of cells (viable count) spotted in each dilution. On SC-ura plates, only cells expressing the *URA3* reporter gene can grow. On 5-FOA plates, only cells that keep the *URA3* reporter gene silenced can grow. Therefore, the amount of growth on 5-FOA is a measure of the level of silencing of a given Tn*7-URA3* insertion.

this experiment, we could only detect ~3 colonies growing on 5-FOA, indicating that all the cells express the *URA3* gene in this strain and that subtelomeric silencing does not spread to ≥31 kb into the chromosome. It is possible that the three colonies found on 5-FOA arose as a result of a mutation in the *URA3* gene.

4. Notes

1. Plasmid DNA made from Promega Wizard minipreps kit works best in our hands. We have tried miniprep kits from several other providers and these do not yield sufficient results. We have not tested midi-prep DNA extraction kits.
2. Glucose should be prepared as a 40% solution and autoclaved separately. After autoclaving the yeast extract and the peptone, add 40% glucose: 50 mL/L to the autoclaved YP media.
3. 100 mM stock should be diluted freshly from a 1 M stock sterilized by autoclaving. The 1 M stock is stable at room temperature for months.
4. Dilute Salmon sperm carrier DNA stock solution to 2 mg/mL with TE. This solution must be heated at 95°C for 5 min and quickly placed on ice before using it.
5. Subtelomeric sequence-specific primers used to diagnose the correct integration at the desired region should be designed in the chromosomal region adjacent to the cloned fragment that contains the *URA3* insertion, ***outside*** of this fragment. A PCR product will only be amplified, if the integration occurred by homologous recombination at the correct place (see Fig. 2).
6. Prepare two solutions: solution A agar and casamino acids in 500 mL water and autoclave. Solution B dissolves ammonium sulfate and the uracil in 500 mL water. Warm solution B to about 38–42°C until the uracil is completely dissolved. Do not overheat you should be able to touch the flask with the inner part of your arm without feeling pain. Add the 5-FOA, stir with magnetic stirrer until 5-FOA is dissolved. Finally, add the YNB w/o amino acids powder and 50 mL of the 40% glucose stock solution. Do not microwave or heat again this solution. When everything is dissolved, filter-sterilize solution B (Millipore 0.22 μm pore). Mix solutions B and A (once it has cooled down to about 45–50°C) carefully, avoiding the formation of bubbles. Quickly pour the plates without making more bubbles and do not flame them. It is important not to overheat 5-FOA, this can result in plates where no cells grow, regardless of whether they express the *URA3* gene (possibly a toxic compound is made if overheated).

7. We add the salt (sodium acetate) to the DNA solution and phenol extract both simultaneously to remove any possible contamination with a DNase in the sodium acetate stock solution.
8. We routinely add 20 μg of glycogen to avoid loosing DNA at the precipitation step.
9. All primers should be designed such that they should start immediately after a T in the template and the last two nucleotides should be either C or G. Primers 1, 3, 4, and 6 should be 20–25 nucleotides long and a Tm ~60°C. The composite primers 2 and 5 should consist of two homologous sections (the 5′ or 3′ flanking regions and the 5′ or 3′ of *URA3*). Each section should be 20–25 bp long and a Tm ~60°C.
10. For this experiment, it is important to always do "Hot Start" (add the polymerase at the end of the first denaturing step).
11. A high-fidelity enzyme should be used. We use Phusion polymerase (a high-fidelity enzyme), if no product is obtained with *Pfu*. A temperature gradient of annealing temperatures and reducing primer concentration can be tried in these cases.
12. Expand long-template PCR system comes with three different buffers. All three of them should be tried at first since it is not possible to accurately predict which buffer will amplify more efficiently the particular fusion product desired.
13. It is important to use high-efficiency electrocompetent cells to increase the number of recovered transformants in *E. coli*, particularly when the target plasmid is large (over 10 kb).
14. We use colony PCR to roughly map the Tn*7*-*URA3* insertions to choose the ones to integrate into the *C. glabrata* genome. Insertions can occur either in the vector or in the insert (subtelomeric sequences). Two PCR mixes can be made to include both combinations of primers to amplify insertions in opposite orientations. The exact insertion point can be identified by sequencing these plasmids using Tn*7* outward primers.
15. Modified Tn*7*-*URA3* transposon usually creates simple insertions into the target molecule with very little, if any, target site specificity. Therefore, an in vitro transposition reaction made as described creates a pool of random insertions into the target molecule (within the insert or in the vector, see Fig. 1a). In addition, when relatively small target molecules are used (<40 kb), almost always only one Tn*7*-*URA3* is inserted due to the phenomenon of transposon immunity that prevents the insertion of more than one element per molecule of target DNA over short distances (9).
16. It is important to use insertions that leave at least 500 bp of *C. glabrata* genomic sequences flanking either end of the

Tn*7*-*URA3* to allow efficient homologous recombination into the genome. In *C. glabrata*, the rate of nonhomologous recombination increases with homologous fragments smaller than 500 bp (20).

17. For the positive control in the PCR mix, plasmid pIC6 can be used using Tn*7* outward primers 1 and 2 and reverse: 5′-TCACACAGGAAACAGCTATGAC-3′). pUC forward should be used with Tn*7* outward 2 (expected size for PCR product is 198 bp) and pUC reverse should be used with Tn*7* outward 1 (expected product is 154 bp).

18. The restriction enzymes to excise the Tn*7*-*URA3* with the flanking subtelomeric sequences should cut ***within*** the *C. glabrata* sequences to generate homologous ends that will recombine at the homologous site in the chromosome with high efficiency. Flanking subtelomeric sequences can also be cloned adding type IV restriction enzymes to the primers so that the enzymes recognize the sequence in the primer but cut outside the recognition site and within the subtelomeric sequence, thereby generating DNA ends perfectly homologous to the target site. Polylinker enzymes that leave even 4 bp sequences at the end of the fragment increase significantly the likelihood of nonhomologous recombination. Care should be taken to insure that the restriction enzyme sites added to the primers do not cut within the fragment (containing the *URA3* insertion) to be recombined.

19. It is important to inactivate the restriction enzyme (by heat inactivation or phenol extraction) before the DNA is transformed into *C. glabrata*. This increases transformation efficiency.

20. The transformation mix can be prepared as a cocktail for several transformations, multiplying by the number of reactions to be made and mixing thoroughly. Pipette 301 μL of this mix into each transformation tube containing the *C. glabrata* cell suspension.

21. Cell pellet forms clumps that have to be dissolved. Do not vortex as cells are very fragile after the lithium acetate and heat shock. They can be resuspended by pipetting up and down gently.

22. To check by PCR, the genomic structure of the subtelomeric region where the Tn*7*-*URA3* was recombined, the primers that anneal in the *C. glabrata* subtelomeric region must be ***outside*** the fragment that was cloned and used to recombine the insertion (see Fig. 2, bottom part). It is important to perform at least the two PCR reactions to probe both junctions of the Tn*7*-*URA3* insertion, since it is somewhat frequent that rearrangements occur only on one side of the

insertion. We also routinely perform a PCR across the insertion using the primer pair that anneals outside the cloned fragment (primers 5 and 6, see Fig. 2).

23. Plates must be relatively dry so that culture spots are absorbed quickly and do not mix with one another. If plates are freshly poured, they can be dried in a hood for 20 min.

Acknowledgments

The authors would like to thank Omar Arroyo-Helguera for excellent technical assistance. This work was supported by fellowship no. 167877 to A.J.R. and grants no. CB-2005-48304 to I.C. and CB-2005-48279 to A.D.L.P.

References

1. Gottschling, D. E., Aparicio, O. M., Billington, B. L., and Zakian, V. A. (1990) Position effect at S. cerevisiae telomeres: reversible repression of Pol II transcription. *Cell* **63**, 751–62.
2. Rusche, L. N., Kirchmaier, A. L., and Rine, J. (2003) The establishment, inheritance, and function of silenced chromatin in Saccharomyces cerevisiae. *Annu Rev Biochem* **72**, 481–516.
3. Tham, W. H., and Zakian, V. A. (2002) Transcriptional silencing at Saccharomyces telomeres: implications for other organisms. *Oncogene* **21**, 512–21.
4. boeke, J. D., LaCroute, F., Fink GR. (1984) A positive selection for mutants lacking orotidine-5'-phosphate decarboxylase activity in yeast: 5-fluoro-orotic acid resistance. *Mol Gen Genet* **197**, 345–346.
5. Pryde, F. E., and Louis, E. J. (1999) Limitations of silencing at native yeast telomeres. *Embo J* **18**, 2538–50.
6. Castano, I., Pan, S. J., Zupancic, M., Hennequin, C., Dujon, B., and Cormack, B. P. (2005) Telomere length control and transcriptional regulation of subtelomeric adhesins in Candida glabrata. *Mol Microbiol* **55**, 1246–58.
7. De Las Penas, A., Pan, S. J., Castano, I., Alder, J., Cregg, R., and Cormack, B. P. (2003) Virulence-related surface glycoproteins in the yeast pathogen Candida glabrata are encoded in subtelomeric clusters and subject to RAP1- and SIR-dependent transcriptional silencing. *Genes Dev* **17**, 2245–58.
8. Rosas-Hernandez, L. L., Juarez-Reyes, A., Arroyo-Helguera, O. E., De Las Penas, A., Pan, S. J., Cormack, B. P., and Castano, I. (2008) yKu70/yKu80 and Rif1 Regulate Silencing Differentially at Telomeres in Candida glabrata. *Eukaryotic Cell* 7, 2168–2178.
9. Biery, M. C., Stewart, F. J., Stellwagen, A. E., Raleigh, E. A., and Craig, N. L. (2000) A simple in vitro Tn7-based transposition system with low target site selectivity for genome and gene analysis. *Nucleic Acids Research* **28**, 1067–1077.
10. Castano, I., Kaur, R., Pan, S., Cregg, R., De Las Penas, A., Guo, N., Biery, M. C., Craig, N. L., and Cormack, B. P. (2003) Tn7-based genome-wide random insertional mutagenesis of Candida glabrata. *Genome Res* **13**, 905–15.
11. Ausubel, F., R. Brent, R. E. Kingston, d. D. Moore, J. G. Seidman, J. A. Smith, and K. Struhl. (2001) *Current Protocols in Molecular Biology.* Wiley & Sons, Inc., New York, NY., New York.
12. Guthrie, C. a. F. G. R. (1991) *Guide to Yeast Genetics and Molecular Biology. Methods in Enzymology,* Vol. **194**, Academic Press, San Diego CA.
13. Dujon, B., Sherman, D., Fischer, G., Durrens, P., Casaregola, S., Lafontaine, I., De Montigny, J., Marck, C., Neuveglise, C., Talla, E., Goffard, N., Frangeul, L., Aigle, M., Anthouard, V., Babour, A., Barbe, V., Barnay, S., Blanchin, S., Beckerich, J. M., Beyne, E., Bleykasten, C., Boisrame, A., Boyer, J., Cattolico, L., Confanioleri, F., De Daruvar, A.,

Despons, L., Fabre, E., Fairhead, C., Ferry-Dumazet, H., Groppi, A., Hantraye, F., Hennequin, C., Jauniaux, N., Joyet, P., Kachouri, R., Kerrest, A., Koszul, R., Lemaire, M., Lesur, I., Ma, L., Muller, H., Nicaud, J. M., Nikolski, M., Oztas, S., Ozier-Kalogeropoulos, O., Pellenz, S., Potier, S., Richard, G. F., Straub, M. L., Suleau, A., Swennen, D., Tekaia, F., Wesolowski-Louvel, M., Westhof, E., Wirth, B., Zeniou-Meyer, M., Zivanovic, I., Bolotin-Fukuhara, M., Thierry, A., Bouchier, C., Caudron, B., Scarpelli, C., Gaillardin, C., Weissenbach, J., Wincker, P., and Souciet, J. L. (2004) Genome evolution in yeasts. *Nature* **430**, 35–44.

14. Gietz, R. D., and Sugino, A. (1988) New yeast-Escherichia coli shuttle vectors constructed with in vitro mutagenized yeast genes lacking six-base pair restriction sites. *Gene* **74**, 527–34.
15. Sikorski, R. S., and Hieter, P. (1989) A system of shuttle vectors and yeast host strains designed for efficient manipulation of DNA in Saccharomyces cerevisiae. *Genetics* **122**, 19–27.
16. Aparicio, O. M., Gottschling, D. E. (1994) Overcoming telomeric silencing: a trans-activator competes to establish gene expression in a cell cycle-dependent way. *Genes & Development* **8**, 1133–1146.
17. Kuwayama, H., Obara, S., Morio, T., Katoh, M., Urushihara, H., and Tanaka, Y. (2002) PCR-mediated generation of a gene disruption construct without the use of DNA ligase and plasmid vectors. *Nucleic Acids Research* **30**, e2.
18. Noble, S. M., and Johnson, A. D. (2005) Strains and strategies for large-scale gene deletion studies of the diploid human fungal pathogen Candida albicans. *Eukaryotic Cell* **4**, 298–309.
19. Gietz, D., St. Jean, A., Woods, R. A., and Schiestl, R. H. (1992) Improved method for high efficiency transformation of intact yeast cells. *Nucleic Acids Res* **20**, 1425.
20. Cormack, B. P., and Falkow, S. (1999) Efficient homologous and illegitimate recombination in the opportunistic yeast pathogen Candida glabrata. *Genetics* **151**, 979–87.

Chapter 15

Morphological and Molecular Genetic Analysis of Epigenetic Switching of the Human Fungal Pathogen *Candida albicans*

Denes Hnisz, Michael Tscherner, and Karl Kuchler

Abstract

Candida albicans is a pleiomorphic fungal pathogen whose morphogenetic plasticity has long been considered as a major virulence factor. In addition to the yeast-filament transition, *C. albicans* cells also have the unique ability to switch between two epigenetic phases referred to as white and opaque. White and opaque cells harbor identical genomes yet they differ in cellular morphologies, gene expression profiles, mating abilities, and virulence properties. The switching process is regulated by a small network of transcription factors and is suggested to be driven by stochastic fluctuations of the regulatory components, which correlates with altered switching frequencies. Traditionally, phase variants have been identified based on cellular morphologies and expression levels of a few marker transcripts, yet it has recently become clear that several other criteria are also essential and relevant, because phase markers are regulated at multiple branching sites of transcriptional circuitry regulating switching. Here, we describe basic methods to discriminate between white and opaque switching variants, based on cellular and macroscopic morphologies, expression levels of phase-specific transcripts, Wor1 protein levels, as well as quantitative mating assays.

Key words: *Candida albicans*, White, Opaque, Phenotypic switching, Morphogenesis

1. Introduction

Candida albicans is a commensal fungus residing on mucosal surfaces of healthy people. However, it can cause opportunistic infections when the immune system of the human host is compromised. *C. albicans*, as several pleiomorphic fungal pathogens, is able to undergo reversible transitions between distinct morphogenetic phases. This morphogenetic plasticity is suggested to improve adaptation of the fungal population to various environmental and host stimuli and is considered a key virulence factor (1). In *C. albicans*, two major morphogenetic processes are considered. First, *C. albicans* can alternate between unicellular, yeast phase growth, and multicellular filamentous growth modes.

Attila Becskei (ed.), *Yeast Genetic Networks: Methods and Protocols*, Methods in Molecular Biology, Vol. 734, DOI 10.1007/978-1-61779-086-7_15, © Springer Science+Business Media, LLC 2011

The yeast-filament switch can be triggered by environmental as well as host cues and is controlled by several signal transduction pathways (2). Second, *C. albicans* possesses the unique ability to form two distinct cell types that are termed white and opaque. These are considered as epigenetic states of cells harboring identical genomes yet white and opaque cells are characterized by different cellular morphologies, gene expression profiles, metabolic preferences, mating abilities, virulence properties, and interactions with host immune cells (3).

White–opaque switching is reversible occurring at a frequency of 10^{-3} to 10^{-4} per generation *in vitro*. Notably, environmental factors are known to alter the frequency of switching. The switching process is tightly linked to the sexual cycle of *C. albicans* and is regulated by a small network of transcription factors (4). *C. albicans* is obligatory diploid and contains a mating type-like locus (*MTL*), which harbors two alleles "**a**" and "α." In *MTL* heterozygous cells, a heterodimeric repressor formed by the gene products of the **a** and α alleles locks cells in the white phase. *MTL* homozygous cells lack an **a**/α repressor and are able switch to the opaque phase. The switching event is regulated by the master regulator *WOR1*. White cells do not express Wor1, and stable, high-level expression of Wor1 is required for the conversion to the opaque phase. The opaque phase is suggested to be heritable for many generations, because Wor1 coordinates three positive feed-back loops to ensure high activity of the *WOR1* locus. In this model, the white phase is considered as the default phase. Cells convert to the opaque phase when stochastic fluctuations of Wor1 levels reach a yet undefined critical threshold, initiating activation of the three feed-back loops (3–5). In addition, histone-modifying enzymes were also found to modulate white–opaque switching, probably by influencing the transcriptional activity at the described four key regulatory loci (6). Since only opaque cells are able to mate with opaque cells of the opposite mating type, and because opaque cells show different interactions with immune cells than white cells, white–opaque switching is thought to have evolved to limit the sexual cycle to certain host niches, as well as to escape immune surveillance *in vivo*.

As a consequence of the stochastic fluctuation model, genetic or biochemical interference with the regulatory circuitry results in changes of the switching frequency between the two phases. Traditionally, white and opaque phases were identified as distinct cellular morphologies on solid media. However, it has recently become clear that although being fundamental, cellular morphology *per se* is not a sufficient parameter to determine the phase, because several phase markers are regulated at multiple branching points of the transcriptional circuitry (6). Here, we describe basic methodologies to discriminate between white and opaque variants, using cellular and macroscopic morphologies, expression levels of phase-specific transcripts, Wor1 protein levels, as well as quantitative mating assays.

2. Materials

2.1. Standard Laboratory Media for C. albicans *Cultivation*

1. YPD medium: 10 g/l Bacto Yeast Extract, 20 g/l BactoPeptone both from BD Biosciences (Franklin Lakes, NJ, USA) and 2% (w/v) glucose. All components are autoclaved; glucose is added separately prior to use. For solid medium, 2% (w/v) agar is added (BD Biosciences, Franklin Lakes, NJ, USA) (7). All components and ready media can be stored at room temperature.
2. SD medium: 1.7 g/l Yeast Nitrogen Base (without amino acids and $(NH_4)_2SO_4$, BD Difco, Franklin Lakes, NJ, USA), 5 g/l $(NH_4)_2SO_4$ and 40 mg/l adenine, 40 mg/l L-arginine, 30 mg/l L-tyrosine, 30 mg/l L-isoleucine, 50 mg/l L-phenylalanine, 100 mg/l L-glutamic acid, 100 mg/l L-aspartic acid, 200 mg/l L-threonine, 400 mg/l L-serine, 150 mg/l L-valine, 50 mg/l L-methionine, and 30 mg/l L-lysine. The pH should be adjusted between 5.5 and 6. After autoclaving, the following components are added in the indicated final concentration: 2% (w/v) glucose added from a 10× autoclaved stock, and 30 mg/l L-histidine, 130 mg/l L-leucine, 20 mg/l uracil, and 40 mg/l L-tryptophan. Of the latter supplements, all stock solutions are prepared as 100× concentrated and are autoclaved, except for L-tryptophan which is filter sterilized (7). For solid medium, 2% (w/v) agar is added. For selective plates, the appropriate amino acid or nucleotide is omitted (see Note 1). All components and ready media can be stored at room temperature, except for the L-tryptophan stock solution which is stored at 4°C.
3. Lee's medium: 5 g/l NaCl, 5 g/l $(NH_4)_2SO_4$, 2.5 g/l K_2HPO_4, 1.3 g/l L-leucine, 1.0 g/l L-lysine, 0.5 g/l L-alanine, 0.5 g/l L-phenylalanine, 0.5 g/l L-proline, 0.5 g/l L-threonine, 0.1 g/l L-methionine, 0.07 g/l L-ornithine, 0.07 g/l L-arginine, 1mg/l D-biotin, 0.1 g/l $ZnSO_4$, 1 g/l $MgSO_4$, and 12.5 g/l glucose. The pH should be adjusted to 6.7–6.8 prior to autoclaving (8). All components and ready media can be stored at room temperature.
4. Phloxin B (Sigma, St. Louis, MO, USA) is dissolved in water at 10 mg/l and sterilized by passing through a 0.22 μm filter (Nalgene, Rochester, NY, USA). The final concentration of the dye is 5 μg/l in solid media, but it cannot be used in liquid media (see Note 2).
5. Microscopy slides and cover slips.

2.2. RNA Extraction and Real-Time qPCR

1. TRI reagent (Molecular Research Center, Cincinnati, OH, USA).
2. Glass beads (acid washed 425–600 μm, Sigma, St. Louis, MO, USA).
3. Phenol:chloroform:isoamyl-alcohol solution (25:24:1).
4. Chloroform.
5. Absolute ethanol.
6. 3 M Na–acetate pH 4.5 (autoclaved).
7. 70% Ethanol.
8. RNase-free distilled water.
9. DNase I (Fermentas, Hanover, MD, USA).
10. RiboLock RNase inhibitor (Fermentas, Hanover, MD, USA).
11. First-strand cDNA synthesis kit (Fermentas, Hanover, MD, USA).
12. Light cycler 480 SYBR Green Master Mix (Roche Applied Science).
13. Primers for *CaWH11* (forward: 5′ CAGAACAATTCAAGGATAAGGTTACTG 3′, reverse: 5′ TTGGAGTCACCAAAAATAGCATCAG 3′), *CaOP4* (forward: 5′ CCTCAAAAGCTGCTACCTC 3′, reverse 5′ GTATCAACAGTTGGAGTAGAAGTAG 3′), and *CaPAT1* (forward: 5′ TTATCGGAATGGTCCTCGTG 3′, reverse: 5′ CCAGAAGAACCATCATCAAC 3′).
14. Primers for quality check after the reverse transcription reaction. *CaTDH3* (forward: 5′ TTTCACCAAACTCGAAGGTGCTC 3′, reverse: 5′ GCAGCACCAGTGGAAGATGG 3′).

2.3. Protein Extraction and Immunoblotting

1. Lysis buffer: 1.85 M NaOH. Store at room temperature. Add 7.5% (v/v) β-mercaptoethanol prior to use.
2. 50% (v/v) Trichloroacetic acid (TCA) solution. Store at 4°C.
3. Sample buffer: 50 mM Tris-HCI pH 6.8, 10% (v/v) glycerol, 2% (w/v) SDS, 0.0005% (w/v) bromophenol blue. Store at room temperature. Add 4% (v/v) β-mercaptoethanol prior to use.
4. 1× TBS-T buffer: 3 g/l Tris–HCl, 8 g/l NaCl, 0.2 g/l KCl (adjust pH to 7.4), 0.1% (v/v) Tween-20. Store at room temperature.
5. Anti-CaWor1 antibodies (commercially not available, developed in the laboratory of Alexander D. Johnson, UCSF, USA (9)). Monoclonal anti-mouse tubulin antibody (DM1A, Sigma, St. Louis, MO, USA).
6. Horseradish peroxidase-coupled secondary antibodies (Merck, Whitehouse Station, NJ, USA).

7. Nitrocellulose membrane (Millipore, Bedford, MA, USA).
8. ECL reagents for immunodetection (Pierce, Rockford, IL, USA).

2.4. Quantitative Mating Assays

1. Mating plates: solid YPD additionally supplemented for the appropriate auxotrophic markers that are used (see Note 3).
2. SD agar plates selective for one or two of the auxotrophic markers used.
3. Filter-paper disks with 6–8 cm diameter, cut out of filter paper (Whatman, Maidstone, UK) sterilized at 180°C for 20 min in a drying oven.
4. Sterile forceps.
5. Sterile Petri dishes ($d = 10$ cm).
6. Sterile 15 ml Falcon tubes.
7. Sterile distilled water.
8. Glass spreader.

3. Methods

Traditionally, there are three standard media used for the propagation of switching variants of *C. albicans*. The nutrient-limited Lee's medium, a synthetic medium with defined components. The SD medium is of standard use for selective purposes in several yeast species. In addition, the nutrient-rich complete YPD medium is often used. Notably, the media choice does influence both the cellular morphology and the switching frequencies. Therefore, standardized conditions for all switching assays are absolutely required for consistent results and correct identification of all phenotypes.

There are several criteria to discriminate between white and opaque *C. albicans* cells. As described in the first report in 1987, white phase cells grow as round, yeast-like cells and form dome-shaped colonies on solid agar, whereas opaque cells display an elongated shape and form flat colonies that can be differentially stained pink on solid medium containing 5 μg/ml Phloxin B (10). Subsequent hybridization experiments also indentified several transcripts that are expressed in a phase-specific manner. For instance, *WH11* is a classic white phase marker, whereas *OP4* is expressed exclusively in opaque cells (11, 12). In 2002, it was also demonstrated that opaque cells are the mating-competent forms of *C. albicans*, which shed new light on the biological significance of the switching system (13). In 2006, the master opaque regulator Wor1 was identified (9, 14). Hence, the correct identification of the epigenetic state requires morphological and expression analyses, as well as functional (mating) assays.

3.1. Stable Propagation of Epigenetic Switching Variants

1. Streak strains from −80°C frozen stocks onto YPD plates and incubate for 3 days at 25°C (see Note 4).
2. Restreak single colonies onto medium of choice (Lee's, SD, and YPD) and incubate for 5 days at 25°C (Lee's and SD) or 3 days at 25°C (YPD).
3. Pick several single colonies (1–5 per genotype or strain tested) and resuspend each in 200 μl sterile water.
4. Measure OD_{600nm} of cell suspensions, dilute in sterile water and spread onto medium of choice (Lee's, SD, and YPD) containing 5 μg/ml Phloxin B at a low density [about 100 colony forming units on a standard Petri dish (d = 10 cm)] (see Notes 5 and 6).
5. Incubate plates for 5 days at 25°C (Lee's and SD) or 3 days at 25°C (YPD) (see Note 7).
6. Microscopic inspection of cells is performed by dropping 5–10 μl of water onto a glass slide, mixing cells of selected colonies with a pipette tip in the drop and covering it with a cover slip. The differences in cellular shape are visible at 40× magnification or higher (Fig. 1a).

3.2. Monitoring Marker Gene Expression

1. Inoculate freshly grown single colonies into 5 ml of liquid medium.
2. Incubate cultures at 25°C with shaking until mid-exponential growth phase (OD_{600nm} of 1–3; see Note 8).
3. Check purity of the cultures under a light microscope by dropping 5 μl of the suspension on a microscopy slide and covering it with a cover slip.
4. Measure OD_{600nm} and transfer around 1 OD_{600nm} (about 10^7 cells) into an Eppendorf tube (see Note 9). Alternatively, cells grown on solid surfaces can also be used; scrape off 1–3 colonies and resuspend in 1 ml sterile water. From this point work on ice.
5. Centrifuge at for 3 min at 3,000 × g at 4°C.
6. Discard supernatant and wash cells with 1 ml ice-cold distilled water.
7. Discard supernatant and resuspend cells in 500 μl TRI reagent.
8. Add 200–300 μl sterile glass beads.
9. Break cells for 45 s at 5 m/s on a FastPrep instrument (MP Biomedicals, Irvine, CA, USA). Alternatively, cells can be also broken by vortex-mixing for five times 1 min intervals with cooling in between.
10. Centrifuge cells for 1 min at 3,000 × g at 4°C.

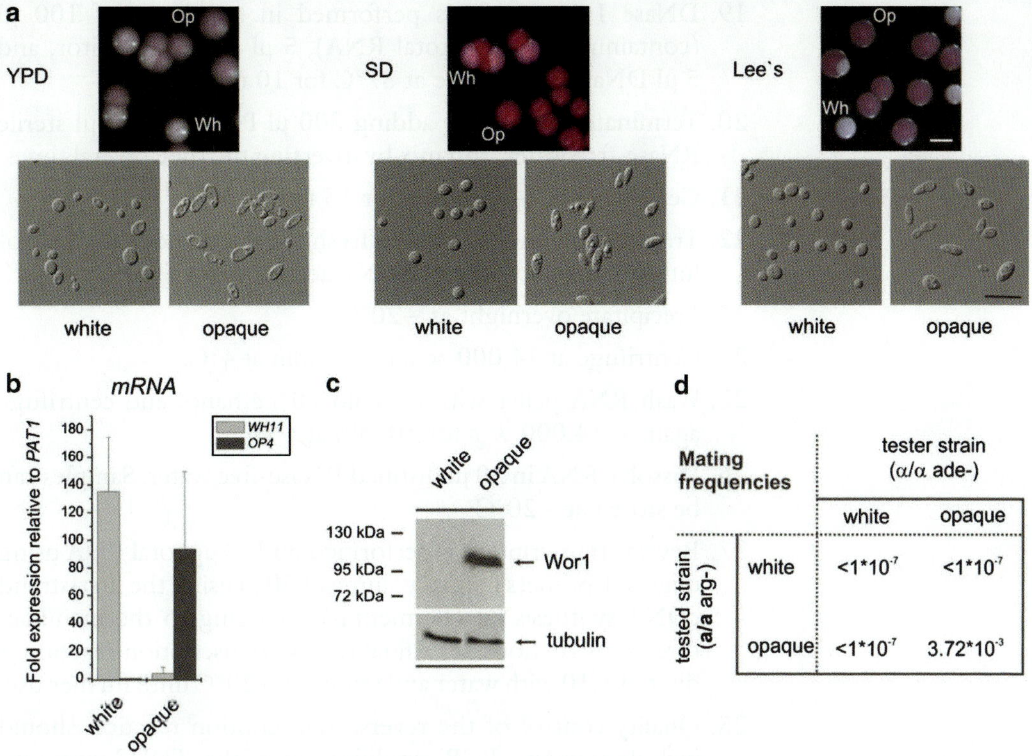

Fig. 1. Morphological, expression, and functional markers of *Candida albicans* white and opaque switching variants. (**a**) Colony and cellular morphologies of white and opaque cells on the standard laboratory media containing 5 μg/ml Phloxin B. Opaque cells display an elongated shape and the colony stains *pink*. Scale bars correspond to 2 μm (*upper panel*) and 5 μm (*lower panel*). (**b**) Quantitative qRT-PCR analysis of phase-specific mRNA transcripts. *WH11* (white-specific) and OP4 (opaque-specific) transcript levels were normalized to the transcript level of *PAT1*. qRT-PCR reactions were performed in triplicates and cDNA isolated from three independent cultures were analyzed. Data are shown as mean + SD. (**c**) Immunoblot analysis of Wor1 expression. A tubulin antibody was used to verify equivalent loading. (**d**) Quantitative mating assays confirm that opaque is the mating competent form of *C. albicans*. In this experiment, a mating type α *ade* tester strain was used to test the mating ability of a mating type **a** *arg* strain.

11. Transfer supernatant (around 300 μl) into a fresh tube, add 500 μl TRI reagent and 160 μl chloroform, and mix well by inverting the tube several times.
12. Centrifuge at 14,000 × *g* for 15 min at 4°C.
13. Transfer aqueous phase to a fresh tube and add 300 μl PCI.
14. Transfer aqueous phase to a fresh tube and add 900 μl absolute ethanol and 30 μl 3 M Na–acetate pH 4.5.
15. Precipitate overnight at −20°C.
16. Centrifuge at 14,000 × *g* for 15 min at 4°C.
17. Wash pellet with ice-cold 70% ethanol and centrifuge at 14,000 × *g* for 10 min at 4°C.
18. Dissolve RNA in 50 μl distilled RNase-free water. Samples can be stored at −20°C.

19. DNase I treatment is performed in a volume of 100 μl (containing 5–10 μg total RNA), 5 μl RNase inhibitor, and 5 μl DNase I. Incubate at 37°C for 10 min.
20. Terminate reaction by adding 300 μl PCI and 200 μl sterile RNase-free water, and mix by inverting the tube several times.
21. Centrifuge at 14,000 × *g* for 15 min at 4°C.
22. Transfer aqueous phase to a fresh tube and add 900 μl absolute ethanol and 30 μl 3 M Na–acetate pH 4.5.
23. Precipitate overnight at −20°C.
24. Centrifuge at 14,000 × *g* for 15 min at 4°C.
25. Wash RNA pellet with ice-cold 70% ethanol and centrifuge again at 14,000 × *g* for 10 min at 4°C.
26. Dissolve RNA in 20 μl distilled RNase-free water. Samples can be stored at −20°C.
27. Reverse transcription is performed on 1–3 μg total RNA using oligo-dT primers in final volume of 20 μl using the first-strand cDNA synthesis kit (Fermentas) according to the manufacturer's instructions. The final reverse transcription reaction is diluted 1:10 with water and stored at −20°C until further use.
28. Quality control of the reverse transcription reaction should include standard PCR amplification of the *TDH3* transcript using the DNase-treated RNA samples as negative controls.
29. Real-time qPCR amplification is performed with the Light Cycler SYBR Green Master Mix (Roche, Basel, Switzerland). Aliquots of 2 μl of cDNAs are added to a total volume of 20 μl. Primers should be used at 300 nM final concentrations. Triplicates of each cDNA sample should be analyzed. Reactions are submitted to cycling using the following conditions: initial denaturation 95°C for 5 min, followed by 40 cycles (each at 95°C for 15 s, 56°C for 15 s, 72°C for 15 s; during these steps, the increase of the fluorescence is recorded), melting curve analysis is done from 60 to 95°C for 30 min. For relative quantification, data are analyzed according to the ΔCt method and are expressed as the fold expression (R) of the gene of interest (GOI) versus the expression of a housekeeping gene as a control (*PAT1*). The equation used is $R = 2^{\Delta Ct}$, where $\Delta Ct = (\Delta Ct_{GOI}t - \Delta Ct_{PAT1}t)$ (Fig. 1b).

3.3. Total Cell Extract Preparation and Immunoblot Analysis of Wor1

1. Inoculate single fresh colonies in 5 ml of liquid medium.
2. Incubate cultures at 25°C in a shaker until mid-exponential growth phase (OD_{600nm} 1–3; see Note 8).
3. Check purity of the cultures under a light microscope by dropping 5 μl of the suspension on a microscopy slide and covering it with a cover slip.

4. Measure OD_{600nm} and transfer around $3OD_{600nm}$ aliquots (3×10^7 cells) into an Eppendorf tube. Alternatively, cells grown on solid surfaces can also be used; scrape off 3–9 colonies and resuspend in 1 ml sterile water. From this point work on ice.
5. Centrifuge at for 3 min at 3,000 × *g* at 4°C.
6. Discard supernatant and wash cells with 1 ml ice-cold distilled water.
7. Discard supernatant and resuspend cells in 1 ml ice-cold distilled water.
8. Add 150 μl lysis buffer and incubate on ice for 10 min.
9. Add 150 μl ice-cold 50% TCA; incubate on ice for at least 10 min.
10. Centrifuge at 14,000 × *g* for 5 min at 4°C.
11. Discard supernatant and centrifuge again at 14,000 × *g* for 5 min at 4°C.
12. Discard supernatant and add 150 μl sample buffer.
13. Resuspend by incubating tubes at 37°C for 10 min. Subsequently, samples can be stored at −20°C.
14. After thawing cell extracts at room temperature for 5 min, separate by SDS–PAGE a 10 μl aliquot on a 10% polyacrylamide gel (0.75 cm) and transfer onto nitrocellulose membrane.
15. Block membrane in 1× TBS-T containing 3% non-fat dry milk for 1–2 h.
16. After a short wash in 1× TBS-T, probe membranes with the primary antibody diluted in 1× TBS-T at 4°C, overnight (see Note 10).
17. Wash membranes three times with 1× TBS-T and incubate membranes with the secondary antibody diluted in 1× TBS-T at room temperature under continuous shaking for 1 h.
18. Wash membranes three times with 1× TBS-T and detect immune complexes using an ECL substrate according to the manufacturer's instructions (Fig. 1c).

3.4. Quantitative Mating Assay

1. Inoculate single colonies of the strains to be tested and the mating tester strains in 5 ml of liquid medium (see Notes 7 and 11).
2. Incubate cultures at 25°C in a shaker until mid-exponential growth phase (OD_{600nm} 1–3; see Note 8).
3. Check purity of the cultures under a light microscope by dropping 5 μl of the suspension on a microscopy slide and covering it with a cover slip (see Note 12).
4. Mix aliquots of about 3×10^7 cells of each mating partner.

5. Place a sterile disk of Whatman filter paper (6–8 cm in diameter) on an YPD plate supplemented for the auxotrophic markers used (see Notes 1 and 3).
6. Centrifuge the mating mixes for 3 min at 3,000 × *g* and discard supernatant so that the residual volume above the pellet is around 300–500 μl.
7. Resuspend cell pellets in the residual medium and gently pipette cell mixtures on the middle of the Whatman filter paper.
8. Incubate mating plates at 25°C for 18 h (see Note 13).
9. Remove filter-paper disks with a sterile forceps into an empty sterile Petri dish.
10. By holding the filter-paper disk in a perpendicular position above the Petri dish wash cells off the disk with 10 ml sterile water.
11. Transfer cell suspension to a 15 ml sterile Falcon tube and vortex briefly.
12. Prepare tenfold dilution series (six dilutions) and plate 150 μl aliquots of each dilution onto four plates: YPD, both single-selective SD and double-selective SD.
13. Incubate scoring plates at 30°C for 2 days.
14. Select appropriate dilutions where the number of colonies per plate is around 100 [on a standard Petri dish plate ($d = 10$ cm)].
15. Score for the number of prototrophic mating products per the total number of the limiting mating partner (the one that gives lower CFU score on the appropriate single selective plate) (Fig. 1d).

4. Notes

1. In several studies, *ura3* mutant *C. albicans* strains have been used as mating tester strains. Since the *CaURA3* gene encodes the orotidine-5′-phosphate decarboxylase enzyme, uridine has to be supplemented to the SD plates instead of the standard uracil.
2. In some reports, Phloxin B is added to solid YPD plates at concentrations as high as 50 μg/ml. However, such a high concentration can already compromise cell viability. Using not more than 5 μg/ml is highly recommended, especially when using nutrient-poor plates media such as Lee's or SD.
3. Although YPD contains low amounts of all essential amino acids and nucleotides, prolonged incubation of auxotrophic

strains (especially nucleotide) on YPD can compromise cell viability. Additionally, in case of using a standard *ade2* mutant as a mating tester strain, the accumulation of a toxic intermediate stains the colonies red and interferes with the visual identification of opaque phase cells on media containing Phloxin B. Therefore, additional supplementation of the mating plates for the auxotrophic markers used is absolutely necessary.

4. Opaque phase cells are unstable at high temperatures. All culturing procedures should be performed at 25°C. Alternatively, room temperature can also be used.
5. One OD_{600nm} corresponds to about 10^7 cells/ml.
6. The density of colonies on plates can influence both colony morphologies and switching frequencies. Therefore, densities should not exceed 100–120 CFUs per a normal Petri plate ($d = 10$ cm). When using bigger plates, scale up estimated numbers of CFUs accordingly.
7. The incubation time effects the growth phase and the purity of colonies. When using nutrient-poor Lee's or SD plates, a 5-day incubation period is recommended. When using YPD plates, a 3-day incubation period is recommended. However, experiments can be initiated using 4- or 2-day-old plates, respectively, although it has to be kept standardized throughout the entire work performed.
8. Generation times are different in the three standard media, which effects the incubation times for liquid cultures. Approximate times of reaching mid-exponential growth phase are YPD: 4–8 h, SD: 8–12 h, and Lee's: 12–16 h. Alternatively, it is possible to freshly dilute an overnight culture to OD_{600nm} 0.1–0.5 and incubate cultures for 2–4 h to reach the mid-exponential growth phase. However, such a prolonged incubation time cannot be used when working with strains that switch phases at significantly elevated frequencies.
9. Using cell numbers exceeding about 5×10^7 cells can saturate the aqueous phase after the centrifugation step, which results in incomplete separation of proteins and DNA from RNA.
10. Adding 5 mM sodium azide to primary antibody dilutions and performing incubations at 4°C will allow to reuse primary antibody dilutions several time, if stored at 4°C. Sodium azide should not be added to the dilutions of horseradish peroxidase-coupled secondary antibodies and washing buffers, since it inhibits the enzyme activity.
11. In case of working with prototrophic clinical isolates, it is possible to use *his–arg–MPA*^{R} mating tester strains which are auxotropic for histidine and arginine and contain an

exogene conferring resistance to mycophenolic acid. In this case, the conjugants are selected on auxotrophic selective media containing mycophenolic acid (15). It is also possible to use dominant markers [such as nourseothricin acetyltransferase (*NAT1*)] instead of auxotrophic markers, in which case media has to be prepared accordingly.

12. The described protocol is optimized for using opaque cultures of the mating tester strain. Purity of the mating tester culture should be at least around 90% based on microscopic inspection.
13. The incubation time effects mating frequencies and needs to be standardized. 18 h are optimal to detect maximum differences in the mating competence of white and opaque phase isolates.

Acknowledgments

We thank Alexander Johnson for generously providing C-terminal anti-Wor1 antibodies. This work was supported by a grant from the Christian Doppler Society to K.K. D.H. and M.T. were supported through the international Vienna Biocenter PhD Program WK001.

References

1. Gow NA, Brown AJ, Odds FC. (2002). Fungal morphogenesis and host invasion. *Curr Opin Microbiol* **5**, 366–71.
2. Whiteway M, Bachewich C. (2007). Morphogenesis in *Candida albicans. Annu Rev Microbiol* **61**, 529–53.
3. Lohse MB, Johnson AD. (2009). White–opaque switching in *Candida albicans. Curr Opin Microbiol* **12**, 650–4.
4. Bennett RJ, Johnson AD. (2002). Mating in *Candida albicans* and the search for a sexual cycle. *Annu Rev Microbiol* **59**, 233–55.
5. Zordan RE, Miller MG, Galgoczy DJ, Tuch BB, Johnson AD. (2007). Interlocking transcriptional feedback loops control white–opaque switching in *Candida albicans. PLoS Biol* **5**, e256.
6. Hnisz D, Schwarzmüller T, Kuchler K. (2009). Transcriptional loops meet chromatin: a dual–layer network controls white–opaque switching in *Candida albicans. Mol Microbiol* **74**, 1–15.
7. Kaiser C, Michaelis S, Mitchell A. (1994). Methods in Yeast Genetics. A Laboratory Course Manual. New York: Cold Spring Harbor Laboratory Press;
8. Lee KL, Buckley HR, Campbell CC. (1975). An amino acid liquid synthetic medium for the development of mycelial and yeast forms of *Candida albicans. Sabouraudia* **13**, 148–53.
9. Zordan RE, Galgoczy DJ, Johnson AD. (2006). Epigenetic properties of white–opaque switching in *Candida albicans* are based on a self–sustaining transcriptional feedback loop. *Proc Natl Acad Sci U S A* **103**, 12807–12.
10. Slutsky B, Staebell M, Anderson J, Risen L, Pfaller M, Soll DR (1987). "White–opaque transition": a second high–frequency switching system in *Candida albicans. J Bacteriol* **169**, 189–97.
11. Srikantha T, Soll DR. (1993). A white–specific gene in the white–opaque switching system of *Candida albicans. Gene* **131**, 53–60.

12. Morrow B, Srikantha T, Anderson J, Soll DR. (1993). Coordinate regulation of two opaque-phase-specific genes during white–opaque switching in *Candida albicans. Infect Immun* **61**, 1823–8.
13. Miller MG, Johnson AD. (2002). White–opaque switching in *Candida albicans* is controlled by mating-type locus homeodomain proteins and allows efficient mating. *Cell* **110**, 293–302.
14. Huang G, Wang H, Chou S, Nie X, Chen J, Liu H. (2006). Bistable expression of *WOR1*, a master regulator of white–opaque switching in *Candida albicans. Proc Natl Acad Sci U S A* **103**, 12813–8.
15. Magee BB, Legrand M, Alarco AM, Raymond M, Magee PT. (2002). Many of the genes required for mating in *Saccharomyces cerevisiae* are also required for mating in *Candida albicans. Mol Microbiol* **46**, 1345–51.

Chapter 16

Quantitation of Cellular Components in *Cryptococcus neoformans* for System Biology Analysis

Arpita Singh, Asfia Qureshi, and Maurizio Del Poeta

Abstract

Methods and procedures in molecular biology used to study fungal pathogenesis have significantly improved during the last decade. In this chapter, we provide step-by-step procedures for performing genetics and biochemical studies in the human pathogenic fungal microorganism *Cryptococcus neoformans* (*Cn*). These methods are employed for studying the pathobiology of *Cn* and for experimental validation of theoretical models of fungal pathogenicity.

Key words: *Cryptococcus neoformans*, Fungal infection, Genetic, Molecular biology, Biochemistry, Sphingolipid, DNA, RNA, Protein, Lipids

1. Introduction

Cryptococcus neoformans (*Cn*) is the causative agent of cryptococcosis, a fungal disease acquired by inhalation of infectious particles from the environment. Cryptococcosis is a relatively frequent disease in immunocompromised subjects and in certain regions of the world such as sub-Saharan Africa in which the estimated number of deaths associated with cryptococcal disease, at half a million per year, is comparable with the number attributed to tuberculosis (1, 2). In the USA, the prevalence of cryptococcosis in HIV positive patients is 5–10%, which is approximately the same as that for meningococcal meningitis (3). Emerging groups at risk include patients suffering from chronic lymphatic leukemia, Hodgkin's disease, chronic myelogenous leukemia, and multiple myeloma (4). The median overall survival of patients with lymphoproliferative disorders affected by cryptococcosis is 2 months, which is significantly shorter than the 9-month

Attila Becskei (ed.), *Yeast Genetic Networks: Methods and Protocols*, Methods in Molecular Biology, Vol. 734, DOI 10.1007/978-1-61779-086-7_16,

median survival of an AIDS patient with cryptococcosis (5). Cryptococcosis is also associated with organ transplantation (6, 7), and was documented in 2.8% of organ transplant recipients with an overall death rate of 42% (8). Some cases of cryptococcosis occur in patients with apparently normal immune function (9–12).

One area of investigation that has significantly improved in the last 2 decades is the molecular biology of this microorganism. The development of molecular epidemiology and phylogeny and molecular technology for clinical diagnosis have significantly helped the clinicians to better manage this life-threatening disease. However, it was the advent of genetics and biochemistry of this microorganism that allowed basic and clinical investigators to address mechanistic questions and study the pathophysiology of cryptococcosis. This was (and still is) an essential step to define fungal features and characteristics necessary for the organism to cause disease (13). These fungal factors can then be exploited for the understanding of fungal pathogenicity and fungal interaction with the host cells and, ultimately, and for the development of new therapeutic strategies. With the rise of its importance as a human pathogen, there has been a concurrent rise in the ability to molecularly study its physiopathology.

In Chapter 9, we provide a mathematical model of the regulation of melanin production by the sphingolipid pathway. In particular, we show that a specific enzyme of the sphingolipid pathway, inositol phosphoryl ceramide synthase 1 (Ipc1), regulates melanin formation in *Cn* through the production of diacylglycerol (DAG) and the consequent activation of protein kinase C 1 (Pkc1). Thus, the downregulation or/and deletion of *IPC1* or/and *PKC1* genes by homologous recombination should produce mutant strains that make less or no melanin. We would expect IPC and DAG lipid measurements to be decreased in the mutant in which Ipc1 is downregulated. Also in this mutant, Pkc1 enzymatic activity should be decreased. This experimental approach is necessary to validate the changes in the network behavior simulated by the mathematical model. Therefore, the deletion of the gene of interest by homologous recombination and confirmation by Southern or/and Northern blot of the isolated genomic DNA or total RNA, respectively, and the analysis of protein and lipid levels regulated by those genes are useful methods to refine and validate the mathematical model.

In this chapter, we provide basic molecular methods for performing genetics and biochemistry studies in *Cn*. These methods can be employed to validate hypotheses and theoretical models of *Cn* pathogenicity or simply to study the pathobiology of this important human pathogen.

2. Materials

2.1. DNA Isolation from Cryptococcus neoformans

1. Yeast Peptone Dextrose (YPD) agar plates and YPD broth (see Note 1).
2. Sterile PBS 1×.
3. 1 M Tris–HCl pH 7.5.
4. 0.5 M, EDTA pH 8.0.
5. 5 M NaCl.
6. 100% Triton X-100.
7. 20% SDS.
8. TENTS: 10 mM Tris–HCl, pH 7.5, 1 mM EDTA, pH 8.0, 100 mM NaCl, 2% Triton X-100, 1% SDS.
9. Acid washed 0.425–600 μm glass beads (SIGMA).
10. Phenol:chloroform:isoamyl alchohol = 25:24:1 (SIGMA).
11. 3 M Sodium acetate (NaOAC).
12. TE buffer, pH 8.0, sterile.

2.2. Biolistic Delivery in Cryptococcus neoformans

1. YPD agar + 1 M Sorbitol plates.
2. YPD agar + Nourseothricin/Hygromycin (100 μg/ml) plates.
3. 0.6 μm Gold beads (BIORAD).
4. MacroCarriers (BIORAD).
5. Rupture Disks, 1,350 psi, (BIORAD).
6. Stopping Screens (BIORAD).
7. 2.5 M $CaCl_2$ sterile.
8. 1.0 M spermidine (filter sterilize) (SIGMA), can be stored at −20°C.
9. 100% Ethanol.
10. Isopropanol.
11. This instruction assumes the use of PDS-1000/He Biolistic Particle Delivery System from BIORAD.

2.3. Southern Hybridization of DNA Extracted from Cryptococcus neoformans

1. Denaturing solution: 1.5 M NaCl, 0.5 M NaOH.
2. Neutralizing Solution: 1 M Tris–HCl pH 8.0, 1.5 M NaCl.
3. Nytran SPC (0.45 μm Nylon Transfer Membrane) (Whatman).
4. Whatman 3MM Blotting Paper.
5. Paper towels, preferably single fold.

6. 20× SSC: 175.3 g of NaCl, 88.2 g of sodium citrate in 800 ml of double distilled water. Adjust pH with NaOH pellets and adjust the total volume to 1 L. Autoclave. Can be stored at room temperature.
7. 20× SSPE : 175.3 g of NaCl, 27.6 g of $NaH_2PO_4 \cdot H_2O$, 7.4 g EDTA in 800 ml of double distilled water. Adjust the pH with NaOH pellets to 7.4. Final volume made up to 1 L. Autoclave. Can be stored at room temperature.
8. 20% SDS.
9. Nonfat dry milk.
10. Prehybridizing solution: 10 ml 20× SSPE, 10 ml 20% SDS, 2 ml 10% nonfat dry milk in a total volume of 40 ml. Can be stored at 4°C with 0.02% sodium azide for 2–3 days.
11. Random Primers DNA labeling system kit (Invitrogen).
12. ^{32}P dCTP (Perkin Elmer).
13. Microspin G-25 column (Amersham Biosciences).
14. Sterile TE buffer, pH 8.0.

2.4. Highly Pure Total RNA Isolation from *Cryptococcus neoformans* (e.g., for Microarray Studies)

1. YPD agar plate and YPD broth.
2. Phosphate Buffered Saline (PBS) 1× sterile.
3. Tri Reagent (Molecular Research Centre).
4. BAN as a phase separation reagent, molecular biology grade (Molecular Research Centre).
5. RNeasy Mini Kit and RNeasy MinElute Cleanup Kit (Qiagen).
6. RNase Zap for removing RNase contamination from external surface (Ambion).
7. RNase/DNase free plastic wares.

2.5. Protein Extraction from *Cryptococcus neoformans*

1. YPD media and agar plates (made from YPD 50 g/L and agar 20 g/L).
2. Buffer for *Cn* cell lysis: 1 ml 1 M Tris–HCl pH 8, 9 ml H_2O, 1.5 ml glycerol (13% v/v), 10 μl CLAP: chymostatin, leupeptin, antipain, and pepstatin A (each at 10 mg/ml in DMSO and stored at −20°C), and 20 μl 100 mM solution phenylmethylsulfonyl fluoride (PMSF) in isopropanol.
3. Glass beads, acid washed, 425 μm (30–40 US sieve) (Sigma).

2.6. Lipid Extraction

1. Mandala lipid extraction buffer: 150 ml ethanol, 150 ml distilled water, 50 ml diethyl ether, 10 ml pyridine, and 180 μl 14.2 N ammonium hydroxide.
2. Use glass tubes for all extraction steps (VWR) fit best in the ThermoSavant SPD2010 SpeedVac system we use).

3. Waters Sep-Pak Classic Silica cartridges (WAT 051900, 690 mg) for analytical scale or WAT036930 200 cc, 5 g cartridges for semipreparative scale lipid isolation and purification.
4. 10″ × 10″ glass tank for thin layer chromatography (TLC).
5. 3MM Whatman chromatography paper (Fisher).
6. TLC chromatography plates (Fisher M5628-5 or 05-713-329 depending on analytical or semipreparative purposes).
7. Soy glucosylceramide standard (Avanti Polar Lipids) made up to 3.5 mM (2.5 μg/μl) in chloroform/methanol (2:1).
8. Prepare 70% H_2SO_4 by adding 14 ml H_2SO_4 slowly to 6 ml water on ice, with mixing. Add 40 mg resorcinol to 20 ml 70% H_2SO_4. Stir well at room temperature with a magnetic stirrer bar. Pour solution into a glass TLC sprayer.

2.7. In Vitro Enzyme Activity Assay

1. NBD-C6-ceramide (Avanti Polar Lipids).
2. Lysis buffer: 25 mM Tris–HCl pH 7.4, 5 mM EDTA, 1 mM PMSF, and CLAP: chymostatin, leupeptin, antipain, and pepstatin A (each at 10 mg/ml in DMSO and stored at −20°C).
3. Silica gel 60 TLC plates (EM Sciences, Fisher).

2.8. Mass Spectrometry of Lipids

1. Commercially available synthetic lipid standards.

3. Methods

3.1. DNA Isolation (14, 15)

1. Inoculate a 10–15 ml YPD broth with a single colony from a fresh YPD agar plate and grow them for 20–24 h at 30°C with constant shaking. Pellet cells from this culture at 1,200 × *g* 4°C for 10 min.
2. Wash with sterile PBS 1× twice and resuspend in 1 ml of sterile double distilled water and transfer to a 2-ml screw cap tube (see Note 2).
3. Pellet cells in a Microcentrifuge for 30 s at 1,200 × *g* at room temperature.
4. Pour off the water; add 0.5 ml of TENTS and vortex at 7–8 speed for three times, 45 s each. This step assumes the use of Vortex Genie 2 from Scientific Industries.
5. Add two cups (1 cup = 400–500 μl) of acid washed 0.45 μm glass beads (see Note 3) and 0.5 ml phenol–chloroform–Isoamyl alcohol (see Note 4).
6. This step assumes the uses of Bead Beater 16 from Scientific Industries .Tubes were vortexed/homogenized in a Bead

Beater three times, 45 s each, with a gap of 45 s on ice, in between each cycle (see Note 5).

7. After homogenizing (or lysing) of the cells, centrifuge the cells for 10 min at 8,000 × *g* at room temperature to separate the cell debris and unbroken cells.
8. Remove the upper aqueous phase which now contains the DNA, to a fresh 1.5-ml Eppendorf and add 1 ml of ethanol 100% and keep at −20°C overnight (see Note 6).
9. Centrifuge the tube at 8,000 × *g* for 30 min at 4°C. Remove the supernatant, dissolve the pellet in 200 μl of TE containing RNase A at a concentration of 100 μg/ml and then incubate at 37°C for 20 min.
10. After incubation, add equal volume of phenol–chloroform–isoamyl alcohol and mix gently by inverting 4–6 times. Centrifuge at 8,000 × *g*, 10 min, 4°C. Remove the aqueous phase and repeat the step with the aqueous phase.
11. Add 20 μl of 3M NaOAC and 400 μl of ethanol (100%) to the final aqueous phase and incubate at −20°C for 30–60 min for complete precipitation.
12. After precipitation, centrifuge the tube at 8,000 × *g* at 4°C for 5 min.
13. Wash the DNA pellet twice with 200 μl of ice-cold 70% ethanol and air dry the pellet (see Note 7).
14. Dissolve the pellet in 30–50 μl of sterile TE gently and store at −20°C.

3.2. Biolistic Delivery in Cryptococcus neoformans *(14, 16)*

1. Spin down 19–20 h grown culture (15 ml) of the recipient strain and throw off 12 ml of the supernatant (see Note 8).
2. Plate 200–250 μl of this cell suspension on prewarmed YPD agar + 1 M sorbitol plates and let them dry for 4–5 h at 30°C (see Note 9). This should include a "non-shot" control plate.
3. During this time of incubation, prepare the shot. For preparing a stock of Gold Beads – 60 mg/ml, 30 mg was weighed out and dissolved in 100% ethanol, vortexed vigorously for 3 min, incubated at room temperature for 15 min and spun for 1 min. Discard the supernatant and suspend the gold beads in 1 ml of sterile water. Incubate or allow the particles to settle down, pellet and discard the supernatant. Add 500 μl of 50% Glycerol to make a final concentration of 60 mg/ml. This stock can be stored in 4°C.
4. Each Shot should be prepared as follows in the same sequence:

   10 μl of 60 mg/ml of gold beads

   1 μl of ≥1 μg/μl of DNA

   10 μl of 2.5 M $CaCl_2$

2 μl of 1.0 M Spermidine

Vortex the mix for 3–5 min and let it settle for 5 min at room temperature. Spin for 20 s and take off the supernatant. Wash the Bead-DNA mix once with 500 μl of 100% ethanol by vortexing and spin down the Bead-DNA. Throw off the supernatant. Finally, resuspend the Bead-DNA in 25 μl of 100% ethanol (see Note 10).

5. The Macrocarriers should be prepared inside a Laminar Hood to prevent contamination. Dip the macrocarriers (one for each shot) in 100% ethanol. Blot off the excess liquid on a sterile wiper and keep in a sterile Petri dish until completely dry.
6. Vortex the Bead-DNA well so that the beads are uniformly coated with the DNA (see Note 11). Spread 10 μl of this mix, first onto the center of the macrocarrier then working outward, within 5 mm to the edge in a slow circular motion. Let it dry. If there is any extra Bead-DNA, it can be added to each macrocarrier in the same fashion (see Note 12).
7. The machine (PDS-1000/He Biolistic Particle Delivery System) should be sterilized with 70% ethanol and dried before shooting. The chamber should be kept closed as much as possible. Open the Helium tank pressure valve and set the pressure regulator at 1,800–2,100 psi.
8. Soak the rupture disk, 1,350 psi in isopropanol, place in the retaining cap and screw the unit onto the gas acceleration tube of the machine with the retaining cap torque wrench (see Note 13).
9. Unscrew the macrocarrier cover lid and place a stopping screen on the stopping screen support. Place the macrocarrier on top (Bead-DNA side up) of the macrocarrier holder, invert and place on the fixed nest. The dried microcarriers should face toward the stopping screen. Screw the macrocarrier cover lid to the assembly until tightened and place this in the top slot inside the chamber.
10. Place the target shelf on the second to bottom shelf (see Note 14). Place the YPD agar + 1 M sorbitol Petri dishes with cells, on this shelf without the lid on.
11. Close the chamber and set the vacuum switch at "VAC" position till the desired vacuum of 28.5–29″ is reached. Hold the vacuum chamber at this level of vacuum by quickly pressing the switch to "HOLD" position and press the "FIRE" switch to bombard the sample into the plate until the rupture disk pops. Vent the chamber and immediately cover the Petri dish with lid and remove it from the chamber.
12. Repeat the shooting until all the macrocarriers coated with Bead-DNA were utilized. All the parts should be cleaned and

surface sterilized with 70% ethanol between two different DNA samples.

13. Incubate the "shot" along with a "non-shot control" plates for 2 h at 30°C (see Note 15).
14. Label Falcon 2054 tubes, one for each of the shot and non-shot plates.
15. Aliquot 1 ml of prewarmed YPD broth onto each plate. Rub the liquid broth across the whole surface of the plate with sterile hockey stick and scrape off the cells. Tilt the plate and pipette the liquid into the labeled Falcon tubes.
16. Plate 200–250 μl of the scraped liquid and spread uniformly onto prewarmed YPD Nourseothricin/Hygromycin plates. Incubate the plates at 30°C for several days.

3.3. Southern Hybridization (17)

1. After taking picture of the Gel (see Note 16), denature in the Denaturing Solution (use fresh) for 1 h at room temperature with constant shaking.
2. Neutralize the gel in Neutralizing Solution for 1–2 h.
3. Wash the gel with double distilled water.
4. Wet the membrane and the 3MM Whatman paper in 2× SSC until complete wet and assemble the transfer. Transfer overnight at room temperature or at least for 16–18 h.
5. Before removing the gel, mark with pencil the wells on the membrane (see Note 17). Keep the membrane on a filter paper presoaked with 6× SSC at room temperature and semi-dry. Auto cross-link for 1 min at 1,200 (μJ × 100; this instruction assumes the use of UV Stratalinker 1800 from Stratagene). The membrane, if not set for hybridization can be stored at 4°C for 2–3 days in a sealed bag.
6. Prehybridize the membrane in prehybridizing solution for 1–2 h at 65°C.
7. Labeling of probes: Spun down the contents of the Random Primer Labeling kit for 30 s in microcentrifuge after thawing. Boil 9 μl of DNA (for probe) for 5 min and cool it down on ice for 1 min. Add to the DNA 1 μl each of dATP, dGTP, dTTP, 2 μl of Random Primers, 5 μl of ^{32}P dCTP, and lastly 1 μl of Klenow. Incubate the mix for 30 min at 37°C. After incubation, add 2 μl of Stop buffer and 20 μl of TE. Snap the tip of a microspin G 25 column and put in a 1.5-ml Eppendorf and spin for 30 s in a centrifuge inside the Laminar Hood and then run the probe through the column (see Note 18). Boil the probe for 5 min and cool it on ice for 1 min. Add 1 ml of 5× SSPE with a syringe

into the probe and transfer it to the hybridizing chamber carefully. Hybridize overnight and wash sequentially with 50 ml of 0.1% SDS in 2× SSC for 20–30 min at 65°C, 50 ml of 0.5% SDS in 0.1× SSC thrice, each for 20–30 min at 65°C.

8. Dry the membrane over a filter paper and saran wrap and tape it on a cassette. Inside the darkroom put the film on top of the membrane and expose the film at −80°C overnight or at the least 4–5 h before developing.

3.4. Isolation of Total RNA (18)

1. Harvest cells (20–24 h) grown in the required media by pelleting down at 1,200 × *g* at 4°C for 10 min (see Note 19).
2. Wash the pelleted cells with sterile PBS twice and spin down at 1,200 × *g*, 4°C for 5 min. Drain out the PBS on a sterile wipe and flash freeze in a dry ice – ethanol bath and set for lyophilization (see Note 20).
3. Aliquot ~100 μl (about 50–75 mg) of lyophilized cells in a 2-ml screw cap tube and grind or smash the cells to powder form with the help of the spatula used to scoop out the lyophilized cells (see Note 21). Add 1–1.25 ml of Tri reagent. Cap the tubes properly and homogenize in Bead Beater 16 with pulses as follows 45 s thrice, 30 s once with a gap of 45 s between each cycle on ice.
4. Incubate the tubes for 10 min at room temperature. Centrifuge for 10 min at 4°C at 8,000 × *g* to pellet the cell debris and unbroken cells.
5. Transfer the supernatant to a fresh tube and add 50–60 μl of BAN (50 μl of BAN/ml of Tri reagent added) and shake vigorously for 20–30 s. Incubate for 5 min at room temperature and centrifuge at 8,000 × *g* at 4°C for 10 min.
6. Transfer the aqueous phase to a new tube and add equal volume of 70% Ethanol and mix gently and properly (see Note 22).
7. Load the aqueous phase (700 μl at a time) onto an RNeasy isolation column and spin for 30 s in a centrifuge at 8,000 × *g* at room temperature. If the volume exceeds 700 μl, the same column can be reloaded until the whole aqueous phase had passed through it.
8. Discard the flow-through and wash the column with RW1 buffer provided with the kit. Discard the flow-through and wash with 500 μl of RPE twice, spin for 30 s at 8,000 × *g* and discard the flow-through. Transfer the RNA isolation column to a new 2-ml collection tube and spin for 2 min at 8,000 × *g* at room temperature.

9. Elute the RNA in 50 μl of RNase free water in a fresh 1.5-ml Eppendorf. Re-elute the residual RNA in another aliquot of 50 μl of RNase free water in the same tube.
10. Concentrate the RNA with the column from RNeasy MinElute Cleanup kit following instructions of the manufacturer. Elute in a final volume of 20 μl of DNase–RNase free water.

3.5. Protein Extraction from Cn

1. These instructions assume the use of a Bead Beater 8.
2. Streak out *Cn* strains of interest (e.g., wt or mutant) onto YPD agar plate and incubate at 30°C for 48 h.
3. Pick a single colony into a 50-ml Corning Centrifuge tube containing 10 ml YPD media and allow to grow at 30°C with shaking for 24 h.
4. Centrifuge the culture 10 min at 1,200 × *g* at ambient temperature (20°C), wash once with doubly distilled water and then resuspend into 7 ml doubly distilled water.
5. Aliquot 1 ml each into 1.5-ml conical tubes with screw caps and centrifuge at 3,500 × *g* for 10 min at 25°C.
6. Meanwhile prepare the lysis buffer.
7. Following centrifugation, discard the supernatant and resuspend each pellet into 200 μl lysis buffer.
8. Add one "cupful" glass beads (see Note 3), then vortex and place on ice.
9. Place each tube into the beadbeater in a 4°C coldroom and beadbeat for 40 s, followed by 1 min on ice. Repeat this four times (see Note 23).
10. Centrifuge each tube at 3,500 × *g* for 12 min at 4°C.
11. Collect the supernatant and carry out Bio-Rad protein assay to determine the amount of protein.

3.6. Lipid Extraction

1. Under sterile conditions, fill 50-ml tube with 9 ml yeast-peptone (YP) and 1 ml 20% glucose. Add a single colony of the strain of interest (in this case *Cn* Gcs1^{REC}) and incubate 48 h at 30°C, 250 rpm.
2. Centrifuge at 1,200 × *g* for 10 min at 4°C. Wash pellet twice with water then resuspend in 9 ml sterile water.
3. Count the cells after appropriate serial dilution and aliquot 5×10^8 cells per tube. Centrifuge 10 min 1,200 × *g* at 4°C. Suction out water carefully (see Note 24).

3.6.1. Mandala Extraction (for Extraction of Inositol-Containing Phospholipids and Phosphatidylcholine) (see Note 25)

4. Add 1.5 ml Mandala extraction buffer (19) to each tube. Vortex and sonicate 20 s each.
5. Incubate at 60°C in a water bath for 15 min, vortex and sonicate for 20 s each then reincubate at 60°C for 15 min.
6. Sonicate 20 s then centrifuge 10 min at 1,200 × *g* at 4°C. Using a glass Pasteur pipette, combine supernatant from two tubes together into a clean tube.
7. Evaporate the solvent in the Speedvac (see Note 25).

3.6.2. Bligh and Dyer Lipid Extraction (for Determination of Neutral Lipids)

8. Following evaporation, add 2 ml methanol and vortex. Sonicate if necessary.
9. Add 1 ml chloroform and vortex. Ensure there is one phase, even if turbid (see Note 26).
10. Incubate the samples at 37°C for 1 h. During this period, vortex each sample twice for 30 s.
11. Centrifuge at 1,200 × *g* for 5 min at room temperature, then transfer the lower phase to a clean tube with a glass Pasteur pipette. Add 1 ml Chloroform and 1 ml water and vortex twice for 30 s each. Recentrifuge samples at 1,200 × *g* for 5 min at room temperature.
12. Once again, using a glass Pasteur pipette transfer lower phase to a clean tube. Up to three tubes can be combined into one to lessen the amount of tubes being handled.
13. Evaporate the solvent in the Speedvac (see Note 25).

3.6.3. Additional Purification Steps (e.g., Isolation of Glucosylceramide Using a Silica Column)

3.6.3.1. Silica Column Purification 1

14. Resuspend the lipids in 1 ml chloroform/acetic acid (99:1).
15. Wash the SepPak cartridges with 15 ml chloroform (see Note 27). Apply sample (in 1 ml) and rinse with 1.5 ml chloroform/acetic acid (99:1). Collect flow-through after 0.5 ml has been allowed to collect into waste.
16. Add 15 ml chloroform/acetic acid (99:1) and collect 5 ml per tube.
17. Add 15 ml acetone and collect 5 ml per tube. Evaporate acetone from these tubes in the SpeedVac then resuspend in minimum amount acetone to combine into one tube. Reevaporate (see Note 28).

3.6.4. Base Hydrolysis

18. Add 0.5 ml chloroform, followed by 0.5 ml 0.6 M KOH in methanol to each sample. Vortex well and leave at room temperature for 1 h.
19. Add 0.325 ml 1 M HCl followed by 0.125 ml distilled water. Vortex well then centrifuge at 1,200 × *g* for 10 min at room temperature. Transfer lower organic phase to a clean tube.
20. Evaporate solvent in the SpeedVac. You should have a small dark brown pellet at this stage (see Note 25).

3.6.5. Silica Column Purification 2

21. Resuspend the pellet in 1 ml chloroform/acetic acid (99:1). Repeat steps 16 and 17.
22. Change eluting solvent to chloroform/methanol (95:5); add 10 ml and collect in two tubes.
23. Change eluting solvent to chloroform/methanol (90:10); add 15 ml and collect into 3 tubes. ***These are the tubes that will contain glucosylceramide, the lipid of interest for this example.*** Evaporate solvent using a SpeedVac. Do not combine the tubes (see Note 29).
24. Wash the column with 15 ml methanol and collect in case needed.

3.6.6. Thin Layer chromatography

25. Prepare a 10′ × 10′ glass TLC tank by adding chloroform/methanol/water (97.5:37.5:6) to a *clean, dry* tank lined with white chromatography paper. Apply a thin layer of vacuum grease around the top lip of the tank to ensure a good seal (see Note 30). Leave until paper is well saturated, usually at least 5 h to overnight.
26. Spot the soy standard onto a TLC plate 1.5 cm from the bottom using a 10 μl pipette, using 1, 2, and 3 μl in three separate lanes (equivalent to 2.5, 5, and 7.5 μg, respectively).
27. Resuspend the dried lipid from step 24 in 30 μl chloroform/methanol (2:1), and spot either 30 μl (analytical) or 5 μl (semipreparative scale) onto the TLC plate into a fourth lane. Allow solvent to evaporate in fume hood (~1–2 min) before placing the TLC plate in the tank.
28. Make sure the TLC tank is tightly closed. Allow the solvent front to migrate up to 1 cm from the top of the plate, before removing the plate from the tank.
29. Dry the TLC plate in the hood at room temperature prior to placing it in another tank containing only iodine crystals to allow visualization of the lipids. Alternatively, the plate can be sprayed with resorcinol in 70% H_2SO_4, and then placed in an oven for 10 min to allow a dark purple color to develop wherever sugar moieties are located on the lipids.

3.7. In Vitro Enzyme Activity Assay

1. This protocol describes the in vitro activity assay of Ipc1 (20) but could be adapted for assaying any enzyme from *Cn*. Ipc1 activity is measured by using the fluorescent ceramide analog NBD-C6-ceramide as substrate and monitoring the formation of NBD-C6-IPC, as described by Fischl et al. (21) with some modifications.
2. Grow wt and mutant *Cn* strains in YPD media in a shaker incubator for 24 h at 30°C. Harvest the cells by centrifugation and wash with sterile distilled water (see Note 24).

3. Resuspend the pellets in lysis buffer, add acid-washed glass beads for a volume equal to 3/4 of the cell suspension and homogenize three times for 45 s, followed by 1 min on ice each time, using the Bead Beater 8.
4. Centrifuge at 2,500 × *g* for 10 min at 4°C, then transfer the supernatant (~100 μl) to a sterile 1.5-ml microcentrifuge tube for protein quantification.
5. Following protein determination, incubate 100 μg protein from the cell lysates for 30 min at 30°C in 50 mM bis-Tris–HCl buffer (pH 6.5) containing 1 mM phosphatidyl inositol, 5 mM Triton X-100, 1 mM $MnCl_2$, 5 mM $MgCl_2$, and 20 μM NBD-C6-ceramide in a final reaction volume of 100 μl.
6. Terminate the reaction by addition of 0.5 ml 0.1 N HCl in methanol.
7. Add 1 ml chloroform and 1.5 ml 1 M $MgCl_2$, mix well and centrifuge at 1,000 × *g* for 10 min to separate the phases.
8. Analyze the chloroform-soluble product, NBD-IPC, by TLC on silica gel 60 plates (EM Science) as described above using chloroform/methanol/water (65:25:4).
9. Identify and quantify NBD-IPC by direct fluorescence using a Molecular Dynamics 840 Storm unit.

3.8. Mass Spectrometry of Lipids (22, 23)

1. This protocol describes MS and MS/MS of Cn glucosylceramide but is applicable to any *Cn* lipid molecule.
2. Following Bligh and Dyer extraction described above under lipid extraction, MS and MS/MS scans of glucosylceramide were carried out on a Thermo Finnigan TSQ7000 triple quadrupole mass spectrometer equipped with electrospray ionization as described in ref. 6.
3. A 31 min method was used with A; water/0.2% formic acid/ 2 mM ammonium formate and B: methanol/0.2% formic acid/1 mM ammonium formate, on a 150 × 3 mm Spectra 3 μm C8SR column (Peeke Scientific) using gradient elution and addition of internal standards.
4. Include multiple reaction monitoring (MRM) for the characteristic product ion *m/z* 276.2.
5. Quantify Cn glucosylceramide using soy glycosylceramide (Avanti Polar lipids) for standard curve generation.
6. Normalize mass spectral data to inorganic phosphate determination.

4. Notes

1. All solutions should be prepared in water which has a resistivity of 18.3 MΩ-cm and total organic content of less than 5 ppb. This water is referred to double distilled water in this text.
2. Each tube should contain ~100 μl of cell pellet.
3. A cup was made by cutting out from the bottom till 0.5 ml marking of a 1.5-Eppendorf tube. Drive 23 gauge BD needle into the cup through the upper part to make a makeshift handle.
4. Tubes should be capped properly and the mouth should be wiped with kimwipes ensuring that tubes seal properly before vortexing.
5. The Vortex should have the single unit assembly during vortexing.
6. The volume of ethanol should be 2–2.5 times the volume of the aqueous phase. Incubating at −20°C at 2 h can also be done, however, the yield may be less.
7. The tube containing the DNA pellet can be covered by Para film, punctured and kept at 4°C to let the ethanol dry off. However, it should not be too much dried.
8. 200–250 μl was to be used from this cell suspension, so this would suffice for 15–12 plates for biolistic delivery. If more number of shots is desired, the culture volume should increase proportionately as during shooting the recipient cell density should be high.
9. The cells should be spread in a monolayer over the plate. To do this, spread the cell suspension with a sterile glass hockey stick in a single direction.
10. 10–15 μl extra ethanol was added to compensate for evaporation. It was always wise to include at least two extra shot when preparing for the Bead-DNA.
11. Spread Bead-DNA mix immediately as they have a tendency to settle down. Best is to spread from a continuously vortexed mix.
12. The macrocarrier should be used for shooting within 1–2 h of its preparation.
13. The rupture disk should not be kept for more than 30–60 s in the isopropanol and excess liquid should be blotted off as this may cause delamination. The rupture disk should also be wet while being loaded as the liquid reduces failure rate of the rupture disk. The retaining cap should be clean for any

residual rupture disk part from previous shooting as this may cause rupture of the disk at a wrong pressure and thereby no delivery of the DNA into the Cryptococcal cells.

14. This distance is the best for delivering DNA into Cryptococcal cells.
15. If not transforming with any selectable marker like Nourseothricin/Hygromycin, these plates, after shooting can be incubated at 30°C directly, for several days.
16. The amount of DNA before restriction enzyme digestion is quantified by agarose gel electrophoresis and the DNA should be completely digested.
17. The total well should be marked with a pencil.
18. The amount of the DNA used as probe should be at least 100 ng. The angle of the Eppendorf with the G25 column after loading of the Probe should be the same as before in the microcentrifuge.
19. Be extremely cautious about RNase contamination. Wipe with RNase ZAP the whole external surface of the working area, pipettes, etc., before starting and change gloves frequently. If Minimal Media (YNB or DMEM) is to be used, it can be supplemented with 50 mM Hepes, 1 M sorbitol, and 10% FCS if required.
20. Lyophilization for a 75–100 ml culture should be at least for 24 h but not more than 48 h.
21. The lyophilized cells in powder form give better yield.
22. Do not let the tip touch into the interphase while transferring the aqueous phase.
23. It is important to go through four cycles on the beadbeater when lysing the *Cn* cells otherwise insufficient protein will be extracted.
24. At this stage, the cell pellet can be frozen at −80°C until ready for extraction.
25. To see the original references on how this protocol was established, see reference by Barbara Hanson (24).
26. The tubes can be left at 4°C overnight if there are time constraints.
27. For analytical scale, use WAT051900; 15 ml is equivalent to a 5 bed volume wash.
28. Dry down other tubes as well in case needed later, then store at −20°C.
29. Try to get as much compound down as possible by rinsing the walls of the glass tube with 9:1 chloroform:methanol.

30. You can add two weights on top to ensure the cover seals well. The weights can be 2 × 250 ml glass bottles filled with water.

Acknowledgments

This work was supported by Grants AI56168 and AI72142 (to M.D.P) and was conducted in a facility constructed with support from the National Institutes of Health, Grant Number C06 RR015455 from the Extramural Research Facilities Program of the National Center for Research Resources. Dr. Maurizio Del Poeta is a Burroughs Wellcome New Investigator in Pathogenesis of Infectious Diseases.

References

1. Harrison, T. S. (2009) The burden of HIV-associated cryptococcal disease. *AIDS* **23**, 531–2.
2. Park, B. J., Wannemuehler, K. A., Marston, B. J., Govender, N., Pappas, P. G., and Chiller, T. M. (2009) Estimation of the current global burden of cryptococcal meningitis among persons living with HIV/AIDS. *AIDS* **23**, 525–30.
3. Hajjeh, R. A., Conn, L. A., Stephens, D. S., Baughman, W., Hamill, R., Graviss, E., Pappas, P. G., Thomas, C., Reingold, A., Rothrock, G., Hutwagner, L. C., Schuchat, A., Brandt, M. E., and Pinner, R. W. (1999) Cryptococcosis: population-based multistate active surveillance and risk factors in human immunodeficiency virus-infected persons. Cryptococcal Active Surveillance Group. *J Infect Dis* **179**, 449–54.
4. Kaplan, M. H., Rosen, P. P., and Armstrong, D. (1977) Cryptococcosis in a cancer hospital: clinical and pathological correlates in forty-six patients. *Cancer* **39**, 2265–74.
5. White, M., Cirrincione, C., Blevins, A., and Armstrong, D. (1992) Cryptococcal meningitis: outcome in patients with AIDS and patients with neoplastic disease. *J Infect Dis* **165**, 960–3.
6. Kohno, S., Varma, A., Kwon-Chung, K. J., and Hara, K. (1994) Epidemiology studies of clinical isolates of Cryptococcus neoformans of Japan by restriction fragment length polymorphism. *Kansenshogaku Zasshi* **68**, 1512–7.
7. Shaariah, W., Morad, Z., and Suleiman, A. B. (1992) Cryptococcosis in renal transplant recipients. *Transplant Proc* **24**, 1898–9.
8. Husain, S., Wagener, M. M., and Singh, N. (2001) Cryptococcus neoformans infection in organ transplant recipients: variables influencing clinical characteristics and outcome. *Emerg Infect Dis* 7, 375–81.
9. Fraser, J. A., Giles, S. S., Wenink, E. C., Geunes-Boyer, S. G., Wright, J. R., Diezmann, S., Allen, A., Stajich, J. E., Dietrich, F. S., Perfect, J. R., and Heitman, J. (2005) Same-sex mating and the origin of the Vancouver Island Cryptococcus gattii outbreak. *Nature* **437**, 1360–4.
10. Seaton, R. A., Verma, N., Naraqi, S., Wembri, J. P., and Warrell, D. A. (1997) Visual loss in immunocompetent patients with *Cryptococcus neoformans* var. *gattii* meningitis. *Trans R Soc Trop Med Hyg* **91**, 44–9.
11. Seaton, R. A., Naraqi, S., Wembri, J. P., and Warrell, D. A. (1996) Predictors of outcome in *Cryptococcus neoformans* var. *gattii* meningitis. *Qjm* **89**, 423–8.
12. Findley, K., Rodriguez-Carres, M., Metin, B., Kroiss, J., Fonseca, A., Vilgalys, R., and Heitman, J. (2009) Phylogeny and phenotypic characterization of pathogenic Cryptococcus species and closely related saprobic taxa in the Tremellales. *Eukaryot Cell* **8**, 353–61.
13. Perfect, J. R. (2005) Cryptococcus neoformans: a sugar-coated killer with designer genes. *FEMS Immunol Med Microbiol* **45**, 395–404.
14. Casadevall, A., and Perfect, J. R. (1998) *Cryptococcus neoformans*, ASM Press, Washington, DC, 381–405.
15. Hull, C. M., and Heitman, J. (2002) Genetics of Cryptococcus neoformans. *Annu Rev Genet* **36**, 557–615.

16. Toffaletti, D. L., Rude, T. H., Johnston, S. A., Durack, D. T., and Perfect, J. R. (1993) Gene transfer in *Cryptococcus neoformans* by use of biolistic delivery of DNA. *J. Bacteriol.* **175**, 1405–11.
17. Sambrook, J., Fritsch, E. F., and Maniatis, T. (1989) Molecular cloning: a laboratory manual, Cold Spring Harbor Laboratory Press, Cold Spring Harbor, New York.
18. Baker, L. G., Specht, C. A., Donlin, M. J., and Lodge, J. K. (2007) Chitosan, the deacetylated form of chitin, is necessary for cell wall integrity in Cryptococcus neoformans. *Eukaryot Cell* **6**, 855–67.
19. Mandala, S. M., Thornton, R. A., Frommer, B. R., and et al. (1995) The discovery of australifungin, a novel inhibitor of sphinganine N-acyltransferase from Sporormiella australis. Producing organism, fermentation, isolation, and biological activity. *J. Antibiot. (Tokyo)* **48**, 349–56.
20. Luberto, C., Toffaletti, D. L., Wills, E. A., Tucker, S. C., Casadevall, A., Perfect, J. R., Hannun, Y. A., and Del Poeta, M. (2001) Roles for inositol-phosphoryl ceramide synthase 1 (*IPC1*) in pathogenesis of *C. neoformans. Genes Dev.* **15**, 201–12.
21. Fischl, A. S., Liu, Y., Browdy, A., and Cremesti, A. E. (2000) Inositolphosphoryl ceramide synthase from yeast. *Methods Enzymol.* **311**, 123–30.
22. Bielawski, J., Pierce, J. S., Snider, J., Rembiesa, B., Szulc, Z. M., and Bielawska, A. (2009) Comprehensive quantitative analysis of bioactive sphingolipids by high-performance liquid chromatography-tandem mass spectrometry. *Methods Mol Biol* **579**, 443–67.
23. Bielawski, J., Szulc, Z. M., Hannun, Y. A., and Bielawska, A. (2006) Simultaneous quantitative analysis of bioactive sphingolipids by high-performance liquid chromatography-tandem mass spectrometry. *Methods* **39**, 82–91.
24. Hanson, B. A., and Lester, R. L. (1980) The extraction of inositol-containing phospholipids and phosphatidylcholine from Saccharomyces cerevisiae and Neurospora crassa. *J Lipid Res* **21**, 309–15.

Index

A

B

Attila Becskei (ed.), *Yeast Genetic Networks: Methods and Protocols*, Methods in Molecular Biology, vol. 734,
DOI 10.1007/978-1-61779-086-7, © Springer Science+Business Media, LLC 2011

C

D

E

F

G

H

I

M

T

U

V

W

X